Lecture Notes in Artificial In

Edited by R. Goebel, J. Siekmann, and '

Subseries of Lecture Notes in Computer Science

Lionel Prevost   Simone Marinai
Friedhelm Schwenker (Eds.)

Artificial Neural Networks in Pattern Recognition

Third IAPR Workshop, ANNPR 2008
Paris, France, July 2-4, 2008
Proceedings

 Springer

Series Editors

Randy Goebel, University of Alberta, Edmonton, Canada
Jörg Siekmann, University of Saarland, Saarbrücken, Germany
Wolfgang Wahlster, DFKI and University of Saarland, Saarbrücken, Germany

Volume Editors

Lionel Prevost
ISIR, Université Pierre et Marie Curie - Paris 6
75252 Paris, France
E-mail: lionel.prevost@upmc.fr

Simone Marinai
Dipartimento di Sistemi e Informatica
Università di Firenze
50139 Firenze, Italy
E-mail: marinai@dsi.unifi.it

Friedhelm Schwenker
Institute of Neural Information Processing
University of Ulm
89069 Ulm, Germany
E-mail: friedhelm.schwenker@uni-ulm.de

Library of Congress Control Number: 2008929582

CR Subject Classification (1998): I.2.6, I.2, I.5, I.4, H.3, F.2.2, J.3

LNCS Sublibrary: SL 7 – Artificial Intelligence

ISSN	0302-9743
ISBN-10	3-540-69938-4 Springer Berlin Heidelberg New York
ISBN-13	978-3-540-69938-5 Springer Berlin Heidelberg New York

Springer is a part of Springer Science+Business Media

springer.com

© Springer-Verlag Berlin Heidelberg 2008
Printed in Germany

Typesetting: Camera-ready by author, data conversion by Scientific Publishing Services, Chennai, India
Printed on acid-free paper   SPIN: 12435340   06/3180   5 4 3 2 1 0

Preface

The Third IAPR TC3 Workshop on Artificial Neural Networks in Pattern Recognition, ANNPR 2008, was held at Pierre and Marie Curie University in Paris (France), July 2–4, 2008. The workshop was organized by the Technical Committee on Neural Networks and Computational Intelligence (TC3) that is one of the 20 TCs of the International Association for Pattern Recognition (IAPR). The scope of TC3 includes computational intelligence approaches, such as fuzzy systems, evolutionary computing and artificial neural networks and their use in various pattern recognition applications. ANNPR 2008 followed the success of the previous workshops: ANNPR 2003 held at the University of Florence (Italy) and ANPPR 2006 held at Reisensburg Castle, University of Ulm (Germany). All the workshops featured a single-track program including both oral sessions and posters with a focus on active participation from every participant.

In recent years, the field of neural networks has matured considerably in both methodology and real-world applications. As reflected in this book, artificial neural networks in pattern recognition combine many ideas from machine learning, advanced statistics, signal and image processing for solving complex real-world pattern recognition problems.

High quality across such a diverse field of research can only be achieved through a rigorous and selective review process. For this workshop, 57 papers were submitted out of which 29 were selected for inclusion in the proceedings. The oral sessions included 18 papers, while 11 contributions were presented as posters. ANNPR 2008 featured research works in the areas of supervised and unsupervised learning, multiple classifier systems, pattern recognition in signal and image processing, and feature selection.

We would like to thank all authors for the effort they put into their submissions, and the Scientific Committee for taking the time to provide high-quality reviews and selecting the best contributions for the final workshop program.

A number of organizations supported ANNPR 2008 including the IAPR, TC3, and the Pierre and Marie Curie University. Last, but not least, we are grateful to Springer for publishing the ANNPR 2008 proceedings in their LNCS/LNAI series.

April 2008

Lionel Prevost
Simone Marinai
Friedhelm Schwenker

Organization

Organization Committee

Lionel Prevost, Université Pierre and Marie Curie - Paris 6, (France)
Simone Marinai, University of Florence, (Italy)

Program Committee

Shigeo Abe
Monica Bianchini
Hervé Bourlard
Horst Bunke
Mohamed Cheriet
Abdel Ennaji
Patrick Gallinari
Marco Gori
Barbara Hammer

Laurent Heutte
Tom Heskes
José Manuel Inesta
Rudolf Kruse
Cheng-Lin Liu
Marco Maggini
Maurice Milgram
Erkki Oja
Günther Palm

Marcello Pelillo
Raul Rojas
Fabio Roli
Robert Sabourin
Friedhelm Schwenker
Edmondo Trentin
Michel Verleysen

Local Arrangements

Pablo Negri, Shehzad Muhammad Hanif, Mohamed Chetouani, Michle Vié

Sponsoring Institutions

International Association for Pattern Recognition (IAPR)
Pierre and Marie Curie University Paris (France)

Table of Contents

Multiple Classifiers

Applications

Feature Selection

Patch Relational Neural Gas – Clustering of Huge Dissimilarity Datasets

Alexander Hasenfuss[1], Barbara Hammer[1], and Fabrice Rossi[2]

[1] Clausthal University of Technology, Department of Informatics,
Clausthal-Zellerfeld, Germany
[2] Projet AxIS, INRIA Rocquencourt, Domaine de Voluceau, Rocquencourt,
B.P. 105, 78153 Le Chesnay Cedex, France

Abstract. Clustering constitutes an ubiquitous problem when dealing with huge data sets for data compression, visualization, or preprocessing. Prototype-based neural methods such as neural gas or the self-organizing map offer an intuitive and fast variant which represents data by means of typical representatives, thereby running in linear time. Recently, an extension of these methods towards relational clustering has been proposed which can handle general non-vectorial data characterized by dissimilarities only, such as alignment or general kernels. This extension, relational neural gas, is directly applicable in important domains such as bioinformatics or text clustering. However, it is quadratic in m both in memory and in time (m being the number of data points). Hence, it is infeasible for huge data sets. In this contribution we introduce an approximate patch version of relational neural gas which relies on the same cost function but it dramatically reduces time and memory requirements. It offers a single pass clustering algorithm for huge data sets, running in constant space and linear time only.

1 Introduction

The presence of huge data sets, often several GB or even TB, poses particular challenges towards standard data clustering and visualization such as neural gas or the self-organizing map [10,12]. At most a single pass over the data is still affordable such that online adaptation which requires several runs over the data is not applicable. At the same time, alternative fast batch optimization cannot be applied due to memory constraints. In recent years, researchers have worked on so-called single pass clustering algorithms which run in a single or few passes over the data and which require only a priorly fixed amount of allocated memory. Popular methods include heuristics such as CURE, STING, and BIRCH [5,16,18] and approximations of k-means clustering as proposed in [4,9]. In addition, dynamic methods such as growing neural gas have been adapted to cope with the scenario of life-long adaptivity, see e.g. [15].

The situation becomes even more complicated if data are non-vectorial and distance-based clustering methods have to be applied, which often display a quadratic time complexity [8]. Although a variety of methods which can directly work with relational data based on general principles such as extensions of the

L. Prevost, S. Marinai, and F. Schwenker (Eds.): ANNPR 2008, LNAI 5064, pp. 1–12, 2008.
© Springer-Verlag Berlin Heidelberg 2008

self-organizing map and neural gas have been proposed [3,7,11], these methods are not suited for huge data sets. For complex metrics such as alignment of DNA strings or complex kernels for text data, it is infeasable to compute all pairs of the distance matrix and at most a small fraction can effectively be addressed. A common challenge today, arising especially in Computational Biology, are huge datasets whose pairwise dissimilarities cannot be hold at once within random-access memory during computation, due to the sheer amount of data.

In this work, we present a new technique based on the Relational Neural Gas approach [7] that is able to handle this situation by a single pass technique based on patches that can be chosen in accordance to the size of the available random-access memory. This results in a linear time and finite memory algorithm for general dissimilarity data which shares the intuitivity and robustness of NG.

2   Neural Gas

Neural Gas (NG), introduced by Martinetz et al. [12], is a vector quantization technique aiming for representing given data $v \in V \subseteq \mathbb{R}^d$ faithfully by prototypes $w_i \in \mathbb{R}^d$, $i = 1, \ldots, n$. For a continuous input distribution given by a probability density function $P(v)$, the cost function minimized by NG is

$$E \sim \frac{1}{2} \sum_{i=1}^{n} \int h_\lambda(k(w_i, v)) \cdot \|v - w_i\|^2 P(v) dv,$$

where $k(w_i, v) = |\{w_j : \|v - w_j\| < \|v - w_i\|\}|$ denotes the rank of neuron w_i arranged according to the distance from data point v. The parameter $\lambda > 0$ controls the neighbourhood range through the exponential function $h_\lambda(t) = \exp(-t/\lambda)$.

Typically, NG is optimized in an online mode using a stochastic gradient descent method. However, for a given discrete training set $\{v_1, v_2, \ldots, v_m\}$ the cost function of NG becomes

$$E(W) \sim \frac{1}{2} \cdot \sum_{i=1}^{n} \sum_{j=1}^{m} h_\lambda(k(w_i, v)) \cdot \|v_j - w_i\|^2 \qquad (1)$$

For this case, an alternative batch optimization technique has been introduced [3]. It, in turn, determines the ranks $k_{ij} = k(w_i, v_j)$ for fixed prototype locations w_i and then determines new prototype locations via the update formula

$$w_i = \sum_j h_\lambda(k_{ij}) \cdot v_j / \sum_j h_\lambda(k_{ij})$$

for the fixed ranks k_{ij}. Batch NG shows the same accuracy and behaviour as NG, whereby its convergence is quadratic instead of linear as for NG.

3   Relational Neural Gas

Relational data do not necessarily originate from an Euclidean vector space, instead only a pairwise dissimilarity measure d_{ij} is given for the underlying datapoints $v_i, v_j \in V$. The only demands made on dissimilarity measures are non-negativity $d_{ij} \geq 0$ and reflexivity $d_{ii} = 0$, so they are not necessarily metric or even symmetric by nature. Obviously, NG cannot directly deal with such data and its original formulation is restricted to vectorial updates.

One way to deal with relational data is Median clustering [3]. This technique restricts prototype locations to given data points, such that distances are well defined in the cost function of NG. Batch optimization can be directly tranferred to this case. However, median clustering has the inherent drawback that only discrete adaptation steps can be performed which can dramatically reduce the representation quality of the clustering.

Relational Neural Gas (RNG) [7] overcomes the problem of discrete adaptation steps by using convex combinations of Euclidean embedded data points as prototypes. For that purpose, we assume that there exists a set of (in general unknown and presumably high dimensional) Euclidean points V such that $d_{ij} = \|v_i - v_j\|$ for all $v_i, v_j \in V$ holds, i.e. we assume there exists an (unknown) isometric embedding into an Euclidean space. The key observation is based on the fact that, under the assumptions made, the squared distances $\|w_i - v_j\|^2$ between (unknown) embedded data points and optimum prototypes can be expressed merely in terms of known distances d_{ij}.

In detail, we express the prototypes as $w_i = \sum_j \alpha_{ij} v_j$ with $\sum_j \alpha_{ij} = 1$. With optimal prototypes, this assumption is necessarily fulfilled. Given a coefficient matrix $(\alpha_{ij}) \in \mathbb{R}^{n \times m}$ and a matrix $\Delta = (d_{ij}^2) \in \mathbb{R}^{m \times m}$ of squared distances, it then holds

$$\|w_i - v_j\|^2 = (\alpha_{i*} \cdot \Delta)_j - \frac{1}{2} \cdot \alpha_{i*} \Delta \alpha_{i*}^T \qquad (2)$$

where $*$ indicates vector indices. Because of this fact, we are able to substitute all terms $\|w_i - v_j\|^2$ in Batch NG by (2) and derive new update rules. For optimum prototype locations given fixed ranks we find

$$\alpha_{ij} = h_\lambda(k_i(v_j)) / \sum_t h_\lambda(k_i(v_t)). \qquad (3)$$

This allows to reformulate the batch optimization schemes in terms of relational data as done in [7].

Note that, if an isometric embedding into Euclidean space exists, this scheme is equivalent to Batch NG and it yields identical results. Otherwise, the consecutive optimization scheme can still be applied. It has been shown in [7] that Relational NG converges for every nonsingular symmetric matrix Δ and it optimizes the relational dual cost function of NG which can be defined solely based on distances Δ.

Relational neural gas displays very robust results in several applications as shown in [7]. Compared to original NG, however, it has the severe drawback that the computation time is $\mathcal{O}(m^2)$, m being the number of data points, and the required space is also quadratic (because of Δ). Thus, this method becomes infeasible for huge data sets. Recently, an intuitive and powerful method has been proposed to extend batch neural gas towards a single pass optimization scheme which can be applied even if the training points do not fit into the main

memory [1]. The key idea is to process data in patches, whereby prototypes serve
as a sufficient statistics of the already processed data. Here we transfer this idea
to relational clustering.

4   Patch Relational Neural Gas

Assume as before that data are given as a dissimilarity matrix $D = (d_{ij})_{i,j=1,...,m}$
with entries $d_{ij} = d(v_i, v_j)$ representing the dissimilarity of the datapoints v_i
and v_j. During processing of Patch Relational NG, n_p patches of fixed size $p = \lfloor m/n_p \rfloor$ are cutted consecutively from the dissimilarity matrix D^1, where every
patch

$$P_i = (d_{st})_{s,t=(i-1)\cdot p+1,...,i\cdot p} \in \mathbb{R}^{p\times p}$$

is a submatrix of D centered around the matrix diagonal.

The idea of the original patch scheme is to add the prototypes from the processing of the former patch P_{i-1} as additional datapoints to the current patch P_i,
forming an extended patch P_i^* which includes the previous points in the form of
a compressed statistics. The additional datapoints – the former prototypes – are
weighted according to the size of their receptive fields, i.e. how many datapoints
do they represent in the former patch. To implement this fact, every datapoint
v_j is equipped with a multiplicity m_j, which is initialized with $m_j = 1$ for data
points from the training set and it is set to the size of the receptive fields for
data points stemming from prototypes. This way, all data are processed without
loss of previous information which is represented by the sufficient statistics. So
far, the method has only been tested for stationary distributions. However, it
can expected that the method works equally well for nonstationary distributions
due to the weighting of already processed information according to the number
of already seen data points. In contrast to dynamic approaches such as [15] the
number of prototypes can be fixed a priori.

Unlike the situation of original Patch NG [1], where prototypes can simply be
converted to datapoints and the inter-patch distances can always be recalculated
using the Euclidean metric, the situation becomes more difficult for relational
clustering. In Relational NG prototypes are expressed as convex combinations of
unknown Euclidean datapoints, only the distances can be calculated. Moreover,
the relational prototypes gained from processing of a patch cannot be simply
converted to datapoints for the next patch. They are defined only on the datapoints of the former patch. To calculate the necessary distances between these
prototypes and the datapoints of the next patch, the distances between former
and next patch must be taken into account, as shown in [7]. But that means
touching all elements of the upper half of the distance matrix at least once during processing of all patches, what foils the idea of the patch scheme to reduce
computation and memory-access costs.

In this contribution, another way is proposed. In between patches not the
relational prototypes itselves but representative datapoints obtained from a so

---

1 The remainder is no further considered here for simplicity. In the practical implementation the remaining datapoints are simply distributed over the first $(M - p \cdot n_p)$
patches.

called k-approximation are used to extend the next patch. As for standard patch clustering, the points are equipped with multiplicities. On each extended patch a modified Relational NG is applied taking into account the multiplicities.

k-Approximation. Assume there are given n relational prototypes by their coefficient matrix $(\alpha_{ij}) \in \mathbb{R}^{n \times m}$ defined on Euclidean datapoints V. These prototypes are taken after convergence of the Relational NG method, i.e. these prototypes are situated at optimal locations.

As can be seen from the update rule (3), after convergence in the limit $\lambda \to 0$ it holds

$$\alpha_{ij} \longrightarrow \begin{cases} 1/|R_i| & : \quad v_j \in R_i \\ 0 & : \quad v_j \notin R_i \end{cases}, \text{ because } \begin{cases} h_\lambda(k_{ij}) = 1 \text{ for } v_j \in R_i \\ h_\lambda(k_{ij}) \to 0 \text{ for } v_j \notin R_i \end{cases},$$

where $R_i = \{v_j \in V : \|w_i - v_j\| \leq \|w_k - v_j\| \text{ for all } k\}$ denotes the receptive field of prototype w_i. That means, in the limit only datapoints from the receptive fields have positive coefficients and equally contribute to the winning prototype that is located in the center of gravity of its receptive field.

A k-approximation of an optimal relational prototype w_i is a subset $R' \subseteq R_i$ with $|R'| = \min\{k, |R_i|\}$ such that $\sum_{r' \in R'} \|w_i - r'\|^2$ is minimized. That means, we choose the k nearest points from the receptive field of a prototype as representatives. If there are less than k points in the receptive field, the whole field is taken. This computation can be done in time $\mathcal{O}(|R_i| \cdot k)$. For a set W of relational prototypes, we refer to the set containing a k-approximation for each relational prototype $w_i \in W$ a k-approximation of W.

These k-approximations in combination with their corresponding coefficients can be interpreted as a convex-combined point in the relational model, defined just over the points of the k-approximation. Therefore, if merged into the next patch, the number of the prototype coefficients remains limited, and the distances of these approximated prototypes to points of the next patch can be calculated using the original equations. This way, only a fraction of the inter-patch distances needs to be considered.

Construction of Extended Patches. Let W_t be a set of optimal relational prototypes gained in a step t. Assume N_t denotes the index set of all points included in the union of a k-approximation of W_t pointing onto elements of the dissimilarity matrix D. The extended patch P_t^* is then characterized by the distance matrix

$$P_t^* = \begin{pmatrix} d(N_{t-1}) & d(N_{t-1}, P_t) \\ \hline d(N_{t-1}, P_t)^T & P_t \end{pmatrix}$$

where

$$d(N_{t-1}) = (d_{uv})_{u,v \,\in\, N_{t-1}} \,\in\, \mathbb{R}^{n_t \times n_t}$$

$$d(N_{t-1}, P_t) = (d_{uv})_{u \,\in\, N_{t-1}, v=(t-1)\cdot p+1,\ldots,t\cdot p} \,\in\, \mathbb{R}^{n_t \times p}$$

denote the inter-distances of points from the k-approximation and the distances between points from the k-approximation and current patch points, respectively. The size n_t is bounded by $|W_t| \cdot k$.

Integrating Multiplicities. The original Relational Neural Gas method has to be modified to handle datapoints v_j equipped with multiplicities m_j which are given by the size of the receptive fields divided by k. Incorporating multiplicities into the cost function yields the update rule

$$\bar{\alpha}_{ij} = \frac{m_j \cdot h_\lambda(k_i(v_j))}{\sum_t m_t \cdot h_\lambda(k_i(v_t))}$$

for prototype coefficients. The computation of distances is not changed.

Patch Relational Neural Gas. Assembling the pieces, we obtain:

Algorithm

Cut the first Patch P_1
Apply Relational NG on $P_1 \longrightarrow$ Relational prototypes W_1
Use k-Approximation on $W_1 \longrightarrow$ Index set N_1
Update Multiplicities m_j according to the receptive fields

Repeat for $t = 2, \ldots, n_p$
    Cut patch P_t
    Construct Extended Patch P_t^* using P_t and index set N_{t-1}
    Apply modified RNG with Multiplicities $\longrightarrow$ Relational prototypes W_t
    Use k-Approximation on $W_t \longrightarrow$ Index set N_t
    Update Multiplicities m_j according to the receptive fields

Return k-approximation of final prototypes N_{n_p}

Complexity. Obviously, the size of extended patches is bounded by the size of the new patch read from the distance matrix and the distances of the at most $k \cdot n$ points representing the n prototypes of the last run by their k approximation. Assume a bounded extended patch size p independent of the number of datapoints, as it would be the case when the patch size is chosen according to memory limitations. The algorithm then works only on $\mathcal{O}(\frac{m}{p} \cdot p^2) = \mathcal{O}(m \cdot p) = \mathcal{O}(m)$ entries of the dissimilarity matrix, compared to $\mathcal{O}(m^2)$ in the original Median NG method. Moreover, the algorithm uses at most $\mathcal{O}(p^2) = const$ entries at a specific point in time.

In case of fixed patch size, also the time complexity is linear, because the Median NG step is $\mathcal{O}(p^2)$ what results in $\mathcal{O}(p^2 \cdot \frac{m}{p}) = \mathcal{O}(p \cdot m) = \mathcal{O}(m)$, an advantage compared to the $\mathcal{O}(m^2)$ time complexity of the original Median NG. Further, the algorithm can be run in a single pass over the data.

These advantages in space and time complexity are obtained by an approximation of the prototypes. As we will see in experiments, this leads only to a small loss in accuracy.

5   Experiments

Practioners often handle huge datasets whose dissimilarities cannot be hold at once within random-access memory due to the sheer amount of data ($\mathcal{O}(m^2)$). At that point, Patch Relational NG comes into play providing a single pass technique based on patches that can be chosen in accordance to the available random-access memory. To show the overall performance of the proposed method, we have chosen some representative dissimilarity datasets. Due to limited computing power and hardware available, the chosen datasets do not represent real-life huge datasets, they should be understood as a proof-of-concept that nevertheless can instantly be transfered to the real problems.

We evaluate the clustering results by means of the classification error for supervised settings, whereby class labels are obtained by posterior labeling of prototypes. Note, however, that the goal of the algorithms is meaningful clustering of data based on a chosen similarity measure and cost function. Hence, the classification error gives only a hint about the quality of the clustering, depending on whether the class labels are compatible to the data clusters and chosen metric or not. We accompany this supervised evaluation be the standard quantization error of the clustering.

For all experiments the initial neighborhood range λ_0 is chosen as $n/2$ with n the number of neurons used. The neighborhood range $\lambda(t)$ is decreased exponentially with the number of adaptation steps t according to $\lambda(t) = \lambda_0 \cdot (0.01/\lambda_0)^{t/t_{\max}}$ (cf. [12]). The value $t_{\max}$ is chosen as the number of epochs.

5.1   Synthetic Dataset

To analyze the relation between the number of patches and the quantization error on one hand, and the effect of k-approximation of relational prototypes on the other hand, an artificial dataset from [3] was taken. It consists of 1250 datapoints in the Euclidean plane gained from three Gaussian clusters.

Effect of k-Approximation. For an empirical study of the effect of k-approximation on the quantization error, we trained 50 neurons with the original Relational NG for 100 epochs, i.e. on average every neuron represents 25 datapoints. On the outcoming relational neurons, k-Approximation for $k = 1, \ldots, 20$ were applied. Figure 1 shows a comparison of the quantization errors yielded with the different approximations to the quantization error gained by the original relational neurons. For each step the average over 10 runs is reported.

As expected, the quantization error decreases with higher numbers k of datapoints used to approximate each relational neuron. Concerning the patch approach, applying a k-approximation to the relational neurons of each patch

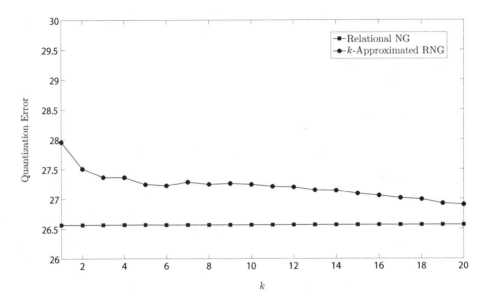

Fig. 1. Quantization error (i.e. $E(W)$ for $\lambda \to \infty$) of original relational neurons compared to different k-approximations on a synthetic dataset

clearly results in a loss of accuracy depending on the choice of parameter k. But as can be seen later on, even with k-approximation the quality of the results is still convincing.

Effect of Patch Sizes. Analyzing the relation between the number of patches chosen and the quantization error, we trained median and relational NG with 20 neurons for 50 epochs. The results presented in figure 2 show the quantization error averaged over 10 runs for each number of patches. As expected, the quantization error increases with the number of patches used. But compared to the Median Patch NG approach the presented Patch Relational NG performs very well with only a small loss even for a larger number of patches used.

5.2   Chicken Pieces Silhouettes Dataset

The task is to classify 446 silhouettes of chicken pieces into the categories wing, back, drumstick, thigh and back, breast. Data silhouettes are represented as a string of the angles of consecutive tangential pieces of length 20, including appropriate scaling. Strings are compared using a (rotation invariant) edit distance, where insertions/deletions cost 60, and the angle difference is taken otherwise.

For training we used 30 neurons. For Patch Median NG the dataset was divided into 4 patches, i.e. a patch size of around 111 datapoints. The results reported in Table 1 are gained from a repeated 10-fold stratified crossvalidation averaged over 100 repetitions and 100 epochs per run. The k-approximation for Patch Relational NG was done with $k = 3$.

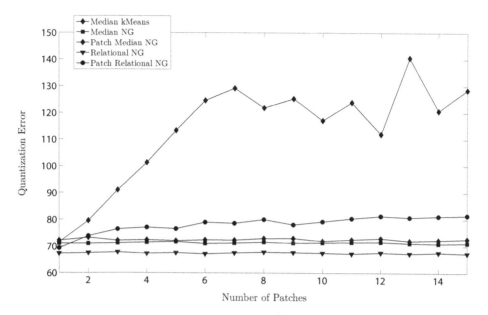

Fig. 2. Quantization error for different patch sizes on a synthetic dataset

Table 1. Classification accuracy on Chicken Pieces Dataset gained from repeated 10-fold stratified crossvalidation over 100 repetitions, four patches were used

Accuracy on Chicken Pieces Dataset

	Relational NG	Patch Relational NG	Median Batch NG	Patch Median NG	Median k-Means
Mean	84.7	85.4	66.4	68.8	72.9
StdDev	1.0	1.1	1.9	2.3	1.7

5.3   Protein Classification

The evolutionary distance of 226 globin proteins is determined by alignment as described in [13]. These samples originate from different protein families: hemoglobin-α, hemoglobin-β, myoglobin, etc. Here, we distinguish five classes as proposed in [6]: HA, HB, MY, GG/GP, and others. Table 2 shows the class distribution of the dataset.

For training we used 20 neurons. For Patch Median NG the dataset was divided into 4 patches, i.e. a patch size of around 57 datapoints. The results reported in Table 3 are gained from a repeated 10-fold stratified crossvalidation averaged over 100 repetitions and 100 epochs per run.

Despite the small size of this dataset – acting more as a proof-of-concept example – the results clearly show a good performance of Patch Median NG.

Table 2. Class Statistics of the Protein Dataset

Class No.	Count	Percentage
HA	72	31.86%
HB	72	31.86%
MY	39	17.26%
GG/GP	30	13.27%
Others	13	5.75%

Table 3. Classification accuracy on Protein Dataset gained from repeated 10-fold stratified crossvalidation over 100 repetitions, four patches were used

Accuracy on Protein Dataset

	Relational NG	Patch Relational NG	Median Batch NG	Patch Median NG	Median k-Means
Mean	92.62	92.61	79.9	77.7	80.6
StdDev	0.92	0.88	1.5	2.4	1.3

Nevertheless, the price of reduced accuracy is obvious, but faster computation and less space requirements are gained in return. The k-approximation for Patch Relational NG was done with $k = 3$.

5.4   Wisconsin Breast Cancer

The Wisconsin breast cancer diagnostic database is a standard benchmark set from clinical proteomics [17]. It consists of 569 data points described by 30 real-valued input features: digitized images of a fine needle aspirate of breast mass are described by characteristics such as form and texture of the cell nuclei present in the image. Data are labeled by two classes, benign and malignant.

Dissimilarities were derived by applying the Cosine Measure

$$d_{cos}(v_i, v_j) = 1 - \frac{v_i \cdot v_j}{\|v_i\|_2 \cdot \|v_j\|_2}.$$

We trained 40 neurons for 100 epochs. As result the accuracy on the test set for a repeated 10-fold stratified crossvalidation averaged over 100 runs is reported. The number of patches chosen for Patch Median NG and Patch Relational NG was 5, i.e. around 114 datapoints per patch. The k-approximation for Patch Relational NG was done with $k = 2$.

Also on this dataset, Patch Relational NG acts merely worse than the original Relational NG. Though, the reduction in accuracy is clearly observable.

5.5   Chromosome Images Dataset

The Copenhagen chromosomes database is a benchmark from cytogenetics. A set of 4200 human chromosomes from 22 classes (the autosomal chromosomes) are represented by the grey levels of their images. These images were transferred

Table 4. Classification accuracy on Wisconsin Breast Cancer Dataset with Cosine Measure gained from repeated 10-fold stratified crossvalidation over 100 repetitions, five patches and a 2-approximation were used

Accuracy on Wisconsin Breast Cancer Dataset

	Relational NG	Patch Relational NG	Median Batch NG	Patch Median NG	Median k-Means
Mean	95.0	94.8	94.7	94.4	94.6
StdDev	0.6	0.7	0.7	0.7	0.7

Table 5. Classification accuracy on Copenhagen Chromosome Image Dataset gained from repeated 2-fold stratified crossvalidation over 10 repetitions, 10 patches and a 3-approximation were used

Accuracy on Copenhagen Chromosome Image Dataset

	Relational NG	Patch Relational NG	Median Batch NG	Patch Median NG	Median k-Means
Mean	89.6	87.0	80.0	67.9	77.1
StdDev	0.6	0.8	1.4	3.1	2.2

to strings representing the profile of the chromosome by the thickness of their silhouettes. Strings were compared using edit distance with substitution costs given by the signed difference of the entries and insertion/deletion costs given by 4.5 [14]. The methods have been trained using 60 neurons for 100 epochs. As result the accuracy on the test set for a repeated 2-fold stratified crossvalidation averaged over 10 runs is reported. The number of patches chosen for Patch Median NG and Patch Relational NG was 10, i.e. around 420 datapoints per patch. The k-approximation for Patch Relational NG was done with $k = 3$.

Also on this dataset, Patch Relational NG acts well. Though, the reduction in accuracy is clearly observable.

6 Summary

In this paper, we proposed a special computation scheme, based on Relational Neural Gas, that allows to process huge dissimilarity datasets by a single pass technique of fixed sized patches. The patch size can be chosen to match the given memory constraints. As explained throughout the paper, the proposed patch version reduces the computation and space complexity with a small loss in accuracy, depending on the patch sizes. We further demonstrated the ability of the proposed method on several representative clustering and classification problems. In all experiments, relational adaptation increased the accuracy of Median clustering.

Note that relational and patch optimization are based on a cost function related to NG such that extensions including semisupervised learning and metric adaptation can directly be transferred to this settings. In future work, the method will be applied to more real-world datasets. The patch scheme also opens a way towards parallelizing the method as demonstrated in [2].

References

1. Alex, N., Hammer, B., Klawonn, F.: Single pass clustering for large data sets. In: WSOM (2007)
2. Alex, N., Hammer, B.: Parallelizing single patch pass clustering (submitted, ESANN 2008)
3. Cottrell, M., Hammer, B., Hasenfuss, A., Villmann, T.: Batch and median neural gas. Neural Networks 19, 762–771 (2006)
4. Guha, S., Mishra, N., Motwani, R., O'Callaghan, L.: Clustering Data Streams. In: IEEE Symposium on Foundations of Computer Science, pp. 359–366 (2000)
5. Guha, S., Rastogi, R., Shim, K.: CURE: an efficient clustering algorithm for large datasets. In: Proceedings of ACM SIGMOD International Conference on Management of Data, pp. 73–84 (1998)
6. Haasdonk, B., Bahlmann, C.: Learning with distance substitution kernels. In: Rasmussen, C.E., Bülthoff, H.H., Schölkopf, B., Giese, M.A. (eds.) DAGM 2004. LNCS, vol. 3175, pp. 220–227. Springer, Heidelberg (2004)
7. Hammer, B., Hasenfuss, A.: Relational Neural Gas. In: Hertzberg, J., Beetz, M., Englert, R. (eds.) KI 2007. LNCS (LNAI), vol. 4667, pp. 190–204. Springer, Heidelberg (2007)
8. Hartigan, J.A.: Clustering Algorithms. Wiley, Chichester (1975)
9. Jin, R., Goswami, A., Agrawal, G.: Fast and Exact Out-of-Core and Distributed K-Means Clustering. Knowledge and Information System (to appear)
10. Kohonen, T.: Self-Organized formation of topologically correct feature maps. Biological Cybernetics 43, 59–69 (1982)
11. Kohonen, T., Somervuo, P.: How to make large self-organizing maps for nonvectorial data. Neural Networks 15, 945–952 (2002)
12. Martinetz, T., Berkovich, S., Schulten, K.: 'Neural gas' network for vector quantization and its application to time series prediction. IEEE Transactions on Neural Networks 4(4), 558–569 (1993)
13. Mevissen, H., Vingron, M.: Quantifying the local reliability of a sequence alignment. Protein Engineering 9, 127–132 (1996)
14. Neuhaus, M., Bunke, H.: Edit distance based kernel functions for structural pattern classification. Pattern Recognition 39(10), 1852–1863 (2006)
15. Prudent, Y., Ennaji, A.: An incremental growing neural gas learns topology. In: IJCNN 2005 (2005)
16. Wang, W., Yang, J., Muntz, R.R.: STING: a statistical information grid approach to spatial data mining. In: Proceedings of the 23rd VLDB Conference, pp. 186–195 (1997)
17. Wolberg, W.H., Street, W.N., Heisey, D.M., Mangasarian, O.L.: Computer-derived nuclear features distinguish malignant from benign breast cytology. Human Pathology 26, 792–796 (1995)
18. Zhang, T., Ramakrishnan, R., Livny, M.: BIRCH: an efficient data clustering method for very large databases. In: Proceedings of the 15th ACM SIGACT-SIGMOD-SIGART Symposium on Principles of Database Systems, pp. 103–114 (1996)

The Block Generative Topographic Mapping

Rodolphe Priam[1], Mohamed Nadif[2], and Gérard Govaert[3]

[1] LMA Poitiers UMR 6086, Université de Poitiers,
86962 Futuroscope Chasseneuil, France
[2] CRIP5 EA N°2517, Université Paris Descartes,
UFR de Mathématiques et Informatique, 75006 Paris, France
[3] Heudiasyc UMR 6599, UTC, BP 20529, 60205 Compiègne, France

Abstract. This paper presents a generative model and its estimation allowing to visualize binary data. Our approach is based on the Bernoulli block mixture model and the probabilistic self-organizing maps. This leads to an efficient variant of Generative Topographic Mapping. The obtained method is parsimonious and relevant on real data.

1 Introduction

Linear methods for exploratory visualization [1] are very powerful and contribute effectively to data analysis every days, but large datasets require new efficient methods. Indeed, the algorithms based on the matricial decomposition become useless for large matrices; moreover, the construction of many maps due to high-dimensionality makes the task of interpretation difficult from the information disseminated on the different maps. Finally a great quantity of data implies a great quantity of information to be synthesized and complex relations between individuals and studied variables. It is then relevant, in this context, to use a self-organizing map (SOM) of Kohonen [2]. SOM is a clustering method with a vicinity constraint on the cluster centers to give a topological sense to the ob-tained final partition. The SOM can be seen like an alternative of the k-means algorithm integrating a topological constraint on the centers. Bishop et al. [3] has re-formulated SOM within a probabilistic setting to give the Generative To-pographic Mapping (GTM). GTM is a method similar to the self-organizing map with constraints of vicinity embedded in a mixture model of gaussian densities. In contrast to SOM, GTM is based on a well-defined criterion; the model im-plements an EM algorithm [4] which guarantees the convergence. Recently, to tackle the visualization of binary data, we have proposed a variant of GTM based on the classical Bernoulli mixture model [5]. The obtained results are encour-aging but when the number of parameters increases with the high-dimensional data, the projection is therefore problematic. To cope with this problem, we propose in this work to use a parsimonious model in order to overcome the high-dimensionality problem.

When the data matrix $\mathbf{x}$ is defined on a set I of objects (rows, observations) and a set J of variables (columns, attributes), the block clustering methods, in contrast to the classical clustering methods, consider the two sets I and J

L. Prevost, S. Marinai, and F. Schwenker (Eds.): ANNPR 2008, LNAI 5064, pp. 13–23, 2008.

simultaneously [6],[7,8],[9]. Recently, these kind of methods were embedded in the mixture approach [10],[11],[12] and a parsimonious model called *Block Latent Model* has been proposed [13,14]. The developed hard and soft algorithms appeared more profitable than the clustering applied separately on I and J [15]. For these reasons, we propose to tackle the problem of visualization of I, by combining the block mixture model and the probabilistic self-organizing maps. This leads to propose a new generative topographic model.

This paper is organized as follows. In Section 2, to give the necessary background of the block clustering approach under the mixture model, we review the block latent model. In Section 3, we focus on the binary data and we propose a *Block Generative Topographic Mapping* based on a block Bernoulli model. In Section 4 devoted to the numerical experiments, we illustrate our method with three binary benchmarks. Finally, the last section summarizes the main points of work and indicates some perspectives.

Hereafter, the partition $\mathbf{z}$ into g clusters of a sample I will be represented by the classification matrix $(z_{ik}, i = 1, \ldots, n, k = 1, \ldots, g)$ where $z_{ik} = 1$ if i belongs to cluster k and 0 otherwise. A similar notation will be used for a partition $\mathbf{w}$ into m clusters of the set J. Moreover, to simplify the notation, the sums and the products relating to rows, columns, row clusters and column clusters will be subscripted respectively by the letters i, j, k and ℓ, without indicating the limits of variation which will be implicit. So, for example, the sum $\sum_i$ stands for $\sum_{i=1}^{n}$, and $\sum_{i,j,k,\ell}$ stands for $\sum_{i=1}^{n} \sum_{j=1}^{d} \sum_{k=1}^{g} \sum_{\ell=1}^{m}$.

2   The Latent Block Model

2.1   Block Clustering

In the following, the $n \times d$ matrix data is defined by $\mathbf{x} = \{(x_{ij}); i \in I \text{ and } \in J\}$ where $x_{ij} \in \{0, 1\}$. The aim of block clustering is to try to summary this matrix by homogeneous blocks. This problem can be studied under the simultaneous partition approach of two sets I and J into g and m clusters respectively. Govaert [7,8] has proposed several algorithms which perform block clustering on contingency tables, binary, continuous and categorical data. These algorithms consist in optimizing a criterion $E(\mathbf{z}, \mathbf{w}, \mathbf{a})$, where $\mathbf{z}$ is a partition of I into g clusters, $\mathbf{w}$ is a partition of J into m clusters and $\mathbf{a}$ is a $g \times m$ matrix which can be viewed as a summary of the data matrix $\mathbf{x}$. A more precise definition of this summary and criterion E will depend on the nature of data. The search of the optimal partitions $\mathbf{z}$ and $\mathbf{w}$ was made using an iterative algorithm. This one is based on the alternated k-means with appropriate metric applied on reduced intermediate $g \times d$ and $n \times m$ matrices. In [13,14], these methods were modeled in the mixture approach. Hard and soft algorithms were then developed. Efficient and scalability are the advantages of these new methods. Next, we review this approach.

2.2   Definition of the Model

Some of the most popular heuristic clustering methods can be viewed as approximate estimations of probability models. For instance, the inertia criterion optimized by the k-means algorithm corresponds to the hypothesis of a population arising from a gaussian mixture. For the classical mixture model, the probability density function (pdf) of a mixture sample $\mathbf{x} = (x_1, \ldots, x_n)$ can be also written [13] $f(\mathbf{x}; \boldsymbol{\theta}) = \sum_{\mathbf{z} \in \mathcal{Z}} p(\mathbf{z}; \boldsymbol{\theta}) f(\mathbf{x}|\mathbf{z}; \boldsymbol{\theta})$ where $\mathcal{Z}$ denotes the set of all possible assignments $\mathbf{z}$ of I into g clusters, $p(\mathbf{z}; \boldsymbol{\theta}) = \prod_{i,k} p_k^{z_{ik}}$ and $f(\mathbf{x}|\mathbf{z}; \boldsymbol{\theta}) = \prod_{i,k} \varphi(x_i; \alpha_k)^{z_{ik}}$. In the context of the block clustering problem, this formulation can be extended to propose a latent block model defined by the following pdf $f(\mathbf{x}; \boldsymbol{\theta}) = \sum_{\mathbf{u} \in U} p(\mathbf{u}; \boldsymbol{\theta}) f(\mathbf{x}|\mathbf{u}; \boldsymbol{\theta})$ where U denotes the set of all possible assignments of $I \times J$, and $\boldsymbol{\theta}$ is the parameter of this mixture model.

In restricting this model to a set of assignments of $I \times J$ defined by a product of assignments of I and J, assumed to be independent, we obtain the following decomposition

$$f(\mathbf{x}; \boldsymbol{\theta}) = \sum_{(\mathbf{z}, \mathbf{w}) \in \mathcal{Z} \times \mathcal{W}} p(\mathbf{z}; \boldsymbol{\theta}) p(\mathbf{w}; \boldsymbol{\theta}) f(\mathbf{x}|\mathbf{z}, \mathbf{w}; \boldsymbol{\theta}),$$

where $\mathcal{Z}$ and $\mathcal{W}$ denote the sets of all possible assignments $\mathbf{z}$ of I and $\mathbf{w}$ of J. Now, as in latent class analysis, the $n \times d$ random variables generating the observed x_{ij} cells are assumed to be independent once $\mathbf{z}$ and $\mathbf{w}$ are fixed; we then have

$$f(\mathbf{x}|\mathbf{z}, \mathbf{w}; \boldsymbol{\theta}) = \prod_{i,j,k,\ell} \varphi(x_{ij}; \alpha_{k\ell})^{z_{ik} w_{j\ell}},$$

where $\varphi(.; \alpha_{k\ell})$ is a pdf defined on the real set $\mathbb{R}$ and $\alpha_{k\ell}$ an unknown parameter. The parameter $\boldsymbol{\theta}$ is formed by $\boldsymbol{\alpha} = (\alpha_{11}, \ldots, \alpha_{gm})$, $\mathbf{p}$ and $\mathbf{q}$; $\mathbf{p} = (p_1, \ldots, p_g)$ and $\mathbf{q} = (q_1, \ldots, q_m)$ are the vectors of probabilities p_k and q_ℓ that a row and a column belong to the kth component and to the ℓth component respectively.

For instance, for binary data, we obtain a Bernoulli latent block model defined by the following pdf

$$f(\mathbf{x}; \boldsymbol{\theta}) = \sum_{(\mathbf{z}, \mathbf{w}) \in \mathcal{Z} \times \mathcal{W}} \prod_{i,k} p_k^{z_{ik}} \prod_{j,\ell} q_\ell^{w_{j\ell}} \prod_{i,j,k,\ell} (\alpha_{k\ell})^{x_{ij}} (1 - \alpha_{k\ell})^{1 - x_{ij}},$$

where $x_{ij} \in \{0, 1\}$, and $\alpha_{k\ell} \in (0, 1)$. Using this block model is dramatically more parsimonious than using a classical mixture model on each set I and J: for instance, with $n = 1000$ objects and $d = 500$ variables and equal class probabilities $p_k = 1/g$ and $q_\ell = 1/m$, if we need to cluster the binary data matrix into $g = 4$ clusters of rows and $m = 3$ clusters of columns, the Bernoulli latent block model will involve the estimation of 12 parameters $\boldsymbol{\alpha} = (\alpha_{k\ell}, k = 1, \ldots, 4, \ell = 1, \ldots, 3)$, instead of $(4 \times 500 + 3 \times 1000)$ parameters with two Bernoulli mixture models applied on I and J separately.

2.3   Estimation of the Parameters

Now we focus on the estimation of an optimal value of $\boldsymbol{\theta}$ by the maximum likelihood approach associated to this block mixture model. For this model, the complete data are taken to be the vector $(\mathbf{x}, \mathbf{z}, \mathbf{w})$ where unobservable vectors $\mathbf{z}$ and $\mathbf{w}$ are the labels; the classification log-likelihood

$$L_C(\mathbf{z}, \mathbf{w}, \boldsymbol{\theta}) = L(\boldsymbol{\theta}; \mathbf{x}, \mathbf{z}, \mathbf{w}) = \log f(\mathbf{x}, \mathbf{z}, \mathbf{w}; \boldsymbol{\theta})$$

can then be written

$$L_C(\mathbf{z}, \mathbf{w}, \boldsymbol{\theta}) = \sum_{i,k} z_{ik} \log p_k + \sum_{j,\ell} w_{j\ell} \log q_\ell + \sum_{i,j,k,\ell} z_{ik} w_{j\ell} \log \varphi(x_{ij}; \alpha_{k\ell}).$$

The EM algorithm [4] maximizes the log-likelihood $L_M(\boldsymbol{\theta})$ w. r. to $\boldsymbol{\theta}$ iteratively by maximizing the conditional expectation of the complete data log-likelihood $L_C(\mathbf{z}, \mathbf{w}, \boldsymbol{\theta})$ w. r. to $\boldsymbol{\theta}$ given a previous current estimate $\boldsymbol{\theta}^{(t)}$ and the observed data $\mathbf{x}$

$$Q(\boldsymbol{\theta}, \boldsymbol{\theta}^{(t)}) = \sum_{i,k} c_{ik}^{(t)} \log p_k + \sum_{j,\ell} d_{j\ell}^{(t)} \log q_\ell + \sum_{i,j,k,\ell} e_{ikj\ell}^{(t)} \log \varphi(x_{ij}; \alpha_{k\ell}), \quad (1)$$

with

$$c_{ik}^{(t)} = P(Z_{ik} = 1 | \boldsymbol{\theta}^{(t)}, \mathbf{X} = \mathbf{x}),$$
$$d_{j\ell}^{(t)} = P(W_{j\ell} = 1 | \boldsymbol{\theta}^{(t)}, \mathbf{X} = \mathbf{x}),$$
$$e_{ikj\ell}^{(t)} = P(Z_{ik} W_{j\ell} = 1 | \boldsymbol{\theta}^{(t)}, \mathbf{X} = \mathbf{x}),$$

where the upper case letters $\mathbf{X}$, Z_{ik} and $W_{j\ell}$ denote the random variables.

Unfortunately, difficulties arise owing to the dependence structure among the variables X_{ij} of the model, and more precisely, to the determination of $e_{ikj\ell}^{(t)}$. To solve this problem a variational approximation by the product $c_{ik}^{(t)} d_{j\ell}^{(t)}$ and a use of the Generalized EM algorithm (GEM) provide a good solution in the clustering and estimation contexts [14].

Next we develop the *Generative Topographic Mapping* which is based on a constrained block Bernoulli mixture whose parameters can be optimized by using a Generalized EM algorithm.

3   Block Generative Topographic Mapping

The Generative Topographic Mapping is a method similar to SOM but based on a constrained gaussian mixture density estimation. The clusters are typically arranged in a regular grid, which is the latent discretized space. The parameters are parameterized as a linear combination of g vectors of h smooth nonlinear basis functions ϕ evaluated on g coordinates of a rectangular grid $\{s_k\}_{k=1}^{k=g}$, so for $k = 1, \cdots, g$ we note

$$\xi_k = \Phi(s_k) = (\phi_1(s_k), \phi_2(s_k), \cdots, \phi_h(s_k))^T,$$

where each basis function ϕ is a kernel-like function,

$$\phi(s_k) = exp\left(-\frac{||s_k - \mu_\phi||^2}{2\nu_\phi^2}\right),$$

with $\mu_\phi \in \mathbb{R}^2$ a mean center and ν_ϕ a standard deviation. More formally, we parameterize the $\alpha_{k\ell}$'s of the block latent model by using the latent space projected into a higher space of h dimensions and we obtain m new h-dimensional unknown vectors noted w_ℓ to be estimated. To keep the dependence on ℓ and k of $\alpha_{k\ell}$, we use the inner product $w_\ell^T \xi_k$ which is then normalized to a probability by the sigmoid function $\sigma(.)$ as a parameter of the Bernoulli pdf. With this formulation, the $g \times m$ matrix $\boldsymbol{\alpha}$ is replaced by the $h \times m$ matrix $\boldsymbol{\Omega} = [w_1|w_2|\cdots|w_m]$. As h is small in practice, as several tens, the model remains parsimonious. In the previous example where the binary data consists of 1000 rows and 500 columns, we end to about several hundred $h \times m$ parameters because h is typically less than 40 and m less than 10. The number of parameters is still less than in the case of a classical mixture approach applied to the both sets separately. Our model has a good foundation to avoid overfitting and its estimation may be less prone to fall into local optima thanks to the small number of parameters: alternative models have a linear increasing of the number of their parameters when the dimension of the data space becomes higher, contrary to the Block GTM. The following figure 1 shows how the discretized plane becomes a non linear space of probability with the constraints of vicinity.

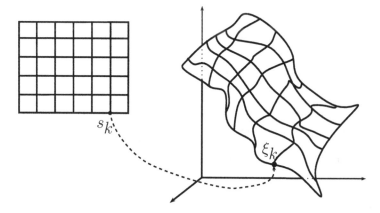

Fig. 1. The graphic illustrates the parameterization of the non linear sigmoid with transformation from a bidimensional Euclidean space to a space of parametric probabilities. In the left the rectangular mesh of the s_k's coordinates is drawn, and in the right the distribution space from φ. Each coordinate of the mesh s_k, $k = 1, \cdots, g$, is mapped in order to become a Bernoulli pdf by writing $\sigma(w_\ell^T \xi_k)$, $\ell = 1, \cdots, m$.

The maximization of the new expression of (1) depending on $\boldsymbol{\Omega}$ can also be performed by the alternated maximization of conditional expectations $Q(\boldsymbol{\theta}, \boldsymbol{\theta}^{(t)}|\mathbf{d})$ and $Q(\boldsymbol{\theta}, \boldsymbol{\theta}^{(t)}|\mathbf{c})$ [14]. When the proportions are supposed equal, the two criteria take the following form

$$Q(\boldsymbol{\theta}, \boldsymbol{\theta}^{(t)}|\mathbf{d}) = \sum_{i,k} c_{ik}^{(t)} \left\{ \sum_\ell u_{i\ell} w_\ell^T \xi_k - d_\ell \log(1 + e^{w_\ell^T \xi_k}) \right\},$$
$$Q(\boldsymbol{\theta}, \boldsymbol{\theta}^{(t)}|\mathbf{c}) = \sum_{j,\ell} d_{j\ell}^{(t)} \left\{ \sum_k v_{jk} w_\ell^T \xi_k - c_k \log(1 + e^{w_\ell^T \xi_k}) \right\},$$

with $u_{i\ell} = \sum_j d^{(t)}_{j\ell} x_{ij}$, $d_\ell = \sum_j d_{j\ell}$ and $c^{(t)}_{ik} \propto \prod_\ell (\sigma(w^T_\ell \xi_k))^{u_{i\ell}} (1 - \sigma(w^T_\ell \xi_k))^{d_\ell - u_{i\ell}}$, and $v_{jk} = \sum_i c^{(t)}_{ik} x_{ij}$, $c_k = \sum_i c_{ik}$ and $d^{(t)}_{j\ell} \propto \prod_k (\sigma(w^T_\ell \xi_k))^{v_{jk}} (1 - \sigma(w^T_\ell \xi_k))^{c_k - v_{jk}}$. A closed form for maximizing these two expectations does not exist yet because of the non linearities from the sigmoid functions, so we use a gradient approach to calculate

$$w^{(t+\frac{1}{2})} = argmax_w \ Q(\boldsymbol{\theta}, \boldsymbol{\theta}^{(t)}|\mathbf{d}) \text{ and } w^{(t+1)} = argmax_w \ Q(\boldsymbol{\theta}, \boldsymbol{\theta}^{(t+\frac{1}{2})}|\mathbf{c}).$$

By derivative of the two criteria, we get the gradient vectors $\mathbf{Q}^{(t)}_u$, $\mathbf{Q}^{(t)}_v$, and the Hessian matrices $\mathbf{H}^{(t)}_u$, $\mathbf{H}^{(t)}_v$. As the Hessian are block diagonal matrices, we are able to increase the log-likelihood at each step of EM, by two consecutive Newton-Raphson ascents for $\ell = 1, \ldots, m$. This leads to the Generalized EM algorithm. If we note $\Phi = [\xi_1|\xi_2|\cdots|\xi_g]^T$ the $g \times h$ matrix of basis functions, each w_ℓ is then expressed as

$$w^{(t+\frac{1}{2})}_\ell = w^{(t)}_\ell + \frac{1}{d_{(\ell)}} \left(\Phi^T G F_\ell \Phi \right)^{-1} \left(\Phi^T C u_\ell - d_{(\ell)} \Phi^T G \alpha_\ell \right),$$
$$w^{(t+1)}_\ell = w^{(t+\frac{1}{2})}_\ell + \frac{1}{d_{(\ell)}} \left(\Phi^T G F_\ell \Phi \right)^{-1} \left(\Phi^T V d_\ell - d_{(\ell)} \Phi^T G \alpha_\ell \right),$$

where $C = (c^{(t)}_{ik})$ is a $g \times n$ matrix of posterior probabilities, $V = (v^{(t)}_{jk})$ a $g \times d$ matrix of sufficient statistics, $G = (c^{(t)}_k)$ and $F_\ell = (\alpha^{(t)}_{k\ell}(1 - \alpha^{(t)}_{k\ell}))$ are $g \times g$ diagonal matrices, $\alpha_\ell = (\alpha^{(t)}_{k\ell})$ a $g \times 1$ vector, $u_\ell = (u^{(t)}_{i\ell})$ a $n \times 1$ vector, $d_\ell = (d^{(t)}_{j\ell})$ a $d \times 1$ vector, and $d_{(\ell)} = d^{(t)}_\ell$ is a scalar.

Finally, for each ℓ, the current parameters $w^{(t)}_\ell \in \mathbb{R}^h$ converges towards the solution $\hat{w}_\ell$. To avoid overfitting and bad numerical solutions, we use a bayesian gaussian prior [16] inducing the bias $-\eta_\ell ||w_\ell||^2/2$ for each w_ℓ. The correction of the estimates is then done by adding $-\eta_\ell w_\ell$ to the gradient and $-\eta_\ell \mathbb{I}_h$ to the diagonal of the Hessian, where $\mathbb{I}_h$ is the h-dimensional identity matrix. The value of the hyperparameters η_ℓ can be manually chosen or estimated.

This Newton-Raphson process in a matrix form sounds like an IRLS [17] step, a crude alternative is a simple gradient with training constant ρ_u and ρ_v instead of the Hessian inverse. Finally, one can notice that the symmetry of the two original mirrored formulas for each side of the matrix is lost because only rows are mapped. Next we illustrate the proposed model on several datasets and demonstrate its good behavior in practice.

4   Numerical Experiments

We experiment our new mapping method on three classical datasets to illustrate the approach. The initialization of the map is done with the help of the first factorial plane from Correspondence Analysis [18], by drawing a mesh over this plane and constructing the initial Bernoulli parameters $\alpha^{(0)}_{k\ell}$ according to this crude clustering.

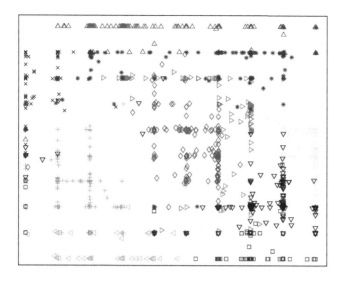

Fig. 2. The Block GTM mapping of the 2000 × 240 image matrix from binarized digits

The first dataset is compound of 2000 binarized images from a database of handwritten digits. For each of the 10 digits '0', '1' and '9', there are 200 images which were digitalized into 240 multi-dimensional vectors, so the data matrix is 2000 × 240 with 10 classes for the row side. No information about class for the column side is provided. The mapping of these data is presented in figure 2 which shows quite good separation of the classes, close to that of the early work of [19,20]. We used a map of size 10 by 10, and 9 nonlinear basis functions plus one intercept and the linear position of the node over the plane, so $h = 12$, and $g = 100$. We choose empirically the value of $m = 20$ as a good number of classes for columns after several manual trials. On the figure 2, the posterior means, $\sum_k \hat{c}_{ik} s_k$, are visualized by a different symbol and color plot for each different class label.

To check more easily the block latent model property and the behavior of the proposed algorithm, two textual datasets are studied with $m = 10$, $g = 81$ and $h = 28$.

The second dataset is compound of 400 selected documents from a textual database of 20000 news. Four newsgroups among the twenty existing ones were kept: "sci.cryp", "sci.space", "sci.med", "soc.religion.christian". For each newsgroup, 100 mails were chosen randomly. The data matrix was then constructed as following. From all the texts, the whole vocabulary of the stemmed words is sought for the entire corpus. Then, a first matrix is constructed with its rows corresponding to texts, and columns corresponding to terms. The value of a cell in this matrix is the number of occurrence of the word in the text. The final list of words is chosen by evaluating mutual information to maximize separation between classes of document thanks to known labels. The final matrix is 400 × 100 with 4 clusters of documents [19,20]. The mapping of this texts on

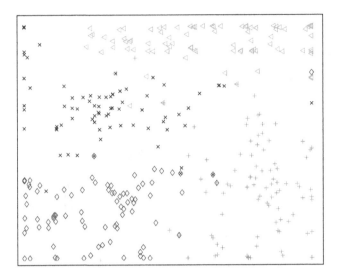

Fig. 3. The Block GTM mapping of the 400×100 textual matrix from four newsgroups

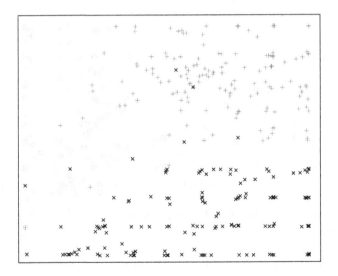

Fig. 4. The Block GTM mapping of the 449×167 textual matrix from three scientific datasets

figure 3 reveals the four topics of discussion which are easily recognizable. The classes are well separated with precise frontiers and we were able to interpret clusters of words too.

The third dataset is a sample of the Classic3 [21] matrix which is a bag of words coding of scientific articles. They come from the three bases Medline, Cisia, Cranfield. We select 450 documents from this file, by randomly drawing

150 documents from each cluster. We select the more frequent words over 30 from all the vocabulary of 4303 terms and we end to a random matrix with approximately 450 rows and 170 columns while discarding the empty rows. One of these matrices was then mapped. Our approach permits us to observe the behavior of the algorithm and we noticed that the solution was stable with only some few outliers badly placed over the plane. The reason is that the clusters are not perfectly separated and that binary coding is not the optimal way to see textual contents. Moreover, we observe quite similar mapping by our binary solution when comparing with the state of art multinomial model-based mapping for the newsgroups file. In future, a contingency table should be modeled to get an even better result in the textual case. Despite this remark, we are still able to visualize the three classes almost perfectly well separated by the non linear mapping in the Figure 4.

5   Conclusion

Considering clustering and visualization within the mixture model approach, we have proposed a new generative self-organizing map for binary data. The proposed Block Generative Topographic Mapping achieves topological organization of the cluster centers basing on a parsimonious block latent model. It counts far fewer parameters than the previously existing models based on a multivariate Bernoulli mixture model [19], a multinomial pLSA [22] or a Bernoulli pLSA [5]. In table 1, when we consider the clustering only on the rows and the proportions p_k and q_ℓ being equal, we report the number of parameters used from the cited models. We note that with our model, the number of parameters increases only with the number of column clusters.

In the visualization context, our variant of GTM gives encouraging results on three applications in two real domains (images and texts). While the linear correspondence analysis is not able to show separately the different clusters over this first plane, our algorithm appears more efficient. Furthermore, the number of parameters of an alternative of the multinomial model in [23,24] for binary data remains the same as for the unconstrained model. So the Block GTM appears clearly as the best candidate to scale for data mining problems.

A first appealing perspective of the model is in domain of textual analysis. Thanks to the clustering of the columns, we are able to map clusters for texts

Table 1. The number of parameters for Bernoulli probabilistic SOMs with $m \ll d$

Model	unconstrained	constrained
Bernoulli mixture model	gd	hd
Bernoulli pLSA	$(n + d)g$	$ng + dh$
Multinomial pLSA	gd	hd
Block latent model	gm	hm

and words together, by evaluating the new heuristic probability that the j-st word belongs to the k-st class, with the following formula

$$d_{jk} \propto \sum_{\ell} d_{j\ell} \alpha_{k\ell}$$

which appears as a crude marginalization over an hidden random variable classifying the columns. The first experiments provide promising results. The vocabulary from each topic appears clearly more probable where each corresponding topic lies on the map. This clustering is very different from the usual one: in the literature, it is usually shown the most probable terms for each cluster of document. A distribution over the map is learned for rows and indirectly for columns, so an original perspective is to construct a non linear *biplot* as [25] by a fully probabilistic and automatic method.

References

1. Lebart, L., Morineau, A., Warwick, K.: Multivariate Descriptive Statistical Analysis. J. Wiley, Chichester (1984)
2. Kohonen, T.: Self-organizing maps. Springer, Heidelberg (1997)
3. Bishop, C.M., Svensén, M., Williams, C.K.I.: Developpements of generative topographic mapping. Neurocomputing 21, 203–224 (1998)
4. Dempster, A., Laird, N., Rubin, D.: Maximum-likelihood from incomplete data via the EM algorithm. J. Royal Statist. Soc. Ser. B. 39, 1–38 (1977)
5. Priam, R., Nadif, M.: Carte auto-organisatrice probabiliste sur données binaires (in french). In: RNTI (EGC 2006 proceedings), pp. 445–456 (2006)
6. Bock, H.: Simultaneous clustering of objects and variables. In: Diday, E. (ed.) Analyse des Données et Informatique, INRIA, pp. 187–203 (1979)
7. Govaert, G.: Classification croisée. In: Thèse d'état, Université Paris 6, France (1983)
8. Govaert, G.: Simultaneous clustering of rows and columns. Control and Cybernetics 24(4), 437–458 (1995)
9. Cottrell, M., Ibbou, S., Letrémy, P.: Som-based algorithms for qualitative variables. Neural Networks 17(89), 1149–1167 (2004)
10. Symons, M.J.: Clustering criteria and multivariate normal mixture. Biometrics 37, 35–43 (1981)
11. McLachlan, G.J., Basford, K.E.: Mixture Models, Inference and applications to clustering. Marcel Dekker, New York (1988)
12. McLachlan, G.J., Peel, D.: Finite Mixture Models. John Wiley and Sons, New York (2000)
13. Govaert, G., Nadif, M.: Clustering with block mixture models. Pattern Recognition 36, 463–473 (2003)
14. Govaert, G., Nadif, M.: An EM algorithm for the block mixture model. IEEE Trans. Pattern Anal. Mach. Intell. 27(4), 643–647 (2005)
15. Govaert, G., Nadif, M.: Block clustering with bernoulli mixture models: Comparison of different approaches. Computational Statistics & Data Analysis 52, 3233–3245 (2008)
16. MacKay, D.J.C.: Bayesian interpolation. Neural Computation 4(3), 415–447 (1992)

17. McCullagh, P., Nelder, J.: Generalized linear models. Chapman and Hall, London (1983)
18. Benzecri, J.P.: Correspondence Analysis Handbook. Dekker, New-York (1992)
19. Girolami, M.: The topographic organization and visualization of binary data using multivariate-bernoulli latent variable models. IEEE Transactions on Neural Networks 20(6), 1367–1374 (2001)
20. Kabán, A., Girolami, M.: A combined latent class and trait model for analysis and visualisation of discrete data. IEEE Trans. Pattern Anal. and Mach. Intell., 859–872 (2001)
21. Dhillon, I.: Co-clustering documents and words using bipartite spectral graph partitioning. In: Seventh ACM SIGKDD Conference, San Francisco, California, USA, pp. 269–274 (2001)
22. Hofmann, T.: Probmap - a probabilistic approach for mapping large document collections. Intell. Data Anal. 4(2), 149–164 (2000)
23. Kaban, A.: A scalable generative topographic mapping for sparse data sequences. In: ITCC 2005: Proceedings of the International Conference on Information Technology: Coding and Computing (ITCC 2005), Washington, DC, USA, vol. I, pp. 51–56. IEEE Computer Society, Los Alamitos (2005)
24. Kaban, A.: Predictive modelling of heterogeneous sequence collections by topographic ordering of histories. Machine Learning 68(1), 63–95 (2007)
25. Priam, R.: CASOM: Som for contingency tables and biplot. In: 5th Workshop on Self-Organizing Maps (WSOM 2005), pp. 379–385 (2005)

Kernel k-Means Clustering Applied to Vector Space Embeddings of Graphs

Kaspar Riesen and Horst Bunke

Institute of Computer Science and Applied Mathematics, University of Bern,
Neubrückstrasse 10, CH-3012 Bern, Switzerland
{riesen,bunke}@iam.unibe.ch

Abstract. In the present paper a novel approach to clustering objects given in terms of graphs is introduced. The proposed method is based on an embedding procedure that maps graphs to an n-dimensional real vector space. The basic idea is to view the edit distance of an input graph g to a number of prototype graphs as a vectorial description of g. Based on the embedded graphs, kernel k-means clustering is applied. In several experiments conducted on different graph data sets we demonstrate the robustness and flexibility of our novel graph clustering approach and compare it with a standard clustering procedure directly applied in the domain of graphs.

1 Introduction

Clustering, a common task in pattern recognition, data mining, machine learning, and related fields, refers to the process of dividing a set of given objects into homogeneous groups. Whereas a large amount of clustering algorithms based on pattern representations in terms of feature vectors have been proposed in the literature (see [1] for a survey), there are only few works where symbolic data structures, and in particular graphs, are used [2]. This is rather surprising since the use of feature vectors implicates two severe limitations. First, as vectors describe a predefined set of features, all vectors in one particular application have to preserve the same length regardless of the size or complexity of the corresponding objects. Furthermore, there is no direct possibility to describe relationships among different parts of an object. However, both constraints can be overcome by graph based object representation [3], as graphs allow us to adapt their size to the complexity of the underlying objects and they also offer a convenient possibility to describe relationships among different parts of an object.

The lack of graph clustering algorithms arises from the fact that there is little mathematical structure in the domain of graphs. For example, computing the sum, the weighted sum, or the product of a pair of entities (which are elementary operations, required in many clustering algorithms) is not possible or not defined in a standardized way in the domain of graphs. However, graph kernels, a relatively novel class of algorithms for pattern recognition, offer an elegant solution to overcome this drawback of graph based representation [4]. Originally, kernel methods have been developed for transforming a given feature space into

L. Prevost, S. Marinai, and F. Schwenker (Eds.): ANNPR 2008, LNAI 5064, pp. 24–35, 2008.

another one of higher dimensionality without computing the transformation explicitly for each individual feature vector. Recently, however, as a fundamental extension, the existence of kernels for symbolic data structures, especially for graphs, has been shown [5].

In the present paper we address the problem of graph clustering by means of kernel k-means clustering. The underlying graph kernel functions are based on a dissimilarity space embedding procedure that has been introduced recently. With a comparison based on four different validation indices we empirically confirm that our novel procedure results in better clusterings when compared to results achieved with an approach directly applied in the domain of graphs.

2   Dissimilarity Space Embedding Graph Kernel

In recent years, kernel methods have become one of the most rapidly emerging sub-fields in intelligent information processing [4]. The fundamental observation in kernel theory is that, given a valid kernel function $\kappa : \mathbb{R}^n \times \mathbb{R}^n \to \mathbb{R}$, there exists a (possibly infinite dimensional) feature space $\mathcal{F}$ endowed with an inner product $\langle ., . \rangle : \mathcal{F} \times \mathcal{F} \to \mathbb{R}$ and a mapping $\phi : \mathbb{R}^n \to \mathcal{F}$ such that $\kappa(\mathbf{x}, \mathbf{x}') = \langle \phi(\mathbf{x}), \phi(\mathbf{x}') \rangle$, for all $\mathbf{x}, \mathbf{x}' \in \mathbb{R}^n$. That is, instead of mapping patterns from $\mathbb{R}^n$ to the feature space $\mathcal{F}$ and computing their scalar product in $\mathcal{F}$, one can simply evaluate the value of the kernel function κ in the original space $\mathbb{R}^n$. This procedure is commonly referred to as *kernel trick*.

What makes kernel theory very interesting is the fact that many algorithms can be kernelized, i.e. reformulated such that only pairwise scalar products rather than explicit vectors are needed[1]. Obviously, by replacing the scalar product by a valid kernel function it is possible to run kernelizable algorithms in a higher dimensional feature vector space $\mathcal{F}$.

Recently, kernel theory has been generalized to the domain of graphs [5]. That is, by means of suitable kernel functions, graphs can be implicitly mapped to vector spaces. Consequently, the whole theory of kernel machines, which has been developed for feature vectors originally, become applicable to graphs. Hence by means of kernel functions one can benefit from both the high representational power of graphs and the rich repository of algorithms available in vector spaces.

Definition 1 (Graph Kernel). *Let $\mathcal{G}$ be a finite or infinite set of graphs, $g_1, g_2 \in \mathcal{G}$, and $\varphi : \mathcal{G} \to \mathbb{R}^n$ a function with $n \in \mathbb{N}$. A graph kernel function is a mapping $\kappa : \mathcal{G} \times \mathcal{G} \to \mathbb{R}$ such that $\kappa(g_1, g_2) = \langle \varphi(g_1), \varphi(g_2) \rangle$.*   □

According to this definition a graph kernel function takes two graphs g_1 and g_2 as arguments and returns a real number that is equal to the result achieved by first mapping the two graphs by a function φ to a vector space $\mathbb{R}^n$ and then computing the scalar product $\langle \varphi(g_1), \varphi(g_2) \rangle$ in $\mathbb{R}^n$. The kernel function $\kappa(g_1, g_2)$ provides us with a shortcut (kernel trick) that eliminates the need for computing $\varphi(.)$ explicitly.

---

[1] Such algorithms together with a kernel function κ are commonly termed *kernel machines*.

The embedding procedure proposed in this paper makes use of graph edit distance. The key idea of graph edit distance is to define the dissimilarity, or distance, of graphs by the minimum amount of distortion that is needed to transform one graph into another. A standard set of distortion operations is given by *insertions*, *deletions*, and *substitutions* of nodes and edges.

Given two graphs, the source graph g_1 and the target graph g_2, the idea of graph edit distance is to delete some nodes and edges from g_1, relabel (substitute) some of the remaining nodes and edges, and insert some nodes and edges in g_2, such that g_1 is finally transformed into g_2. A sequence of edit operations $e_1, \ldots, e_k$ that transform g_1 into g_2 is called an *edit path* between g_1 and g_2. In order to find the most suitable edit path out of all possible edit paths, one introduces a cost for each edit operation, measuring the strength of the corresponding operation. The idea of such cost functions is to define whether or not an edit operation represents a strong modification of the graph. Consequently, the *edit distance* of two graphs is defined by the minimum cost edit path between two graphs. The edit distance of graphs can be computed, for example, by a tree search algorithm [6] or by faster, suboptimal methods which have been proposed recently (e.g. [7]).

The idea underlying our graph embedding method was originally developed for the problem of embedding sets of feature vectors in a dissimilarity space [8]. In this paper we use the extension of this method to the domain of graphs proposed in [9] for the problem of classification and apply it to clustering. Assume we have available a set of graphs $\mathcal{G} = \{g_1, \ldots, g_N\}$ and a graph dissimilarity measure $d(g_i, g_j)$ (in our case graph edit distance). After having selected a set $\mathcal{P} = \{p_1, \ldots, p_n\}$ of $n \leq N$ prototypes from $\mathcal{G}$, we compute the dissimilarity of a given graph $g \in \mathcal{G}$ to each prototype $p \in \mathcal{P}$. This leads to n dissimilarities, $d_1 = d(g, p_1), \ldots, d_n = d(g, p_n)$, which can be arranged in an n-dimensional vector $(d_1, \ldots, d_n)$. In this way we can transform any graph from $\mathcal{G}$ into a vector of real numbers.

Definition 2 (Graph Embedding). *If $\mathcal{G} = \{g_1, \ldots, g_N\}$ is a set of graphs and $P = \{p_1, \ldots, p_n\} \subseteq \mathcal{G}$ is a set of prototype graphs, the mapping $\varphi_n^P : \mathcal{G} \to \mathbb{R}^n$ is defined as the function*

$$\varphi_n^P(g) \mapsto (d(g, p_1), \ldots, d(g, p_n)),$$

where $d(g, p_i)$ is the graph edit distance between graph g and the i-th prototype.

Regarding the graph embedding procedure, the importance of the prototype set $\mathcal{P} = \{p_1, \ldots, p_n\} \subseteq \mathcal{G}$ is obvious. Recently, different prototype selectors have been proposed in the literature which we adopt in this work [8,9]. Note that $\mathcal{P}$ can be an arbitrary set of graphs in principle. However, for the sake of convenience we always use subsets of $\mathcal{G}$ which are obtained by means of prototype selection methods.

Since the computation of graph edit distance is exponential in the number of nodes for general graphs, the complexity of this graph embedding is exponential as well. However, one can use efficient approximation algorithms for graph edit

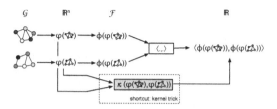

Fig. 1. Graph kernel trick illustrated

distance (e.g. [7] with cubic time complexity). Consequently, given n predefined prototypes the embedding of one particular graph is established by means of n distance computations with polynomial time.

Clearly, the graph embedding procedure described above provides a foundation for a novel class of graph kernels. Based on the mapping φ_n^P, one can define a valid graph kernel κ by computing the standard scalar product of two graph maps in the resulting vector space

$$\kappa(g_i, g_j) = \langle \varphi_n^P(g_i), \varphi_n^P(g_j) \rangle$$

Note that, in contrast to some other kernel methods, the approach proposed in this paper results in an explicit embedding of the considered graphs in a vector space. Hence, not only scalar products, but individual graph maps are available in the target space. We observe that this approach is more powerful than conventional graph kernels as not only kernel machines, but also other non-kernelizable algorithms can be applied to the resulting vector representation. Furthermore, based on the resulting graph maps standard kernel functions for feature vectors in $\mathbb{R}^n$ can be applied, mapping the vector space embedded graphs implicitly into a higher dimensional feature space $\mathcal{F}$. An example is the RBF kernel.

$$\kappa_{RBF}(g_i, g_j) = exp\left(-\gamma ||\varphi_n^P(g_i) - \varphi_n^P(g_j)||^2\right) \text{ , with } \gamma > 0.$$

Obviously, in every kernel machine the scalar product can be replaced by $\kappa(g_i, g_j)$ such that these algorithms can be applied to objects originally given in terms of graphs. In Fig. 1 this procedure is schematically illustrated.

3   Kernel k-Means Clustering

The k-means algorithm [10] is one of the most popular clustering algorithms in pattern recognition and related areas. Let us assume that N objects $O = \{o_1, \ldots, o_N\}$ are given. Starting with an initial set of k cluster centers, i.e. a set of $k < N$ objects $M_k = \{o_{(1)}, \ldots, o_{(k)}\} \subset O$, we assign each of the N objects to the closest cluster centers. Based on this clustering, the cluster centers are recomputed. The two preceding steps, i.e. the assignment of objects to the nearest cluster center and the recomputation of the centers, are repeated until a predefined termination criterion is met (e.g. no reassignment of objects from one cluster to another has taken place during the last iteration).

The initialization of k-means is commonly done with a random selection of k objects. However, in the present paper a deterministic procedure is applied. The set of initial cluster centers M_k is constructed by iteratively retrieving the median of set O minus the objects already selected. The median of set O is the object $\mathbf{o} \in O$ that minimizes the sum of distances to all other objects in O, i.e. $median = argmin_{\mathbf{o}_1 \in O} \sum_{\mathbf{o}_2 \in O} d(\mathbf{o}_1, \mathbf{o}_2)$. Obviously, this procedure initializes k-means with objects situated in, or near, the center of the set O.

K-means clustering makes use of the squared error criterion as an objective function. Formally, the k-means algorithms finds k clusters $C_1, \ldots, C_k$ such that the objective function

$$f\left(\{C_j\}_{j=1}^k\right) = \sum_{j=1}^k \sum_{\mathbf{o}_i \in C_j} d(\mathbf{o}_i, \mathbf{m}_j)$$

is minimized. In this formula, the j-th cluster is denoted by C_j, a clustering by $\{C_j\}_{j=1}^k$, d is an appropriate distance function, and $\mathbf{m}_j$ refers to the mean of cluster C_j.

Note that the objects $\mathbf{o}_i$ can either be graphs or vectors. If the objects are given in terms of graphs, the distance function d is given by the graph edit distance and the mean $\mathbf{m}_j$ of the j-th cluster is defined as the set median graph ($\mathbf{m}_j = argmin_{g_1 \in C_j} \sum_{g_2 \in C_j} d(g_1, g_2)$). In the remainder of the present paper we denote k-means applied to graphs as k-medians. If the objects are given in terms of feature vectors[2], the dissimilarity function d is defined as the Euclidean distance, and $\mathbf{m}_j$ is the mean vector of C_j (i.e. $\mathbf{m}_j = \frac{1}{|C_j|} \sum_{\mathbf{x}_i \in C_j} \mathbf{x}_i$).

A well known drawback of k-means clustering is that the individual clusters C_j need to be spherical in order to achieve satisfactory results. (This drawback directly follows from the minimization of the squared error.) Now let us assume that a function ϕ is given, mapping n-dimensional vectors $\mathbf{x}_i$ into a higher dimensional feature space $\mathcal{F}$. Applying k-means clustering in the resulting feature space $\mathcal{F}$, i.e. finding spherical clusters C_j in $\mathcal{F}$, corresponds to finding (possibly) non-spherical clusters in the original vector space $\mathbb{R}^n$. Hence, this clustering procedure is much more powerful than the conventional k-means algorithm.

The objective function for k-means clustering in the higher dimensional feature space $\mathcal{F}$ can be written as the minimization of

$$f\left(\{C_j\}_{j=1}^k\right) = \sum_{j=1}^k \sum_{\mathbf{x}_i \in C_j} \|(\phi(\mathbf{x}_i) - \mathbf{m}_j\|, \text{ where } \mathbf{m}_j = \frac{\sum_{\mathbf{x}_i \in C_j} \phi(\mathbf{x}_i)}{|C_j|}$$

In fact, it turns out that the k-means algorithm can be written as a kernel machine, i.e. can be reformulated in terms of pairwise scalar products only.

---

[2] Note that the vectors $\mathbf{x}_i$ in the present paper are embedded graphs by means of the embeddding function $\varphi_n^{\mathcal{P}}(.)$

Formally, the squared Euclidean distance $||(\phi(\mathbf{x}) - \mathbf{m}||^2$ between vector $\mathbf{x}$ and the mean $\mathbf{m}$ of cluster C can be rewritten as

$$||\phi(\mathbf{x}) - \mathbf{m}||^2 = \langle \phi(\mathbf{x}), \phi(\mathbf{x}) \rangle + \frac{1}{n^2} \sum_{\mathbf{x}_i \in C} \sum_{\mathbf{x}_j \in C} \langle \phi(\mathbf{x}_i), \phi(\mathbf{x}_j) \rangle - \frac{2}{n} \sum_{\mathbf{x}_i \in C} \langle \phi(\mathbf{x}), \phi(\mathbf{x}_i) \rangle$$

Obviously, now we can replace the scalar products $\langle .,. \rangle$ with a valid vector kernel function κ to represent the scalar product in an implicit feature space $\mathcal{F}$ without explicitly computing the transformation $\phi : \mathbb{R}^n \to \mathcal{F}$. That is, one can apply k-means in an implicitly existing feature space $\mathcal{F}$. The resulting procedure is commonly referred to as kernel k-means clustering.

4 Clustering Validation

In order to compare the novel kernel k-means clustering algorithm proposed in this paper with the conventional k-medians algorithm in the graph domain, we use four different validation indices, viz. DUNN [11], C [12], RAND [13], and the BIPARTITE index. Whereas the two former indices (DUNN and C) do not need any ground truth information, the latter ones (RAND and BIPARTITE) are defined with respect to the class memberships of the underlying objects. Note that all indices can be applied to both graphs and vectors.

4.1 Dunn Index

We define the distance between two clusters C and C' as $d(C, C') = min\{d(\mathbf{o}_i, \mathbf{o}_j) \mid \mathbf{o}_i \in C, \mathbf{o}_j \in C'\}$. The diameter of a cluster C is given by $\varnothing(C) = max\{d(\mathbf{o}_i, \mathbf{o}_j) \mid \mathbf{o}_i, \mathbf{o}_j \in C\}$ and accordingly the maximum diameter of all k clusters is defined by $\varnothing_{max} = max\{\varnothing(C_i) \mid 1 \leqslant i \leqslant k\}$. DUNN index measures the ratio of the minimum distance of two different clusters and the maximum diameter of a cluster. Formally,

$$\text{DUNN} = min\left\{ \frac{d(C_i, C_j)}{\varnothing_{max}} \mid 1 \leqslant i < j \leqslant K \right\}$$

DUNN is considered to be positively-correlated such that higher values indicate higher clustering quality.

4.2 C Index

One defines

$$c(\mathbf{o}_i, \mathbf{o}_j) = \begin{cases} 1 & \text{if } \mathbf{o}_i \text{ and } \mathbf{o}_j \text{ belong to the same cluster} \\ 0 & \text{else} \end{cases}$$

Furthermore, Γ is defined by the sum of all distances of objects belonging to the same cluster, and the number of pairs of objects in the same cluster is denoted by a. Formally,

$$\Gamma = \sum_{i=1}^{n-1} \sum_{j=i+1}^{n} d(\mathbf{o}_i, \mathbf{o}_j) c(\mathbf{o}_i, \mathbf{o}_j) \qquad a = \sum_{i=1}^{n-1} \sum_{j=i+1}^{n} c(\mathbf{o}_i, \mathbf{o}_j)$$

With $min(max)$ we denote the sum of the a smallest (largest) distances $d(\mathbf{o}_i, \mathbf{o}_j)$ where $\mathbf{o}_i \neq \mathbf{o}_j$. The C index is then defined as

$$C = \frac{\Gamma - min}{max - min}$$

Obviously, the numerator of the C index measures how many pairs of objects of the a nearest neighboring pairs belong to the same cluster. The denominator is a scale factor ensuring that $0 \leq C \leq 1$. The smaller the C index value is, the more frequently do pairs with a small distance belong to the same cluster, i.e. the higher is the clustering quality.

4.3   Rand Index

For computing the RAND index we regard all pairs of objects $(\mathbf{o}_i, \mathbf{o}_j)$ with $\mathbf{o}_i \neq \mathbf{o}_j$. We denote the number of pairs $(\mathbf{o}_i, \mathbf{o}_j)$ belonging to the same class and to the same cluster with N_{11}, whereas N_{00} denotes the number of pairs that neither belong to the same class nor to the same cluster. The number of pairs belonging to the same class but not to the same cluster is denoted by N_{10}, and conversely N_{01} represents the number of pairs belonging to different classes but to the same cluster. The RAND index is defined by

$$\text{RAND} = \frac{N_{11} + N_{00}}{N_{11} + N_{00} + N_{01} + N_{10}}$$

RAND index measures the consistency of a given clustering, and therefore higher values indicate better clusterings.

4.4   Bipartite Index

In order to compute the BIPARTITE index, we first define the confusion matrix $\mathbf{M}$. Assume a clustering with k clusters $(C_1, \ldots, C_k)$, and a set of l classes $(\Omega_1, \ldots, \Omega_l)$ are given. Note that $\cup_{i=1}^{k} C_i = \cup_{i=1}^{l} \Omega_i$, i.e. the elements underlying the clustering and the classes are identical. Moreover, we define $k = l$, i.e. the number of clusters is equal to the number of classes[3]. The $k \times k$ confusion matrix is defined by

$$\mathbf{M} = \begin{bmatrix} m_{1,1} & \cdots & m_{1,k} \\ \vdots & \ddots & \vdots \\ m_{k,1} & \cdots & m_{k,k} \end{bmatrix}$$

where $m_{i,j}$ represents the number of elements from class Ω_j occuring in cluster C_i. The problem to be solved with this confusion matrix is to find an optimal assignment of the k clusters to the k classes. Such an optimal assignment maximizes the sum of the corresponding cluster-class values $m_{i,j}$. Formally, one has to find a permuation p of the integers $1, 2, \ldots, k$ maximizing $\sum_{i=1}^{k} m_{ip_i}$. Let p be

---

[3] For the sake of convenience we use k to denote both the number of clusters and the number of classes from now on.

the optimal permuation. The BIPARTITE index (BP index for short) is defined as

$$BP = \frac{\sum_{i=1}^{k} m_{ip_i}}{N}$$

Note that BP gives us the maximum possible classification accuracy of the given clustering. The computation of the BP index can be efficiently accomplished by means of Munkres' algorithm [14].

Note that other validation indices could be also used. However, we feel that a validation based on the four indices proposed covers the different aspects of cluster quality evaluation quite well and we leave an analysis involving additional indices to future work.

5   Experimental Results

The intention of the experimental evaluation is to empirically investigate whether kernel k-means based on the proposed graph embedding outperforms the standard k-medians clustering algorithm in the original graph domain. For our novel approach the most widely used RBF kernel $\kappa_{RBF}(g_i, g_j)$ as defined in Section 2 is used. Class information is available for all of our graph data sets. Therefore the number of clusters k is defined for each data set as the number of classes in the underlying graph set.

5.1   Databases

For our experimental evaluation, six data sets with quite different characteristics are used. The data sets vary with respect to graph size, edge density, type of labels for the nodes and edges, and meaning of the underlying objects. Lacking space we give a short description of the data only. For a more thorough description we refer to [9] where the same data sets are used for the task of classification.

The first database used in the experiments consists of graphs representing distorted letter drawings out of 15 classes (Letter). Next we apply the proposed method to the problem of image clustering, i.e. we use graphs representing images out of two categories (*cups, cars*) from the COIL-100 database [15] (COIL). The third data set is given by graphs representing fingerprint images of the NIST-4 database [16] out of the four classes *arch, left, right*, and *whorl* (Fingerprint). The fourth set is given by the Enzyme data set. The graphs are constructed from the Protein Data Bank [17] and labeled with their corresponding enzyme class labels (*EC 1,..., EC 6*) (Enzymes). The fifth graph set is constructed from the AIDS Antiviral Screen Database of Active Compounds [18]. Graphs from this database belong to two classes (*active, inactive*), and represent molecules with activity against HIV or not (AIDS). The last data set consists of graphs representing webpages [19] that originate from 20 different categories (*Business, Health, Politics, ...*) (Webgraphs).

Each of our graph sets is divided into two disjoint subsets, viz. validation and test set. The validation set is used to determine those meta parameters of the clustering algorithm which cannot be directly inferred from the specific application. For k-medians clustering in the original graph domain only the cost function for graph edit distance has to be validated. For our novel approach, however, there are three additional parameters to tune, viz. the prototype selection method (PS), the number of prototypes n (dimensionality of the vector space $\mathbb{R}^n$), and the parameter γ in the RBF kernel. For each of the four validation indices, these three meta parameters are optimized individually on the validation set. Thereafter the best performing parameter combination is applied to the independent test set.

5.2   Results and Discussion

In Table 1 the clustering validation indices for both the k-medians clustering in the original graph domain (GD) and kernel k-means in the embedding vector space (VS) are given for all test data sets. Regarding the DUNN index we observe that the clustering based on the vector space embedded graphs outperforms the clustering in the original graph domain in three out of six cases, i.e. the clusterings in the vector space are not necessarily better than the clusterings in the original graph domain. This finding can be explained by the fact that DUNN's index is very instable in presence of outliers since only two distances are considered. Regarding the other indices the superiority of our novel approach compared to the reference system becomes obvious. Kernel k-means outperforms the k-medians clustering under C, RAND, and BP on all data sets. That is, with the novel procedure pairs with small distances are more frequently in the same cluster, and the clusterings in the embedding space are more accurate and consistent according to the ground truth.

At first glance the meta parameters to be tuned in our novel approach seem to be a kind of drawback. However, since the k-means algorithm is able to find spherical clusters only, these meta parameters establish a powerful possibility to optimize the underlying vector space embedding with respect to a specific validation index. In Fig. 2 clusterings on the COIL database are illustrated as an example. Note that in these illustrations the different colors (black and white) refer to the cluster assignment, while the different shapes (circle and diamond) reflect the class membership of the respective points.

In Fig. 2 (a) the original graphs and the clustering found by the reference system (k-medians) are projected onto the plane by multidimensional scaling

Table 1. Clustering results on the data sets in the graph domain (GD) and the vector space (VS). Bold numbers indicate superior performance over the other system.

Data Set	Dunn GD	Dunn VS	C GD	C VS	Rand GD	Rand VS	BP GD	BP VS
Letter	0.016	**0.157**	0.419	**0.026**	87.30	**90.76**	22.90	**46.27**
COIL	0.132	**0.142**	0.377	**0.053**	69.01	**76.63**	81.11	**86.67**
Fingerprint	**0.209**	0.057	0.094	**0.017**	32.20	**77.66**	45.20	**68.40**
Enzymes	0.027	**0.036**	0.591	**0.021**	49.23	**72.37**	22.00	**27.00**
AIDS	**0.054**	0.044	0.015	**0.000**	83.06	**83.17**	90.67	**90.73**
Webgraphs	**0.064**	0.042	0.111	**0.016**	88.51	**90.08**	53.72	**53.97**

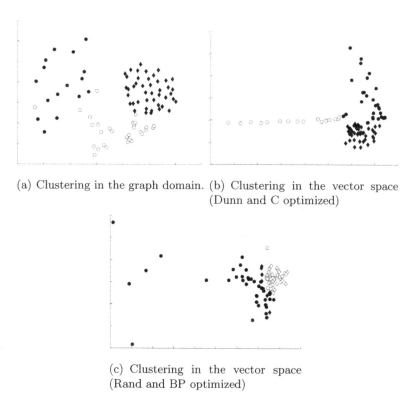

(a) Clustering in the graph domain. (b) Clustering in the vector space
(Dunn and C optimized)

(c) Clustering in the vector space
(Rand and BP optimized)

Fig. 2. Different clusterings of the same data. (Different shapes refer to different classes while different colors refer to different clusters.)

(MDS) [20]. In Fig. 2 (b) and (c) the same graphs processed by different vector space embeddings and kernel k-means clustering in the embedding space are displayed. Fig. 2 (b) shows the vector space embedded graphs and the clustering which is optimized with respect to the validation indices DUNN and C[4]. In this case the underlying vector space is optimized such that the resulting clusters are more compact and better separable than the clusters achieved in the original graph domain (Fig. 2 (a)). In other words, the two clusters of white and black elements can be much better approximated by non-overlapping ellipses than in Fig. 2 (a). Note that for both indices DUNN and C the class membership is not taken into account but only the size and shape of the clusters. In Fig. 2 (c) the embedding and clustering are optimized with respect to RAND and BP. In this case the embedding and clustering are optimized such that spherical clusters are able to seperate the data with a high degree of consistency and accuracy according to ground truth. This can also be seen in Fig. 2 (c) where the shape and color, i.e. class- and cluster membership, correspond to each other more often than in Fig. 2 (a).

---

[4] Coincidentally, for both indices the best performing parameter values are the same.

Summarizing, the embedding process lends itself to a methodology for adjusting a given data distribution such that the clustering algorithm becomes able to achieve good results according to a specific validation criterion. This explains the findings reported in Table 1.

6   Conclusions

In the present paper we propose a procedure for embedding graphs in an n-dimensional vector space by means of prototype selection and graph edit distance. Based on this vector space embedding of graphs, a standard kernel function for feature vectors – the RBF kernel – is applied. This leads to an implicit embedding of the graph maps in a feature space $\mathcal{F}$. These implicit embeddings are then fed into a kernel k-means clustering algorithm. The power of our novel approach is threefold. First, it makes the k-means clustering algorithm available to the graph domain. Because of the lack of suitable procedures for computing the mean of a graph population, only k-medians algorithm has been traditionally applied to graphs. Secondly, by means of the embedding procedure we gain the possibility to apply kernel k-means clustering to data that are not spherically structured, as implicitly assumed by the k-means clustering algorithm. Thirdly, by means of the embedding parameters, the resulting vector space can be adjusted with respect to a specific validation index. The applicability and performance of our novel approach is tested on six different graph sets with four clustering validation indices. According to the DUNN index our novel approach outperforms the reference system in three out of six cases. The other indices C, RAND, and BP indicate that our novel approach outperforms the reference system even on all data sets.

In future work we will study additional clustering validation indices and performance measures, for example, cluster stability [21]. Moreover, other clustering algorithms applicable to vector space embedded graphs will be investigated.

Acknowledgements

This work has been supported by the Swiss National Science Foundation (Project 200021-113198/1).

References

1. Jain, A., Murty, M., Flynn, P.: Data clustering: A review. ACM Computing Surveys 31(3), 264–323 (1999)
2. Englert, R., Glantz, R.: Towards the clustering of graphs. In: Kropatsch, W., Jolion, J. (eds.) Proc.2nd Int.Workshop on Graph Based Representations in Pattern Recognition, pp. 125–133 (2000)
3. Conte, D., Foggia, P., Sansone, C., Vento, M.: Thirty years of graph matching in pattern recognition. Int.Journal of Pattern Recognition and Artificial Intelligence 18(3), 265–298 (2004)

4. Schölkopf, B., Smola, A.: Learning with Kernels. MIT Press, Cambridge (2002)
5. Gärtner, T.: A survey of kernels for structured data. SIGKDD Explorations 5(1), 49–58 (2003)
6. Bunke, H., Allermann, G.: Inexact graph matching for structural pattern recognition. Pattern Recognition Letters 1, 245–253 (1983)
7. Riesen, K., Neuhaus, M., Bunke, H.: Bipartite graph matching for computing the edit distance of graphs. In: Escolano, F., Vento, M. (eds.) GbRPR. LNCS, vol. 4538, pp. 1–12. Springer, Heidelberg (2007)
8. Duin, R., Pekalska, E.: The Dissimilarity Representations for Pattern Recognition: Foundations and Applications. World Scientific, Singapore (2005)
9. Riesen, K., Neuhaus, M., Bunke, H.: Graph embedding in vector spaces by means of prototype selection. In: Escolano, F., Vento, M. (eds.) GbRPR. LNCS, vol. 4538, pp. 383–393. Springer, Heidelberg (2007)
10. MacQueen, J.: Some methods for classification and analysis of multivariant observations. In: Proc. 5th. Berkeley Symp., vol. 1, pp. 281–297. University of California Press 1 (1966)
11. Dunn, J.: Well-separated clusters and optimal fuzzy partitions. Journal of Cybernetics 4, 95–104 (1974)
12. Hubert, L., Schultz, J.: Quadratic assignment as a general data analysis strategy. British Journal of Mathematical and Statistical Psychology 29, 190–241 (1976)
13. Rand, W.: Objective criteria for the evaluation of clustering methods. Journal of the American Statistical Association 66(336), 846–850 (1971)
14. Munkres, J.: Algorithms for the assignment and transportation problems. Journal of the Society for Industrial and Applied Mathematics 5, 32–38 (1957)
15. Nene, S., Nayar, S., Murase, H.: Columbia Object Image Library: COIL-100. Technical report, Department of Computer Science, Columbia University, New York (1996)
16. Watson, C., Wilson, C.: NIST Special Database 4, Fingerprint Database. National Institute of Standards and Technology (1992)
17. Berman, H., Westbrook, J., Feng, Z., Gilliland, G., Bhat, T., Weissig, H., Shidyalov, I., Bourne, P.: The protein data bank. Nucleic Acids Research 28, 235–242 (2000)
18. DTP, DTP.: AIDS antiviral screen (2004), http://dtp.nci.nih.gov/docs/aids/aids_data.html
19. Schenker, A., Bunke, H., Last, M., Kandel, A.: Graph-Theoretic Techniques for Web Content Mining. World Scientific, Singapore (2005)
20. Cox, T., Cox, M.: Multidimensional Scaling. Chapman and Hall, Boca Raton (1994)
21. Kuncheva, L., Vetrov, D.: Evaluation of stability of k-means cluster ensembles with respect to random initialization. IEEE Transactions on Pattern Analysis and Machine Intelligence 28(11), 1798–1808 (2006)

Probabilistic Models Based
on the Π-Sigmoid Distribution

Anastasios Alivanoglou and Aristidis Likas

Department of Computer Science, University of Ioannina
GR 45110, Ioannina, Greece
{aalivano,arly}@cs.uoi.gr

Abstract. Mixture models constitute a popular type of probabilistic neural networks which model the density of a dataset using a convex combination of statistical distributions, with the Gaussian distribution being the one most commonly used. In this work we propose a new probability density function, called the Π-sigmoid, from its ability to form the shape of the letter "Π" by appropriately combining two sigmoid functions. We demonstrate its modeling properties and the different shapes that can take for particular values of its parameters. We then present the Π-sigmoid mixture model and propose a maximum likelihood estimation method to estimate the parameters of such a mixture model using the Generalized Expectation Maximization algorithm. We assess the performance of the proposed method using synthetic datasets and also on image segmentation and illustrate its superiority over Gaussian mixture models.

Keywords: Probabilistic neural networks, mixture models, Π-sigmoid distribution, orthogonal clustering, image segmentation.

1 Introduction

Gaussian mixture models (GMM) are a valuable statistical tool for modeling densities. They are flexible enough to approximate any given density with high accuracy, and in addition they can be interpreted as a soft clustering solution. Thus, they have been widely used in both supervised and unsupervised learning, and have been extensively studied, e.g. [3]. They can be trained through a convenient EM procedure [4] that yields maximum likelihood estimates for the parameters of the mixture. However, it exhibits some weaknesses, the most notable being its lack of interpretability, since it provides spherical (or ellipsoidal in the most general case) shaped clusters that are inherently hard to be understood by humans. It is widely acknowledged that humans prefer solutions in the form of *rectangular shaped* clusters which are directly interpretable. Another weakness of the GMM approach is that it is not efficient when used to model data that are uniformly distributed in some regions.

With the aim to adequately treat the above issues, in this work we propose a new probability distribution called the Π-sigmoid (Πs) distribution. This distribution is obtained as the difference of two translated sigmoid functions and is flexible enough

L. Prevost, S. Marinai, and F. Schwenker (Eds.): ANNPR 2008, LNAI 5064, pp. 36–43, 2008.

to approximate data distributions ranging from Gaussian to uniform depending on the slope of the sigmoids. We also propose a mixture model with Π-sigmoid distributions called Π-sigmoid mixture model (ΠsMM) and show that it is capable of providing probabilistic clustering solutions for the case of rectangular-shaped clusters that are straightforward to transform into an interpretable set of rules. We present a maximum likelihood technique to estimate the parameters of ΠsMM using the Generalized EM algorithm [3,4]. As experimental results indicate, due to its flexibility to approximate both the rectangular uniform and bell-shaped distributions, the ΠsMM provides superior solutions compared to GMMs when the data are not Gaussian.

2 The Π-Sigmoid Distribution

The one-dimensional Π-sigmoid distribution is computed as the difference between two logistic sigmoid functions with the same slope. The logistic sigmoid with slope λ is given by:

$$\sigma(x) = \frac{1}{1 + e^{-\lambda x}} \tag{1}$$

The Π-sigmoid pdf with parameters a, b, λ (with $b > a$) is defined by subtracting two *translated* sigmoids:

$$\Pi s(x) = \left(\frac{1}{b-a}\right)\left[\frac{1}{1 + e^{-\lambda(x-a)}} - \frac{1}{1 + e^{-\lambda(x-b)}}\right], \; b > a, \; \lambda > 0 \tag{2}$$

The term $1/(b-a)$ is the normalization constant to ensure that the integral of $\Pi s(x)$ is unit. It is interesting to note that the integral $\int[\sigma(x-a) - \sigma(x-b)]\,dx = b - a$ independently of λ. Figure 1 describes two translated sigmoids and the resulting Π-sigmoid distribution.

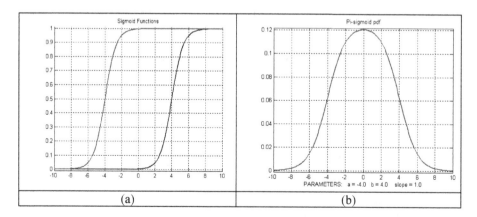

Fig. 1. The one-dimensional Π-sigmoid distribution (b) obtained as the difference of two translated sigmoid functions (a)

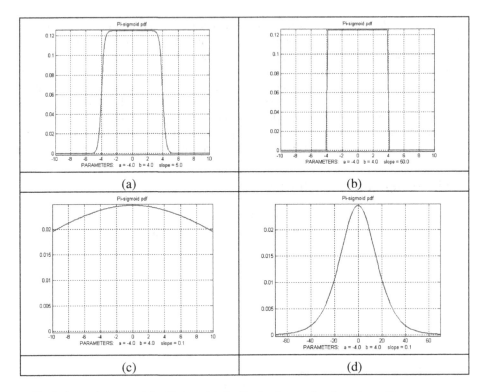

Fig. 2. Several shapes of the Π-sigmoid distribution by varying the values of its parameters

In Fig.2 we present several shapes of the Π-sigmoid distribution by varying the values of its parameters. It is clear that both uniform (Fig. 2b) and bell-shaped distributed (Fig. 2d) can be adequately approximated.

The *multidimensional Π-sigmoid distribution* is obtained under the assumption of independence along each dimension. More specifically, for a vector $x=(x_1, x_2,\ldots, x_D)^T$:

$$\Pi s(x) = \prod_{d=1}^{D} \Pi s(x_d) = \prod_{d=1}^{D} \frac{\dfrac{1}{1+e^{-\lambda_d(x_d-a_d)}} - \dfrac{1}{1+e^{-\lambda_d(x_d-b_d)}}}{b_d - a_d} \tag{3}$$

with $b_d > a_d$ and $\lambda_d > 0$. Fig. 3 illustrates how a two-dimensional Π-sigmoid distribution that approximates a uniform distribution on rectangular domain.

2.1 Maximum Likelihood Estimation

Suppose we are given a dataset $X = \{x^1,\ldots, x^N\}$, $x^i \in R^D$, to be modeled by a Π-sigmoid distribution. The parameters of the distribution can be estimated by maximizing the likelihood of the dataset X:

$$L(X;A,B,\Lambda) = \sum_{i=1}^{N} \log \Pi s(x^i) = \sum_{i=1}^{N} \sum_{d=1}^{D} \log \left(\frac{\dfrac{1}{1+e^{-\lambda_d(x_d^i-a_d)}} - \dfrac{1}{1+e^{-\lambda_d(x_d^i-b_d)}}}{b_d - a_d} \right) \quad (4)$$

with respect to the set of parameters $A = \{a_d\}, B = \{b_d\}, \Lambda = \{\lambda_d\}$, $d=1,...D$.

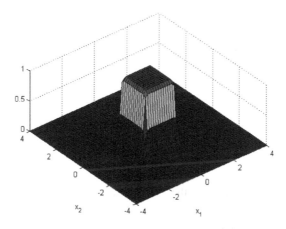

Fig. 3. A two-dimensional Π-sigmoid probability density function

The maximum likelihood solution cannot be obtained in closed form. However, since the gradient of the likelihood with respect to the parameters can be computed, gradient-based maximization methods can be employed (for example the simple gradient ascent or the more sophisticated quasi-Newton methods such as the BFGS).

3 The Π-Sigmoid Mixture Model

Using the proposed Π-sigmoid distribution, a mixture model can be defined called ΠsMM (Π-sigmoid Mixture Model) can be defined as follows

$$p(x) = \sum_{k=1}^{K} \pi_k \Pi s(x; A_k, B_k, \Lambda_k) \quad (5)$$

where K is the number of Π-sigmoid components, $A_k = \{a_{kd}\}, B_k = \{b_{kd}\}, \Lambda_k = \{\lambda_{kd}\}$ are the parameters of k-th component and the mixing weights π_k satisfy the constraints:

$$\pi_k \geq 0, \ \sum_{k=1}^{K} \pi_k = 1.$$

Given a dataset $X = \{x^1, ..., x^N\}$, $x^i \in R^D$ the parameters of the ΠsMM can be estimated through maximum likelihood using the EM algorithm as is the also the case with GMMs. The EM algorithm is an iterative approach involving two steps at each iteration. The *E-step* is the same in all mixture models and computes the posterior probability that x^i belongs to component k:

$$P(k \mid x^i) = \frac{\pi_k \Pi s(x^i; A_k, B_k, \Lambda_k)}{\sum_{j=1}^{K} \pi_j \Pi s(x^i; A_j, B_j, \Lambda_j)} \tag{6}$$

The *M-step* requires the maximization of the complete likelihood L_c (eq. (7)) with respect to the parameters of the ΠsMM model.

$$L_c = \sum_{i=1}^{N} \sum_{k=1}^{K} P(k \mid x^i) \log[\pi_k \Pi s(x^i; A_k, B_k, \Lambda_k)] \tag{7}$$

For the parameters π_k the update equation is the same for all mixture models:

$$\pi_k = \frac{1}{N} \sum_{i=1}^{N} P(k \mid x^i) \tag{8}$$

In contrast to the GMM case, the M-step does not lead to closed form update equations for the parameters A_k, B_k, Λ_k of the Π-sigmoid components. Thus we resort to the GEM (generalized EM) algorithm [3,4], which suggests to update the model parameters so that obtain higher (not necessarily maximum) values of the complete likelihood are obtained. In this work a few updates of each parameter θ along the direction of the gradient $\partial L_c / \partial \theta$ are computed. The GEM algorithm ensures that, at each iteration, the parameters θ are updated so that the likelihood increases, until a local maximum is reached.

It is well known that the EM algorithm is very sensitive to the initialization of the model parameters. To deal with this issue, we first apply the Greedy-EM algorithm [2] to fit a GMM with K components (and diagonal covariance matrices) on the dataset. Then the parameters of the k-th ΠsMM component are initialized using the obtained parameters (mean and variance) of the k-th GMM component. The values of λ are always initialized to 1.5, although the performance is not sensitive to this choice.

4 Experimental Results

We have compared the modeling capabilities of ΠsMM against the widely used GMM. First artificial datasets were considered with a) only uniform rectangular, (Fig. 4a) b) only Gaussian (Fig. 4b) and c) mixed (uniform and Gaussian) clusters (Fig. 4c). We considered the case D=2, K=4 (as in Fig. 4) and D=5, K=4. Each dataset contained 5000 data points used for training and 5000 used for testing. It must be emphasized that we are not interested in data clustering, but in building accurate models of the density of the given datasets. In this spirit, the obtained mixture models

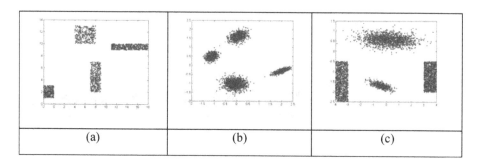

Fig. 4. Three types of artificial datasets: (a) rectangular uniform, (b) Gaussian, (c) mixed

Table 1. Test set likelihood values for three types of datasets (Gaussian, uniform, mixed) for the cases with D=2, K=4 and D=5, K=4

	D=2, K=4			D=5, K=4		
	Uniform	Gaussian	Mixed	Uniform	Gaussian	Mixed
ΠsMM	-8572	3802	-11558	-10167	-8168	-46616
GMM	-11118	3836	-12653	-8856	-8352	-49044

were compared in terms of the likelihood on the test set which constitutes the most reliable measure to compare statistical models.

The results are summarized in Table 1. In the case of uniform clusters the ΠsMM was highly superior to GMM, in the case of Gaussian clusters the GMM was only slightly superior, while in the case of mixed clusters the performance of ΠsMM was much better. The inability of GMMs to efficiently approximate the uniform distribution is an old and well-known problem and the proposed ΠsMM model provides an efficient solution to modeling uniformly distributed data, while at the same time is able to adequately model Gaussian data.

Since GMMs have been successfully employed for image modeling and segmentation using pixel intensity information, we tested the performance of the proposed ΠsMM model on the image segmentation task and compare against GMM. We considered 256x256 grey level images and for each image the intensities of 5000 randomly selected pixels were used as the training set to fit a GMM and a ΠsMM with the same number of components (K=5). After training, the remaining pixels were assigned to the component (cluster) with the maximum posterior probability (eq (8)). The segmentation results for two example images are shown in Fig. 5.

It must be noted that the likelihood values on the large set of pixels not used for training (test set) of the PsMM and GMM were -28229 and -30185, respectively, for the top image and -23487 and -26576 for the bottom image of Fig. 5, also indicating that the ΠsMM provides superior statistical models of the images compared to GMM. This superiority is also confirmed from visual inspection of the segmented images (which is more clear in the top row images).

| Original image | Segmentation with ΠsMM | Segmentation with GMM |

Fig. 5. Segmentation results for two natural images (first column) using ΠsMM (second column) and GMM (third column)

5   Conclusions

We have proposed a new probability density function (Π-sigmoid) defined as the difference of two translated logistic sigmoids. Depending on the slope value of the sigmoids, the shape of the distribution may vary from bell-shaped to uniform allowing the flexibility to model a variety of datasets from Gaussian to uniform. We have also presented the Π-sigmoid mixture model (ΠsMM) and show how to estimate its parameters under the maximum likelihood framework using the Generalized EM algorithm. Experimental comparison with the Gaussian Mixture Models indicate that ΠsMM is more flexible than GMM providing solutions of higher likelihood. Also a notable characteristic of ΠsMM is its ability to accurately identify rectangular shaped clusters, which constitutes a well-known weakness of GMMs.

It must be noted that another probabilistic model cabable of identifying rectangular shaped clusters has been proposed in [1], called mixture of rectangles. In that model a component distribution is a uniform distribution with a Gaussian tail and it is difficult to train, due to the inability to define the gradient of the likelihood with respect to the parameters. Thus one has to resort to line search optimization methods to perform training of the model.

Our current work is focused on extending the Π-sigmoid distribution with the aim to describe rotated rectangles. The main issue to be addressed is how to develop an efficient training algorithm to adjust the additional parameters defining the rotation matrix. Another important issue is to develop a methodology to estimate the number of mixture components in ΠsMM, based on recent methods developed in the context of GMMs.

Acknowledgement. Information dissemination of this work was supported by the European Union in the framework of the project "Support of Computer Science Studies in the University of Ioannina" of the "Operational Program for Education and Initial Vocational Training" of the 3rd Community Support Framework of the Hellenic Ministry of Education, funded by national sources and by the European Social Fund (ESF).

References

[1] Pelleg, D., Moore, A.: Mixture of rectangles: Interpretable soft clustering. In: Proc. ICML (2001)
[2] Vlassis, N., Likas, A.: A greedy EM algorithm for Gaussian mixture learning. Neural Processing Letters 15, 77–87 (2002)
[3] McLachlan, J.G., Krishnan, T.: Finite Mixture Models. Wiley, Chichester (2000)
[4] McLachlan, J.G., Krishnan, T.: The EM algorithm and extensions. Marcel Dekker, New York (1997)

How Robust Is a Probabilistic Neural VLSI System Against Environmental Noise

C.C. Lu, C.C. Li, and H. Chen

The Dept. of Electrical Engineering,
The National Tsing-Hua University, Hsin-Chu, Taiwan 30013
hchen@ee.nthu.edu.tw

Abstract. Implementing probabilistic models in the Very-Large-Scale-Integration (VLSI) has been attractive to implantable biomedical devices for improving sensor fusion and power management. However, implantable devices are normally exposed to noisy environments which can introduce non-negligible computational errors and hinder optimal modelling on-chip. While the probabilistic model called the Continuous Restricted Boltzmann Machine (CRBM) has been shown useful and realised as a VLSI system with noise-induced stochastic behaviour, this paper investigates the suggestion that the stochastic behaviour in VLSI could enhance the tolerance against the interferences of environmental noise. The behavioural simulation of the CRBM system is used to examine the system's performance in the presence of environmental noise. Furthermore, the possibility of using environmental noise to induce stochasticity in VLSI for computation is investigated.

1 Introduction

In the development of implantable devices [1][2] and bioelectrical interfaces [3][4], exposing electronic systems to the noisy environment becomes inevitable. Although noisy data could be transmitted wirelessly out of implanted devices and processed by sophisticated algorithms, transmitting all raw data is power-consuming, and is unfavourable for long-term monitoring. Therefore, an intelligent embedded system which is robust against noise and able to extract useful information from high dimensional, noisy biomedical signals becomes essential. The Continuous Restricted Boltzmann Machine (CRBM) is a probabilistic model both useful in classifying biomedical data and amenable to the VLSI implementation [5]. The usefulness comes from the use of noise-induced stochasticity to represent natural variability in data. The VLSI implementation further explores the utility of noise-induced, continuous-valued stochastic behaviour in VLSI circuits [5]. This leads to the suggestion that stochastic behaviour in VLSI could be useful for discouraging environmental noise and computation errors. Therefore, based on the well-defined software-hardware mapping derived in [5] this paper use behavioural simulation to examines the maximum external noise that the CRBM system can tolerate to model both artificial and real biomedical (ECG) data. The possibility of using environmental noise to replace on-chip noise generators is also investigated.

L. Prevost, S. Marinai, and F. Schwenker (Eds.): ANNPR 2008, LNAI 5064, pp. 44–53, 2008.

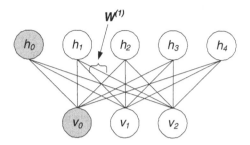

Fig. 1. The architecture of a CRBM model with two visible and four hidden neurons. v_0 and h_0 represent biasing unit with invariant outputs $v_0 = h_0 = 1$.

2 The CRBM Model

The CRBM model consists of one visible and one hidden layers of continuous-valued, stochastic neurons with inter-layer connections, as shown in Fig.1. Circle v_i represents visible neurons while circle h_j represents hidden neurons. The number of visible neurons corresponds to the dimensions of modeled data, while that of hidden neurons is chosen according to data complexity. Let s_i denotes the state of neuron v_i or h_j, and w_{ij} represents the bi-directional connection between v_i and h_j. The stochastic behaviour of a neuron s_i is described by [5]

$$s_i = \varphi_i\left(a_i \cdot \left(\Sigma_j w_{ij} \cdot s_j + N_i(\sigma, 0)\right)\right) \tag{1}$$

where $N_i(\sigma, 0)$ represents a zero-mean Gaussian noise with variance σ^2, and $\varphi_i(\cdot)$ a sigmoid function with asymptotes at ± 1 and slope controlled by a_i. As a *generative* model, the CRBM learns to "regenerate" training data distributions in its visible neurons. Testing data can then be categorised according to the responses of hidden neurons [5]. The training algorithm implemented in the CRBM system is defined by the following equation [6]

$$\triangle\lambda = \eta_\lambda \cdot \left(\langle s_i \cdot s_j \rangle_4 - \langle \hat{s}_i \cdot \hat{s}_j \rangle_4\right) \tag{2}$$

where λ represent parameters w_{ij} or a_i, η_λ the updating rate, $\hat{s}_i$ and $\hat{s}_j$ the one-step Gibbs-sampled states. $\langle \cdot \rangle_4$ stands for taking the expectation over four training data. For parameter a_i, the training algorithm is the same with Eq.(2) but simply replace s_j and $\hat{s}_j$ by s_i and $\hat{s}_i$, respectively.

The modelled distribution of a trained CRBM is obtained by initializing visible neurons with random values, and then Gibbs sampling hidden and visible neurons alternatively for multiple steps. The N-th step samples of visible neurons are called the *N-step reconstruction*, and it approximates the modelled distribution when N is large. The similarity between *N-step reconstruction* and training data indicates how well training data is modelled.

3   The Robustness Against Environmental Noise

Following a brief introduction to the architecture of the CRBM system, this section investigates the influences of environmental noise on the performance of the CRBM system.

3.1   The CRBM System

The prototype CRBM system containing two visible and four hidden neurons has been demonstrated able to reconstruct two-dimensional artificial data with noise-induced stochastic behaviour in VLSI [5]. Fig.2 shows the modular diagram of the CRBM systems excluding its learning circuits. The CRBM neurons mainly comprise of multipliers to calculate the products $(w_{ij} \cdot s_j)$ in Eq.(1), and sigmoid circuits with $\{a_i\}$ controlling

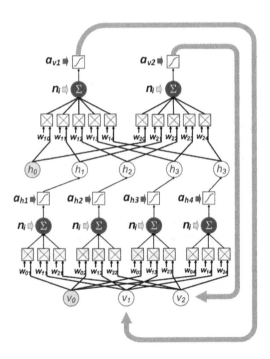

Fig. 2. The modular diagram of the CRBM system with two visible and four hidden neurons

Table 1. The mapping of parameter values between software simulation and hardware implementation

	Matlab	VLSI(V)
s_i	[-1.0, 1.0]	[1.5, 3.5]
w_{ij}	[-2.5, 2.5]	[0.0, 5.0]
a_i	[0.5, 9.0]	[1.0, 3.0]

the slope of φ_i. On the other hand, multi-channel, uncorrelated noise $\{n_i\}$ are injected into the neurons to make the outputs, $\{v_i\}$ and $\{h_i\}$, probabilistic. Parameters $\{w_{ij}\}$ and $\{a_i\}$ are stored as voltages across capacitors, and are adaptable by on-chip learning circuits. The learning circuits can also refresh $\{w_{ij}\}$ and $\{a_i\}$ to specific values after training. Table 1 summarises the mapping for all parameters between software simulation and VLSI implementation, which has been proved useful for simulating the effects of non-ideal training offsets on the performance of the CRBM system [5].

In an implantable device containing digital-signal-processing circuits, multi-purpose sensors, and wireless transceivers, a VLSI system unavoidably suffers from various environmental noise including substrate noise, sensory noise, and electromagnetic interferences. As these interferences mainly affect the precision of voltage signals in VLSI, voltage-represented s_i, w_{ij}, and a_i in the CRBM system are expected to experience serious effects, while the influence on current-mode learning circuits are assumed to be negligible. Therefore, the influence of environmental noise on the CRBM system was simulated by replacing s_i, w_{ij}, and a_i in Eq.(1) and Eq.(2) by the following equations

$$w' = w + n_w$$

$$a' = a + n_a \tag{3}$$

$$s' = s + n_s$$

where n_s, n_w, and n_a represent zero-mean, uncorrelated noise with either Gaussian or Uniform distributions.

3.2 Modelling Artificial Data in the Presence of Environmental Noise

To illustrate the characteristics of the CRBM, as well as to identify a quantitative index for how well the CRBM models a dataset, the CRBM with two visible and four hidden neurons was first trained to model the artificial data in Fig.3(a) in the absence of environmental noise. The training data contains one elliptic and one circular clusters of 1000 Gaussian-distributed data points. With $\sigma = 0.2$, $\eta_w = 0.02$, $\eta_a = 0.2$, and after 15,000 training epochs, the CRBM regenerated the 20-step reconstruction of 1000 data points as shown in Fig.3(b), indicating that the CRBM has modelled data. While visual comparison between Fig.3(a) and (b) can hardly tell how well the data is modelled, the following index is employed to measure the similarity quantitatively.

Let $P^T(\mathbf{v})$ and $P^M(\mathbf{v})$ represent the probability distribution of training data and that modelled by the CRBM, respectively. The *Kulback-Leibler (KL) Divergence* defined as Eq.(4) [7] measures the difference between $P^T(\mathbf{v})$ and $P^M(\mathbf{v})$.

$$G = \Sigma_\mathbf{v} P^T(\mathbf{v}) log \frac{P^T(\mathbf{v})}{P^M(\mathbf{v})} \tag{4}$$

where $\mathbf{v}$ denotes the subset of visible space, and G equals zero when $P^T(\mathbf{v}) = P^M(\mathbf{v})$. As explicit equations for describing the modelled distribution, $P^M(\mathbf{v})$, are normally intractable, $P^T(\mathbf{v})$ and $P^M(\mathbf{v})$ were statically-estimated by dividing the two-dimensional space into 10x10 square grids, counting the number of data points in each grid, and

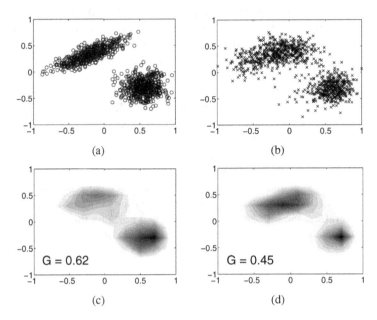

Fig. 3. (a)Artificial training data. (b)The 20-step reconstruction with 1000 points generated by the CRBM trained after 15,000 epochs. (c)The statistical density after 12,500 epochs. (d)The statistical density corresponding to (b).

normalising the counts with respect to the total number of data points. Fig.3(c)-(d) shows the statistical density of the 20-step reconstructions generated by the CRBM after 12,500 and 15,000 training epochs, respectively. The G values calculated according to Eq.(4) are also shown at the bottom-left corners. This result indicates the probability distribution of training data is modelled with negligible errors as G is smaller than 0.45(Fig.3(d)).

Based on the criterion of $G \leq 0.45$, the maximum environmental noise, both Gaussian- and uniformly-distributed, the CRBM can tolerate to model the artificial data are identified and summarised in Table 2. The maximum tolerable noise levels are expressed in terms of voltages based on the mapping in Table 1, and obviously, the tolerable levels are much greater than the noise levels existing in contemporary VLSI technologies. The first three rows in Table 2 show the tolerance identified when only one type of parameters experiences noise. Parameter $\{a_i\}$ has slightly smaller tolerance than $\{w_{ij}\}$ and $\{s_i\}$ because the mapping for $\{a_i\}$ has a largest ratio between the numerical and voltage ranges. This leads $\{a_i\}$ to experience largest numerical errors in the existence of the same noise levels. The forth row in Table 2 shows the tolerance when noise exists in all parameters, the more realistic case. The tolerance is not seriously degraded, indicating that the training algorithm of the CRBM system can compensate for noise-induced errors among parameters and maintain a satisfactory tolerance. Finally, Table 2 reveals that the tolerance against Gaussian-distributed noise is much better than

Table 2. The maximum gaussian- and uniformly-distributed noise tolerable by the CRBM system during modelling artificial data

	Gaussian-distributed noise	Uniformly-distributed noise
n_w	[-0.27V, 0.27V]	[-0.18V, 0.18V]
n_a	[-0.16V, 0.16V]	[-0.08V, 0.08V]
n_s	[-0.18V, 0.18V]	[-0.06V, 0.06V]
n_w, n_a, n_s	[-0.13V, 0.13V]	[-0.04V, 0.04V]

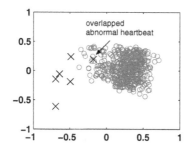

Fig. 4. The projection of 500 ECG training data to its first two principle components. The projection of the five abnormal ECGs in the dataset are denoted by black crosses.

that against uniformly-distributed noise. This is attributed to the fact that the training data and the noise incorporated in the CRBM neurons are Gaussian-distributed. From another point of view, if the distribution of environmental noise is known, training the CRBM system with noise inputs $\{n_i\}$ modified to have the known distribution can enhance the tolerance against a specific type of noise.

3.3 Modelling Biomedical Data in the Presence of Environmental Noise

The tolerable environmental noise for modelling high-dimensional, real-world data was examined in the context of recognising electrocardiograms (ECG), extracted from the MIT-BIH database as in [6]. The training dataset contains 500 heartbeats with only 5 abnormal heartbeats. The testing dataset contains 1700 heartbeats with 27 abnormal heartbeats. Each heartbeat is sampled as a 65-dimemsional datum, and Fig.4 shows the projection of the training dataset onto its first two principle components. Although the dimension reduction makes the quantitative index G remain applicable, pilot simulation showed that modelling training data satisfactorily does not guarantee the detection of abnormal heartbeats with 100% accuracy. This is because the distributions of normal and abnormal heartbeats overlap with each other, as shown in Fig.4. Therefore, detecting abnormal heartbeats with 100% accuracy was used as a stricter criterion for identifying the tolerable noise during modelling ECG data.

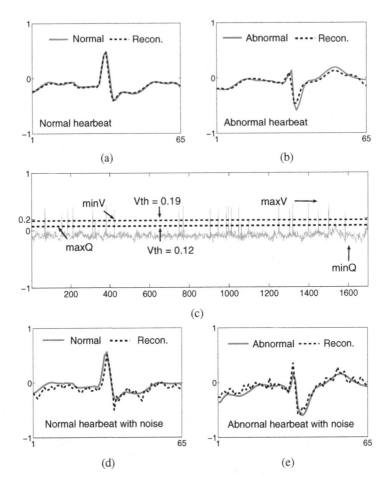

Fig. 5. (a) (b) Heartbeat signals sampled from training data (solid) and reconstruction generated by the trained CRBM (dashed) without noise. (c) Corresponding response of hidden neuron to 1700 testing data. (d)(e) Heartbeat signals and reconstruction generated by the trained CRBM with uniformly-distributed external noise in the range [-0.01, 0.01].

Fig.5(a)(b) shows heartbeat signals reconstructed by a CRBM system trained without noise. Fig.5(c) shows the response of hidden neuron h_2 to 1700 testing data $\{\mathbf{d}\}$, calculated according to Eq.(5). The abnormal heartbeats can be detected with 100% accuracy by setting any threshold between minV and maxQ.

$$h_2 = \varphi_2(a_2 \cdot (\mathbf{w}^{(2)} \cdot \mathbf{d})) \tag{5}$$

With uniformly-distributed noise ranging between -0.01V and 0.01V, the trained CRBM system was able to reconstruct both normal and abnormal ECG signal satisfactorily, as shown in Fig.5(d)(e). Comparison between Fig.5(a)(b) and Fig.5(d)(e) indicates that the influence of environmental noise injection introduce extra fluctuations in the waveform.

Table 3. The maximum gaussian- and uniformly-distributed noise tolerable by the CRBM system to detect abnormal ECGs reliably

	Gaussian-distributed noise	Uniformly-distributed noise
n_w	[-0.39V, 0.39V]	[-0.29V, 0.29V]
n_a	[-0.15V, 0.15V]	[-0.11V, 0.11V]
n_s	[-0.09V, 0.09V]	[-0.05V, 0.05V]
n_w, n_a, n_s	[-0.03V, 0.03V]	[-0.03V, 0.03V]

Table 3 summarises the maximum environmental noise the CRBM can tolerate to model and to detect abnormal ECGs with 100% accuracy. The tolerable noise levels in Table 3 are mostly smaller than those in Table 2 because the ECG data require more sophisticated modelling. On the contrary, parameter $\{w_{ij}\}$ has comparable tolerance in both modelling tasks because the presence of n_w simply distorts the weight vectors and results in fluctuated reconstructions like those in Fig.5(d)(e), while these effects do not impede the CRBM from modelling the distinguishable features between abnormal and normal ECGs. This also explains the relatively smaller tolerance against n_s, i.e. the noise in $\{s_i\}$, as n_s does distort the features of ECGs, making it difficult to identify any distinguishable feature between normal and abnormal ECGs.

4   Noise-Enhanced Robustness in the CRBM System

By reducing the standard deviation of the noise injected into CRBM neurons to $\sigma = 0.1$ in Eq.(1), the maximum environmental noise the CRBM can tolerate to model the artificial data in Fig.3(a) is summarised in Table 4. Comparison between Table 2 and Table 4 indicates that reducing $\sigma = 0.1$ helps to enhance the robustness against environmental noise, as part of environmental noise is incorporated to compensate for the reduction of the"internal noise" n_i, which is essential for inducing stochasticity for modelling the variability of training data. Therefore, as shown by Table 4, the robustness against environment noise, especially for uniformly-distributed noise, is improved significantly. The worst tolerable level is still greater than 110mV, corresponding to a signal-to-noise ratio less than 20 for a CRBM system. The advantage of incorporating noise-induced stochasticity in VLSI is clearly demonstrated.

Furthermore, it is interesting to investigate whether the internal noise n_i could be completely replaced by environmental noise to induce stochasticity for computation. By substituting $s_j + n_s$ for s_j and setting $N_i(\sigma, 0) = 0$ in Eq.(1), the term $\sum w_{ij} \cdot (s_j + n_s)$ becomes a random variable n_i' with mean value $\sum w_{ij} \cdot s_j$ and a variance given as

$$var(n_i') = \Sigma_j w_{ij}^2 \cdot var(n_s) \qquad (6)$$

If n_i' has the same variance as n_i, the stochasticity induced by n_s should have the same level as that induced by n_i in Fig.2. The CRBM trained on the artificial data was found

Table 4. The tolerable interference of external gaussian- and uniformity-distributed noise with reduced variations of on-chip noise generator to model artificial data

	Gaussian-distributed noise	Uniformy-distributed noise
n_w	[-0.46V, 0.46V]	[-0.46V, 0.46V]
n_a	[-0.33V, 0.33V]	[-0.14V, 0.14V]
n_s	[-0.28V, 0.28V]	[-0.18V, 0.18V]
n_w, n_a, n_s	[-0.16V, 0.16V]	[-0.11V, 0.11V]

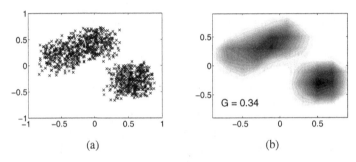

(a)   (b)

Fig. 6. (a) The 20-step reconstruction with 1000 points to model artificial in Fig.3(a) generated by the CRBM trained after 15,000 epochs. (b) The statistical density corresponding to (a).

to have an average value between 0.8 and 1 for $\Sigma_j w_{ij}^2$. According to Eq.6, a uniformly-distributed noise n_s ranging between [-0.3V, 0.3V] will yield an equivalent noise n_i' with a variance of 0.04. Fig.6 shows the results of training the CRBM to model the artificial data in Fig.3(a) without n_i, but with uniformly-distributed noise n_s in the range [-0.3V, 0.3V]. Fig.6(a) and (b) depicts the 20-step reconstruction and its statistical density, respectively. The corresponding G value is 0.34, revealing that the trained CRBM has used the stochasticity induced by environmental noise in $\{s_i\}$ to model data optimally. This supports the suggestion in [8] that intrinsic noise of MOSFETs could be used rather than suppressed to achieve robust computation, based on algorithms like the CRBM. This is especially important when the VLSI technology moves towards the deep-sub-micron era.

5   Conclusion

The behavioural simulation of the CRBM system demonstrates that the CRBM system has satisfactory robustness against environmental noise, confirming the potential of using the CRBM system as an intelligent system in implantable devices. As the promising performance mainly comes from the incorporation of noise-induced stochasticity, the robustness can be further enhanced if the distribution of environmental noise is known and incorporated during training, or by reducing the internal noise of the CRBM system.

It is also demonstrated that environmental noise can be used to induce the stochasticity essential for the CRBM system to model data optimally. In other words, the robustness of the CRBM system can be optimised by training the CRBM to model data with stochasticity induced by environmental noise the CRBM system is exposed to. All these concepts will be further examined by hardware testing with the CRBM system.

References

1. Tong, B.T., Johannessen, E.A., Lei, W., Astaras, A., Ahmadian, M., Murray, A.F., Cooper, J.M., Beaumont, S.P., Flynn, B.W., Cumming, D.R.S.: Toward a miniature wireless integrated multisensor microsystem for industrial and biomedical applications. Sensors Journal 2(6), 628–635 (2002)
2. Johannessen, E.A., Lei, W., Li, C., Tong, B.T., Ahmadian, M., Astaras, A., Reid, S.W.J., Yam, P.S., Murray, A.F., Flynn, B.W.A., Beaumont, S.P.A., Cumming, D.R.S., Cooper, J.M.A.: Implementation of multichannel sensors for remote biomedical measurements in a microsystems format. IEEE Transactions on Biomedical Engineering 51(3), 525–535 (2004)
3. Mingui, S., Mickle, M., Wei, L., Qiang, L., Sclabassi, R.J.: Data communication between brain implants and computer. IEEE Transactions on Neural Systems and Rehabilitation Engineering 11(2), 189–192 (2003)
4. Nicolelis, M.A.L.: Actions from thoughts. Nature 409, 403–407 (2001)
5. Hsin, C., Fleury, P.C.D., Murray, A.F.: Continuous-valued probabilistic behavior in a VLSI generative model. IEEE Transactions on Neural Networks 17(3), 755–770 (2006)
6. Chen, H., Murray, A.F.: Continuous restricted Boltzmann machine with an implementable training algorithm. IEE Proceedings-Vision Image and Signal Processing 150(3), 153–158 (2003)
7. Hinton, G.E., Sejnowski, T.J.: Learning and Relearning in Boltzmann Machine. In: Parallel Distributed Processing: Explorations in the Microstructure of Cognition, pp. 283–317. MIT, Cambridge (1986)
8. Hamid, N.H., Murray, A.F., Laurenson, D., Roy, S., Binjie, C.: Probabilistic computing with future deep sub-micrometer devices: a modelling approach. In: IEEE International Symposium on Circuits and Systems, pp. 2510–2513 (2005)

Sparse Least Squares Support Vector Machines by Forward Selection Based on Linear Discriminant Analysis

Shigeo Abe

Graduate School of Engineering
Kobe University
Rokkodai, Nada, Kobe, Japan
abe@kobe-u.ac.jp
http://www2.eedept.kobe-u.ac.jp/~abe

Abstract. In our previous work, we have developed sparse least squares support vector machines (sparse LS SVMs) trained in the reduced empirical feature space, spanned by the independent training data selected by the Cholesky factorization. In this paper, we propose selecting the independent training data by forward selection based on linear discriminant analysis in the empirical feature space. Namely, starting from the empty set, we add a training datum that maximally separates two classes in the empirical feature space. To calculate the separability in the empirical feature space we use linear discriminant analysis (LDA), which is equivalent to kernel discriminant analysis in the feature space. If the matrix associated with the LDA is singular, we consider that the datum does not contribute to the class separation and permanently delete it from the candidates of addition. We stop the addition of data when the objective function of LDA does not increase more than the prescribed value. By computer experiments for two-class and multi-class problems we show that in most cases we can reduce the number of support vectors more than with the previous method.

1 Introduction

A least squares support vector machine (LS SVM) [1] is a variant of a regular SVM [2]. One of the advantages of LS SVMs over SVMs is that we only need to solve a set of linear equations instead of a quadratic programming program. But the major disadvantage of LS SVMs is that all the training data become support vectors instead of sparse support vectors for SVMs. To solve this problem, in [1], support vectors with small absolute values of the associated dual variables are pruned and the LS SVM is retrained using the reduced training data set. This process is iterated until sufficient sparsity is realized. In [3], LS SVMs are reformulated using the kernel expansion of the square of Euclidian norm of the weight vector in the decision function. But the above pruning method is used to reduce support vectors. Because the training data are reduced during pruning, information for the deleted training data is lost for the trained LS

L. Prevost, S. Marinai, and F. Schwenker (Eds.): ANNPR 2008, LNAI 5064, pp. 54–65, 2008.

SVM. To overcome this problem, in [4], independent data in the feature space are selected from the training data, and using the selected training data the solution is obtained by the least squares method using all the training data. Along the line of kernel expansion, there are some approaches to realize sparse kernel expansion by forward selection of basis vectors based on some criterion such as least squares errors [5,6,7]. Based on the concept of the empirical feature space [8], which is closely related to kernel expansion, in [9] a sparse LS SVM is developed restricting the dimension of the empirical feature space by the Cholesky factorization.

Instead of the Cholesky factorization used in [9], in this paper we propose using forward selection based on the class separability calculated by the linear discriminant analysis (LDA) in the empirical feature space. Namely, starting from the empty set, we add training datum, one at a time, that maximally separates two classes in the empirical feature space. To calculate the separability in the empirical feature space we use LDA in the empirical feature space. Linear discriminant analysis in the empirical feature space is equivalent to kernel discriminant analysis (KDA) in the feature space, but since the variables in the empirical feature space are treated explicitly, the calculation of LDA is much faster. If the matrix associated with LDA is singular, we consider that the datum does not contribute to the class separation and permanently delete it from the candidates of addition. In addition to this, using the incremental calculation of LDA, we speed up forward selection. We stop the addition of data when the objective function of LDA does not increase more than the prescribed value. We evaluate the proposed method using two-class and multi-class problems.

In Section 2, we summarize sparse LS SVMs trained in the empirical feature space, and in Section 3 we discuss forward selection of independent variables based on LDA. In Section 4, we evaluate the validity of the proposed method by computer experiments.

2   Sparse Least Squares Support Vector Machines

Let the M training data pairs be $(\mathbf{x}_1, y_1), \ldots, (\mathbf{x}_M, y_M)$, where $\mathbf{x}_i$ and y_i are the m-dimensional input vector and the associated class label, and $y_i = 1$ and -1 if $\mathbf{x}_i$ belongs to Classes 1 and 2, respectively. In training LS SVMs in the empirical feature space we need to transform input variables into variables in the empirical feature space. To speed up generating the empirical feature space we select independent training data that span the empirical feature space [9]. Let the $N\,(\leq M)$ training data $\mathbf{x}_{i_1}, \ldots, \mathbf{x}_{i_N}$ be independent in the empirical feature space, where $\mathbf{x}_{i_j} \in \{\mathbf{x}_1, \ldots, \mathbf{x}_M\}$ and $j = 1, \ldots, N$. Then, we use the following mapping function: $\mathbf{h}(\mathbf{x}) = (H(\mathbf{x}_{i_1}, \mathbf{x}), \ldots, H(\mathbf{x}_{i_N}, \mathbf{x}))^T$, where $H(\mathbf{x}, \mathbf{x}')$ is a kernel. By this formulation, $\mathbf{x}_{i_1}, \ldots, \mathbf{x}_{i_N}$ become support vectors. Thus, support vectors do not change even if the margin parameter changes. And the number of support vectors is the number of selected independent training data that span the empirical feature space. Then reducing N without deteriorating the generalization ability we can realize sparse LS SVMs.

The LS SVM in the empirical feature space is trained by solving

$$\text{minimize} \quad Q(\mathbf{v}, \boldsymbol{\xi}, b) = \frac{1}{2} \mathbf{v}^T \mathbf{v} + \frac{C}{2} \sum_{i=1}^{M} \xi_i^2 \tag{1}$$

$$\text{subject to} \quad \mathbf{v}^T \mathbf{h}(\mathbf{x}_i) + b = y_i - \xi_i \quad \text{for} \quad i = 1, \dots, M, \tag{2}$$

where $\mathbf{v}$ is the N-dimensional vector, b is the bias term, ξ_i is the slack variable for $\mathbf{x}_i$, and C is the margin parameter.

Substituting (2) into (1) and minimizing the resultant objective function, we obtain

$$b = \frac{1}{M} \sum_{i=1}^{M} (y_i - \mathbf{v}^T \mathbf{h}(\mathbf{x}_i)), \tag{3}$$

$$\left(\frac{1}{C} + \sum_{i=1}^{M} \mathbf{h}(\mathbf{x}_i) \mathbf{h}^T(\mathbf{x}_i) - \frac{1}{M} \sum_{i,j=1}^{M} \mathbf{h}(\mathbf{x}_i) \mathbf{h}^T(\mathbf{x}_j) \right) \mathbf{v}$$

$$= \sum_{i=1}^{M} y_i \mathbf{h}(\mathbf{x}_i) - \frac{1}{M} \sum_{i,j=1}^{M} y_i \mathbf{h}(\mathbf{x}_j). \tag{4}$$

We call the LS SVM obtained by solving (4) and (3) primal LS SVM if N is the same as the dimension of the mapped training data in the feature space. If N is smaller, the primal LS SVM is called sparse LS SVM. We call the LS SVM in dual form [1] dual LS SVM.

3   Selection of Independent Data

3.1   Idea

In [9], independent data are selected by Cholesky factorization of the kernel matrix. During factorization, if the argument of the square root associated with the diagonal element is smaller than the prescribed threshold value, we delete the associated row and column and continue decomposing the matrix. By increasing the threshold value, we can increase the sparsity of the LS SVM.

So long as we select the independent data that span the empirical feature space, different sets of independent data do not affect the generalization ability of the LS SVM, because the different sets span the same empirical feature space.

But the different sets of the independent data for the reduced empirical feature space span different reduced empirical feature spaces. Thus, the processing order of the training data affects the generalization ability of the LS SVM. Therefore, selection may be inefficient if the empirical feature space is reduced.

To overcome this problem, we consider selecting independent data that maximally separate two classes using LDA calculated in the reduced empirical feature space. In the following first we summarize the selection of independent data by the Cholesky factorization and then discuss our proposed method based on LDA.

3.2 Selection of Independent Data by Cholesky Factorization

Let the kernel matrix $H = \{H(\mathbf{x}_i, \mathbf{x}_j)\}$ $(i, j = 1, \ldots, M)$ be positive definite. Then H is decomposed by the Cholesky factorization into $H = LL^T$, where L is the regular lower triangular matrix and each element L_{ij} is given by

$$L_{op} = \frac{H_{op} - \sum\limits_{n=1}^{p-1} L_{pn} L_{on}}{L_{pp}} \qquad \text{for} \quad o = 1, \ldots, M, \quad p = 1, \ldots, o-1, \qquad (5)$$

$$L_{aa} = \sqrt{H_{aa} - \sum\limits_{n=1}^{a-1} L_{an}^2} \qquad \text{for} \quad a = 1, 2, \ldots, M. \qquad (6)$$

Here, $H_{ij} = H(\mathbf{x}_i, \mathbf{x}_j)$.

Then during the Cholesky factorization, if the argument of the square root associated with the diagonal element is smaller than the prescribed value η_C ($>$ 0):

$$H_{aa} - \sum\limits_{n=1}^{a-1} L_{an}^2 \leq \eta_C, \qquad (7)$$

we delete the associated row and column and continue decomposing the matrix. The training data that are not deleted in the Cholesky factorization are independent.

The above Cholesky factorization can be done incrementally [10,11]. Namely, instead of calculating the full kernel matrix in advance, if (7) is not satisfied, we overwrite the ath column and row with those newly calculated using the previously selected data and $\mathbf{x}_{a+1}$. Thus the dimension of L is the number of selected training data, not the number of training data.

To increase sparsity of LS SVMs, we increase the value of η_C. The optimal value is determined by cross-validation. We call thus trained LS SVMs sparse LS SVMs by Cholesky factorization, sparse LS SVMs (C) for short.

3.3 Linear Discriminant Analysis in the Empirical Feature Space

We formulate linear discriminant analysis in the empirical feature space, which is equivalent to kernel discriminant analysis in the feature space. To make notations simpler, we redefine the training data: Let the sets of m-dimensional data belonging to Class i ($i = 1, 2$) be $\{\mathbf{x}_1^i, \ldots, \mathbf{x}_{M_i}^i\}$, where M_i is the number of data belonging to Class i. Now we find the N-dimensional vector $\mathbf{w}$ in which the two classes are separated maximally in the direction of $\mathbf{w}$ in the empirical feature space.

The projection of $\mathbf{h}(\mathbf{x})$ on $\mathbf{w}$ is $\mathbf{w}^T \mathbf{h}(\mathbf{x})/\|\mathbf{w}\|$. In the following we assume that $\|\mathbf{w}\| = 1$. We find such $\mathbf{w}$ that maximizes the difference of the centers and minimizes the variance of the projected data.

The square difference of the centers of the projected data, d^2, is

$$d^2 = (\mathbf{w}^T(\mathbf{c}_1 - \mathbf{c}_2))^2 = \mathbf{w}^T(\mathbf{c}_1 - \mathbf{c}_2)(\mathbf{c}_1 - \mathbf{c}_2)^T \mathbf{w}, \qquad (8)$$

where $\mathbf{c}_i$ are the centers of class i data:

$$\mathbf{c}_i = \frac{1}{M_i} \sum_{j=1}^{M_i} \mathbf{h}(\mathbf{x}_j^i) \qquad \text{for} \quad i = 1, 2. \tag{9}$$

We define $Q_B = (\mathbf{c}_1 - \mathbf{c}_2)(\mathbf{c}_1 - \mathbf{c}_2)^T$ and call Q_B the between-class scatter matrix.

The variance of the projected data is $s^2 = \mathbf{w}^T Q_W \mathbf{w}$, where

$$Q_W = \frac{1}{M} \sum_{j=1}^{M} \mathbf{h}(\mathbf{x}_j) \mathbf{h}(\mathbf{x}_j)^T - \mathbf{c}\,\mathbf{c}^T, \quad \mathbf{c} = \frac{1}{M} \sum_{j=1}^{M} \mathbf{h}(\mathbf{x}_j) = \frac{M_1 \mathbf{c}_1 + M_2 \mathbf{c}_2}{M_1 + M_2}. \tag{10}$$

We call Q_W the within-class scatter matrix.

Now, we want to maximize

$$J(\mathbf{w}) = \frac{d^2}{s^2} = \frac{\mathbf{w}^T Q_B \mathbf{w}}{\mathbf{w}^T Q_W \mathbf{w}}. \tag{11}$$

Taking the partial derivative of (11) with respect to $\mathbf{w}$ and equating the resulting equation to zero, we obtain the following generalized eigenvalue problem:

$$Q_B \mathbf{w} = \lambda Q_W \mathbf{w}, \tag{12}$$

where λ is a generalized eigenvalue.

Substituting

$$Q_W \mathbf{w} = \mathbf{c}_1 - \mathbf{c}_2 \tag{13}$$

into the left-hand side of (12), we obtain $(\mathbf{w}^T Q_W \mathbf{w}) Q_W \mathbf{w} = \lambda Q_W \mathbf{w}$. Thus, by letting $\lambda = \mathbf{w}^T Q_W \mathbf{w}$, (13) is a solution of (12).

If Q_W is positive definite, the optimum $\mathbf{w}$, $\mathbf{w}_{\text{opt}}$, is given by

$$\mathbf{w}_{\text{opt}} = Q_W^{-1} (\mathbf{c}_1 - \mathbf{c}_2). \tag{14}$$

If Q_W is positive semi-definite, i.e., singular, one way to overcome singularity is to add positive values to the diagonal elements [12]:

$$\mathbf{w}_{\text{opt}} = (Q_W + \varepsilon I)^{-1} (\mathbf{c}_1 - \mathbf{c}_2), \tag{15}$$

where ε is a small positive parameter.

Assuming that Q_W is positive definite, we substitute (14) into (11) and obtain

$$J(\mathbf{w}_{\text{opt}}) = (\mathbf{c}_1 - \mathbf{c}_2)^T \mathbf{w}_{\text{opt}}. \tag{16}$$

Linear discriminant analysis in the empirical feature space discussed above is equivalent to kernel discriminant analysis in the feature space, but since we can explicitly treat the variables in the empirical feature space, the calculation is much simpler.

3.4 Forward Selection

Starting from an empty set we add one datum at a time that maximizes (11) if the datum is added. Let the set of selected data indices be S^k and the set of remaining data indices be T^k, where k denotes that k data points are selected. Initially $S^0 = \phi$ and $T^0 = \{1, \ldots, M\}$. Let S_j^k denote that $\mathbf{x}_j$ for $j \in T^k$ is temporarily added to S^k. Let $\mathbf{h}^{k,j}(\mathbf{x})$ be the mapping function with $\mathbf{x}_j$ temporarily added to the selected data with indices in S^k:

$$\mathbf{h}^{k,j}(\mathbf{x}) = (H(\mathbf{x}_{i_1}, \mathbf{x}), \ldots, H(\mathbf{x}_{i_k}, \mathbf{x}), H(\mathbf{x}_j, \mathbf{x}))^T, \tag{17}$$

where $S^k = \{i_1, \ldots, i_k\}$. And let $J_{\mathrm{opt}}^{k,j}$ be the optimum value of the objective function with the mapping function $\mathbf{h}^{k,j}(\mathbf{x})$. Then we calculate

$$j_{\mathrm{opt}} = \arg_j J_{\mathrm{opt}}^{k,j} \qquad \text{for} \quad j \in T^k \tag{18}$$

and if the addition of $\mathbf{x}_{j_{\mathrm{opt}}}$ results in a sufficient increase in the objective function:

$$\left(J_{\mathrm{opt}}^{k,j_{\mathrm{opt}}} - J_{\mathrm{opt}}^k \right) / J_{\mathrm{opt}}^{k,j_{\mathrm{opt}}} \geq \eta_{\mathrm{L}}, \tag{19}$$

where η_{L} is a positive parameter, we increment k by 1 and add j_{opt} to S^k and delete it from T^k. If the above equation does not hold we stop forward selection. We must notice that $J_{\mathrm{opt}}^{k,j}$ is non-decreasing for the addition of data [13]. Thus the left-hand side of (19) is non-negative.

If the addition of a datum results in the singularity of $Q_{\mathbf{w}}^{k,j}$, where $Q_{\mathbf{w}}^{k,j}$ is the within-class scatter matrix evaluated using the data with $S^{k,j}$ indices, we consider the datum does not give useful information in addition to the already selected data. Thus, instead of adding a small value we do not consider this datum for a candidate of addition. This is equivalent to calculating the pseudo-inverse of $Q_{\mathbf{w}}^{k,j}$.

The necessary and sufficient condition of a matrix being positive definite is that all the principal minors are positive. And notice that the exchange of two rows and then the exchange of the associated two columns do not change the singularity of the matrix. Thus, if $\mathbf{x}_j$ causes the singularity of $Q_{\mathbf{w}}^{k,j}$, later addition will always cause singularity of the matrix. Namely, we can delete j from T^k permanently. If there are many training data that cause singularity of the matrix, forward selection becomes efficient.

Thus the procedure of independent data selection is as follows.

1. Set $S^0 = \phi$, $T^0 = \{1, \ldots, M\}$, and $k = 0$. Calculate j_{opt} given by (18) and set $S^1 = \{j_{\mathrm{opt}}\}$, $T^1 = T^0 - \{j_{\mathrm{opt}}\}$, and $k = 1$.
2. If for some $j \in T^k$, $Q_{\mathbf{w}}^{k,j}$ is singular, permanently delete j from T^k and calculate j_{opt} given by (18). If (19) is satisfied, go to Step 3. Otherwise terminate the algorithm.
3. Set $S^{k+1} = S^k \cup \{j_{\mathrm{opt}}\}$ and $T^{k+1} = T^k - \{j_{\mathrm{opt}}\}$. Increment k by 1 and go to Step 2.

Keeping the Cholesky factorization of $Q_{\mathbf{w}}^k$, the Cholesky factorization of $Q_{\mathbf{w}}^{k,j}$ is done incrementally; namely, using the factorization of $Q_{\mathbf{w}}^k$, the factorization of $Q_{\mathbf{w}}^{k,j}$ is obtained by calculating the $(k+1)$st diagonal element and column elements. This accelerates the calculation of the inverse of the within-class scatter matrix.

We call thus trained sparse LS SVM sparse LS SVM by forward selection, sparse LS SVM (L) for short.

4   Performance Evaluation

We compared the generalization ability and sparsity of primal, sparse, and dual LS SVMs using two groups of data sets: (1) two-class data sets [14,15] and (2) multi-class data sets [11,16]. We also evaluated regular SVMs to compare sparsity.

We normalized the input ranges into $[0, 1]$ and used RBF kernels. For the primal LS SVM we set $\eta_C = 10^{-9}$ and for the primal and dual LS SVMs, we determined the parameters C and γ by fivefold cross-validation; the value of C was selected from among $\{1, 10, 50, 100, 500, 1000, 2000, 3000, 5000, 8000, 10000, 50000, 100000\}$, the value of γ from among $\{0.1, 0.5, 1, 5, 10, 15\}$. For the sparse LS SVMs, we used the value of γ determined for the primal LS SVM, and determined the value of η_C from $\{10^{-4}, 10^{-3}, 10^{-2}, 0.05, 0.2\}$ and the value of η_L from $\{10^{-4}, 10^{-3}\}$ by fivefold cross-validation. We measured the computation time using a workstation (3.6GHz, 2GB memory, Linux operating system).

Then we compared the sparse LS SVM (L) with the methods discussed in [5]. We used diagonal Mahalanobis kernels [17] since in [5] the training inputs were converted into those with zero means and unit variances. We selected the value of the scale factor δ from $[0.1, 0.5, 1.0, 1.5, 2.0]$ and the value of C from $[1, 10, 50, 100, 500, 1000]$. We determined these values setting $\eta_L = 10^{-2}$ and then for the determined values we selected, from $i \times 10^{-4}$ $(i = 1, \ldots, 9)$, the largest value of η_L that realize the generalization ability comparable with that in [5].

4.1   Evaluation for Two-Class Problems

The two-class classification problems have 100 or 20 training data sets and their corresponding test data sets. We determined the parameter values by fivefold cross-validation for the first five training data sets.

Table 1 lists the determined parameters. In the table "Sparse (C)" and "Sparse (L)" denote the sparse LS SVM by the Cholesky factorization and that by forward selection proposed in this paper, respectively. The values of γ are not always the same for primal and dual problems. In most cases the values of η_L were 10^{-4} and were more stable than those of η_C. The table also lists the parameter values for the SVM. The values of γ are similar for the LS SVM and SVM.

Table 2 shows the average classification errors and standard deviations. Excluding the SVM, we statistically analyzed the average and standard deviations with the significance level of 0.05. Numerals in italic show that they are statistically inferior. Primal and sparse solutions for the ringnorm problem, primal

Table 1. Parameter setting for two-class problems

Data	Primal		Sparse (C)		Sparse (L)		Dual		SVM	
	γ	C	C	η_C	C	η_L	γ	C	γ	C
Banana	10	10^5	10^6	10^{-4}	10^5	10^{-4}	10	500	15	100
B. Cancer	0.5	500	1000	10^{-4}	1000	10^{-4}	0.5	10	1	10
Diabetes	1	500	10^4	10^{-2}	2000	10^{-3}	10	1	10	1
German	1	100	50	10^{-3}	100	10^{-4}	0.5	50	5	1
Heart	0.1	100	50	10^{-4}	1000	10^{-4}	0.1	10	0.1	50
Image	10	10^6	10^7	10^{-4}	10^8	10^{-4}	10	3000	10	1000
Ringnorm	0.1	1	10	10^{-3}	50	10^{-4}	10	1	15	1
F. Solar	0.5	100	500	10^{-3}	500	10^{-4}	0.1	100	1	1
Splice	5	100	100	0.2	500	10^{-4}	5	50	10	10
Thyroid	10	500	1000	10^{-4}	10^5	10^{-4}	10	50	5	1000
Titanic	5	100	100	10^{-2}	10	10^{-3}	0.5	500	10	10
Twonorm	0.1	1	100	10^{-5}	3000	10^{-3}	0.1	10	1	1
Waveform	10	10	1	5×10^{-2}	10	10^{-4}	10	1	5	10

Table 2. Comparison of the average classification errors (%) and the standard deviations of the errors

Data	Primal	Sparse (C)	Sparse (L)	Dual	SVM
Banana	11.0 ± 0.55	10.9 ± 0.54	11.2 ± 0.63	10.7 ± 0.52	10.4 ± 0.46
B. Cancer	25.5 ± 4.3	25.7 ± 4.2	25.5 ± 4.3	25.7 ± 4.5	25.6 ± 4.5
Diabetes	23.0 ± 1.8	23.0 ± 1.7	23.1 ± 1.8	23.2 ± 1.7	23.4 ± 1.7
German	23.6 ± 2.0	23.8 ± 2.1	23.7 ± 2.0	23.3 ± 2.1	23.8 ± 2.1
Heart	16.4 ± 3.4	16.4 ± 3.4	16.3 ± 3.1	16.0 ± 3.4	16.1 ± 3.1
Image	2.89 ± 0.37	2.85 ± 0.38	3.08 ± 0.38	2.66 ± 0.37	2.84 ± 0.50
Ringnorm	6.02 ± 3.0	7.56 ± 6.3	6.20 ± 2.4	4.08 ± 0.58	2.64 ± 0.35
F. Solar	33.3 ± 1.5	33.3 ± 1.5	33.3 ± 1.6	33.3 ± 1.6	32.3 ± 1.8
Splice	11.2 ± 0.48	11.2 ± 0.61	11.4 ± 0.53	11.3 ± 0.51	10.8 ± 0.71
Thyroid	5.59 ± 2.6	5.60 ± 2.7	5.15 ± 2.7	4.84 ± 2.5	4.05 ± 2.3
Titanic	22.5 ± 0.94	22.5 ± 0.94	22.7 ± 0.85	22.5 ± 0.97	22.4 ± 1.0
Twonorm	3.91 ± 1.9	2.10 ± 0.63	2.68 ± 0.22	1.90 ± 0.61	2.02 ± 0.64
Waveform	9.76 ± 0.38	9.68 ± 0.32	9.74 ± 0.38	14.9 ± 0.98	10.3 ± 0.40

solutions for the twonorm problem and dual solutions for the waveform problem show significantly inferior performance. The inferior solutions arose because of imprecise model selection [9].

Comparing the results of the LS SVMs with the SVM, the SVM showed the better generalization ability for some problems than LS SVMs. Or LS SVMs are not robust for parameter changes.

Table 3 lists the number of support vectors. The smallest number of support vectors in LS SVMs is shown in boldface. The number of support vectors for primal solutions is the number of training data at most. The numbers of support vectors for the sparse LS SVMs are, in general, much smaller than those for the primal and dual LS SVM. This tendency is evident especially for the sparse LS SVM (L); except for three problems it performed best.

Table 3. Comparison of support vectors

Data	Primal	Sparse (C)	Sparse (L)	Dual	SVM
Banana	93	42	**33**	400	173
B. Cancer	187	101	**67**	200	118
Diabetes	447	**22**	34	468	268
German	700	386	**351**	700	416
Heart	170	68	**34**	170	74
Image	1215	476	**198**	1300	149
Ringnorm	400	21	**10**	400	131
F. Solar	82	**16**	37	666	522
Splice	977	921	**619**	1000	749
Thyroid	140	70	**55**	140	13
Titanic	11	11	**9**	150	113
Twonorm	400	169	**8**	400	193
Waveform	400	**233**	313	400	114

Table 4. Comparison of computation time in seconds

Data	Primal	Sparse (C)	Sparse (L)	Dual	SVM
Banana	9.4	2.1	**1.4**	1.0	0.4
B. Cancer	6.2	1.8	**1.0**	0.05	0.3
Diabetes	283	**0.9**	2.0	0.4	1.3
German	1550	**467**	822	1.4	7.5
Heart	3.7	0.6	**0.2**	0.4	0.09
Image	16181	2489	**1123**	10.7	1.5
Ringnorm	168	0.9	**0.6**	1.8	0.9
F. Solar	20	**4.8**	7.5	1.2	7.8
Splice	6170	**5503**	6845	5.6	13
Thyroid	1.7	0.4	**0.3**	0.02	0.01
Titanic	**0.04**	**0.04**	0.05	0.11	0.2
Twonorm	167	30	**0.6**	1.9	1.4
Waveform	167	**58**	110	1.3	0.6

In calculating the classification errors listed in Table 2, we measured the computation time of training and testing for the 100 or 20 data sets in a classification problem. Then we calculated the average computation time for a training data set and its associated test data set. Table 4 lists the results. As a reference we include the computation time for the SVM, which was trained by the primal-dual interior-point methods combined with the decomposition technique. Training of primal problems is slow especially for diabetes, german, image, and splice problems. Except for german, image, splice, and waveform computation time for sparse LS SVMs (L) is the shortest; speed-up was mostly caused by the frequent matrix singularity. But for those four problems, matrix singularity was rare and because of relatively large training data, forward selection took time.

4.2 Evaluation for Multi-class Problems

Each multi-class problem has one training data set and one test data set. We used fuzzy pairwise LS SVMs with minimum operators [11] to resolve unclassifiable regions. For comparisons we also used fuzzy pairwise SVMs.

Table 5. Parameter setting for multi-class problems

Data	Sparse (C)		Sparse (L)		Dual		SVM	
	C	η_C	C	η_L	C	γ	γ	C
Iris	10^6	10^{-4}	10^6	10^{-4}	0.1	500	0.1	5000
Numeral	10^6	10^{-4}	10^7	10^{-3}	0.1	100	5	10
Thyroid (M)	10^8	10^{-3}	10^9	10^{-4}	10	10^6	5	10^5
Blood cell	10^7	10^{-4}	10^7	10^{-4}	1	7000	10	1000
H-50	10^8	10^{-3}	10^5	10^{-4}	5	7000	10	500
H-13	10^9	10^{-4}	10^9	10^{-4}	1	10^8	10	3000
H-105	3000	10^{-2}	2000	10^{-3}	10	50	10	50

Table 6. Comparison of classification errors (%) and the numbers of support vectors

Data	Sparse (C)		Sparse (L)		Dual		SVM	
	ERs	SVs	ERs	SVs	ERs	SVs	ERs	SVs
Iris	2.67	8	4.00	**4**	2.67	50	5.33	10
Numeral	**0.73**	18	**0.73**	**4**	0.85	162	0.61	8
Thyroid (M)	4.87	354	4.96	**160**	**4.52**	2515	2.71	75
Blood cell	5.68	141	6.13	**25**	**5.65**	516	7.16	21
H-50	0.80	148	**0.74**	**37**	0.80	236	0.74	16
H-13	0.49	39	0.48	**23**	**0.14**	441	0.37	10
H-105	**0**	259	0.02	**15**	**0**	441	0	26

Table 7. Comparison of computation time in seconds

Data	Sparse (C)	Sparse (L)	Dual	SVM
Iris	**0.005**	0.007	0.02	0.01
Numeral	2.9	**1.6**	9.8	2.0
Thyroid (M)	39525	**27487**	524	9.1
Blood cell	364	**181**	210	9.6
H-50	12127	**1652**	2589	128
H-13	2387	**1585**	5108	368
H-105	112175	**1176**	14184	358

Table 5 lists the parameters determined by cross-validation. The values of C of sparse solutions are larger than those of dual solutions. The dual LS SVM and the SVM selected similar γ values, but unlike two-class problems, they selected the similar C values.

Table 6 lists the classification errors and the average number of support vectors per class pair. Excluding those of SVMs, the smallest errors and the support vectors are shown in boldface. Including SVMs, the difference of the classification errors is small. The number of support vectors for the sparse LS SVM (L) is the smallest among LS SVMs and sometimes smaller than those of SVMs. Therefore, sufficient sparsity is realized by the sparse LS SVMs (L).

Table 7 lists the computation time of training and testing. Between sparse LS SVMs, the shorter computation time is shown in boldface. Except for the iris problem, computation time of sparse LS SVMs (L) is shorter. Comparing with the dual LS SVM and the SVM, the dual LS SVM is slower because of the slow Cholesky factorization for large matrices.

4.3   Comparison with Other Methods

We used the two-class ringnorm data set with 3000 training data and 4400 test data and the 6-class satimage data set with 4435 training data and 2000 test data used in [5]. Since training took time, we used pairwise LS SVMs instead of one-against-all LS SVMs in [5]. Table 8 shows the parameter setting for the sparse LS SVM (L) and the comparison results. The results other than Sparse (L) are from [5]. We measured the computation time by a 3.0GHz Windows machine with 2Gbyte memory, while in [5] the training time was measured by a 2.4GHz Windows machine with 1Gbyte memory. The "SVs" row denotes the total number of distinct support vectors and the "Rate" row denotes the recognition rate for the test data set. For the satimage data set we could not measure time but it took long time. From the table, the numbers of support vectors for Sparse (L) are smallest with comparable recognition rates. The training time for the ringnorm data set is comparable, but for the satimage data set, it was very slow because we did not use the subset selection method used in [5].

Table 8. Parameter setting and comparison with other methods

Data		Parm	Sparse (L)		Term	Sparse (L)	PFSALS-SVM	KMP	SGGP
Ringnorm	δ	1.5		SVs	**6**	661	77	74
	C	1		Rate	**98.64**	98.52	98.11	98.27
	η_L	10^{-3}		Time	9.9	7.1	7.5	50
Satimage	δ	1.5		SVs	1581	1726	—	—
	C	100		Rate	92.05	**92.25**	—	—
	η_L	2×10^{-4}		Time	—	50	—	—

5   Conclusions

In this paper we proposed sparse LS SVMs by forward selection of independent data based on linear discriminant analysis in the empirical feature space. Namely, starting from the empty set, we add the training data that maximally separate two-classes in the empirical feature space. We measure the class separability by the objective function of linear discriminant analysis. To speed up forward selection, we exclude data that cause matrix singularity from later addition and use the incremental calculation in calculating the inverse of the within-class scatter matrix. For most of the two-class and multi-class problems tested, sparsity of the solutions was increased drastically compared to the method using the Cholesky factorization in reducing the dimension of the empirical feature space and for most of the problems training of LS SVMs by forward selection was the fastest. For comparison with other methods, sparsity of the solutions was the highest.

References

1. Suykens, J.A.K., Van Gestel, T., De Brabanter, J., De Moor, B., Vandewalle, J.: Least Squares Support Vector Machines. World Scientific Publishing, Singapore (2002)
2. Vapnik, V.N.: Statistical Learning Theory. John Wiley & Sons, Chichester (1998)
3. Cawley, G.C., Talbot, N.L.C.: Improved sparse least-squares support vector machines. Neurocomputing 48, 1025–1031 (2002)
4. Valyon, J., Horváth, G.: A sparse least squares support vector machine classifier. In: Proc. IJCNN 2004, vol. 1, pp. 543–548 (2004)
5. Jiao, L., Bo, L., Wang, L.: Fast sparse approximation for least squares support vector machine. IEEE Trans. Neural Networks 18(3), 685–697 (2007)
6. Smola, A.J., Bartlett, P.L.: Sparse greedy Gaussian process regression. Advances in Neural Information Processing Systems 13, 619–625 (2001)
7. Vincent, P., Bengio, Y.: Kernel matching pursuit. Machine Learning 48(1-3), 165–187 (2002)
8. Xiong, H., Swamy, M.N.S., Ahmad, M.O.: Optimizing the kernel in the empirical feature space. IEEE Trans. Neural Networks 16(2), 460–474 (2005)
9. Abe, S.: Sparse least squares support vector training in the reduced empirical feature space. Pattern Analysis and Applications 10(3), 203–214 (2007)
10. Kaieda, K., Abe, S.: KPCA-based training of a kernel fuzzy classifier with ellipsoidal regions. International Journal of Approximate Reasoning 37(3), 145–253 (2004)
11. Abe, S.: Support Vector Machines for Pattern Classification. Springer, Heidelberg (2005)
12. Mika, S., Rätsch, G., Weston, J., Schölkopf, B., Müller, K.-R.: Fisher discriminant analysis with kernels. In: Proc. NNSP 1999, pp. 41–48 (1999)
13. Ashihara, M., Abe, S.: Feature selection based on kernel discriminant analysis. In: Kollias, S., Stafylopatis, A., Duch, W., Oja, E. (eds.) ICANN 2006. LNCS, vol. 4132, pp. 282–291. Springer, Heidelberg (2006)
14. Rätsch, G., Onoda, T., Müller, K.-R.: Soft margins for AdaBoost. Machine Learning 42(3), 287–320 (2001)
15. http://ida.first.fraunhofer.de/projects/bench/benchmarks.htm
16. ftp://ftp.ics.uci.edu/pub/machine-learning-databases/
17. Abe, S.: Training of support vector machines with Mahalanobis kernels. In: Duch, W., Kacprzyk, J., Oja, E., Zadrożny, S. (eds.) ICANN 2005. LNCS, vol. 3697, pp. 571–576. Springer, Heidelberg (2005)

Supervised Incremental Learning with the Fuzzy ARTMAP Neural Network

Jean-François Connolly, Eric Granger, and Robert Sabourin

Laboratoire d'imagerie, de vision et d'intelligence artificielle
Dépt. de génie de la production automatisée
École de technologie supérieure
1100 rue Notre-Dame Ouest,
Montreal, Quebec,
Canada, H3C 1K3*
jfconnolly@livia.etsmtl.ca, eric.granger@etsmtl.ca,
robert.sabourin@etsmtl.ca

Abstract. Automatic pattern classifiers that allow for on-line incremental learning can adapt internal class models efficiently in response to new information without retraining from the start using all training data and without being subject to catastrophic forgeting. In this paper, the performance of the fuzzy ARTMAP neural network for supervised incremental learning is compared to that of supervised batch learning. An experimental protocole is presented to assess this network's potential for incremental learning of new blocks of training data, in terms of generalization error and resource requirements, using several synthetic pattern recognition problems. The advantages and drawbacks of training fuzzy ARTMAP incrementally are assessed for different data block sizes and data set structures. Overall results indicate that error rate of fuzzy ARTMAP is significantly higher when it is trained through incremental learning than through batch learning. As the size of training blocs decreases, the error rate acheived through incremental learning grows, but provides a more compact network using fewer training epochs. In the cases where the class distributions overlap, incremental learning shows signs of over-training. With a growing numbers of training patterns, the error rate grows while the compression reaches a plateau.

1 Introduction

The performance of statistical and neural pattern classifiers depends heavily on the availability of representative training data. The collection and analysis of such data is expensive and time consuming in many practical applications. Training data may, therefore, be incomplete in one of several ways. In an environments where class distributions remain fixed, these include a limited number of training observations, missing components of the input observations, missing class labels during training, and missing classes (*i.e.*, some classes that were not present in the training data set may be encountered during operations) [7].

* This research was supported in part by the Natural Sciences and Engineering Research Council of Canada.

L. Prevost, S. Marinai, and F. Schwenker (Eds.): ANNPR 2008, LNAI 5064, pp. 66–77, 2008.

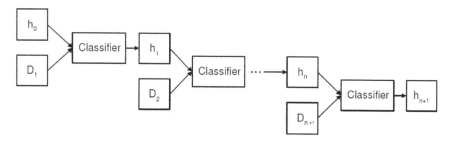

Fig. 1. A generic incremental learning scenario where blocks of data are used to update the classifier in an incremental fashion over time. Let $D_1, D_2, ..., D_{n+1}$ be the blocks of training data available to the classifier at discrete instants in time $t_1, t_2, ..., t_{n+1}$. The classifier starts with initial hypothesis h_0 which constitutes the prior knowledge of the domain. Thus, h_0 gets updated to h_1 on the basis of D_1, and h_1 gets updated to h_2 on the basis of data D_2, and so forth [5].

Given an environment where class distributions are fixed, and in which training data is incomplete, a critical feature of future automatic classification systems is the ability to update their class models incrementally during operational phases in order to adapt to novelty encountered in the environment [5] [9]. As new information becomes available, internal class models should be refined, and new ones should be created on the fly, without having to retrain from the start using all the cumulative training data.

For instance, in many practical applications, additional training data may be acquired from the environment at some point in time after the classification system has originally been trained and deployed for operations (see Fig. 1). Assume that this data is characterized and labeled by a domain expert, and may contain observations belonging to classes that are not present in previous training data, and classes may have a wide range of distributions. It may be too costly or not feasible to accumulate and store all the data used thus far for supervised training, and to retrain a classifier using all the cumulative data[1]. In this case, it may only be feasible to update the system through *supervised incremental learning*.

Assuming that new training data becomes available, incremental learning provides the means to efficiently maintain an accurate and up-to-date class models. Another advantage of incremental learning is the lower convergence time and memory complexity required to update a classifier. Indeed, temporary storage of the new data is only required during training, and training is only performed with the new data. Strategies adopted for incremental learning will depend on the application – the nature of training data, the environment, performance constraints, etc. Regardless of the context, updating a pattern classification system in an incremental fashion raises several technical issues.

In particular, accommodating new training data may corrupt the classifier's previously acquired knowledge structure and compromise its ability to achieve a high level of generalization during future operations. The *stability-plasticity dilemma* [1] refers to

---

[1] The vast majority of statistical and neural pattern classifiers proposed in literature can only perform *supervised batch learning* to learn new data. They must accumulate and store all training data in memory, and retrain from the start using all previously-accumulated training data.

the problem of learning new information incrementally, yet overcoming the problem of catastrophic forgetting.

This paper focuses on techniques that are suitable for supervised incremental learning in an environment where class distributions remain fixed over time. According to Polikar [12], an incremental learning algorithm should:

1. allow to learn additional information from new data,
2. not require access to the previous training data,
3. preserve previously acquired knowledge, and
4. accommodate new classes that may be introduced with new data.

In literature, some promising pattern classification algorithms have been reported for supervised incremental learning in environments where distributions are fixed. For example, the ARTMAP [2] and Growing Self-Organizing [6] families of neural network classifiers, have been designed with the inherent ability to perform incremental learning. In addition, some well-known pattern classifiers, such as the Support Vector Machine (SVM), and the Multi-Layer Perceptron (MLP) and Radial Basis Function (RBF) neural networks have been adapted to perform incremental learning [10] [11] [13]. Finally, some high-level architectures, based on well-known pattern classifiers, e.g., Ensemble of Classifiers, have also been proposed [12].

In this paper, the performance of the fuzzy ARTMAP [4] neural network is characterize for supervised incremental learning of new blocks of training data in an environment where class distributions are fixed. Fuzzy ARTMAP is the most popular ARTMAP network. While its incremental learning capabilities are often cited in literature, to the authors knowledge, these capabilities have not been assessed. An experimental protocole has thus been defined such that the impact on performance of learning a new block of training data incrementally, after each network has previously been trained, is assessed for different types of synthetic pattern recognition problems. The first type of problem consists of data with overlapping class distributions, whereas the second type involves data with complex decision boundaries but no overlap. With this protocole, the advantages and drawbacks of the ARTMAP architectures are discussed for incremental learning data using different data block sizes, and using different data set structures (overlap, dispersion, etc.).

In the next section, fuzzy ARTMAP is briefly reviewed. Then, the experimental protocol, performance measures and synthetic data sets, used for proof-of-concept computer simulations, are described in Section 3. Finally, experimental results are presented and discussed in Section 4.

2   ARTMAP Neural Networks

ARTMAP refers to a family of neural network architectures based on Adaptive Resonance Theory (ART) [1] that is capable of fast, stable, on-line, unsupervised or supervised, incremental learning, classification, and prediction [2]. A key feature of the ARTMAP networks is their unique solution to the stability - plasticity dilemma.

Several ARTMAP networks have been proposed in order to improve the performance of these architectures. Members of the ARTMAP family can be broadly divided according to their internal matching process, which depends on either deterministic or probabilistic category activation. The deterministic type consists of networks such as fuzzy ARTMAP, ART-EMAP, ARTMAP-IC, default ARTMAP, simplified ARTMAP, distributed ARTMAP, etc., and represent each class using one or more fuzzy set hyper-rectangles. In contrast, the probabilistic type consists of networks such as PRO-BART, PFAM, MLANS, Gaussian ARTMAP, ellipsoid ARTMAP, boosted ARTMAP, μARTMAP, etc., and represent each class using one or more probability density functions.

This paper focuses on the popular fuzzy ARTMAP neural network [4]. It integrates the fuzzy ART [3] in order to process both analog and binary-valued input patterns to the original ARTMAP architecture [2]. The rest of this section provides a brief description of fuzzy ARTMAP.

2.1 Fuzzy ARTMAP

The fuzzy ART neural network consists of two fully connected layers of nodes: a $2M$ node input layer F_1 to accomodate complement-coded input patterns, and an N node competitive layer, F_2. A set of real-valued weights $\mathbf{W} = \{w_{ij} \in [0,1] : i = 1, 2, ..., 2M; j = 1, 2, ..., N\}$ is associated with the F_1-to-F_2 layer connections. The F_2 layer is connected, through learned associative links, to an L node map field F_{ab}, where L is the number of classes in the output space. A set of binary weights $\mathbf{W}^{ab} = \{w_{jk}^{ab} \in \{0,1\} : j = 1, 2, ..., N; k = 1, 2, ..., L\}$ is associated with the F_2-to-F_{ab} connections. Each F_2 node $j = 1, ..., N$ corresponds to a category that learns a proto-type vector $\mathbf{w}_j = (w_{1j}, w_{2j}, ..., w_{2Mj})$, and is associate with one of the output classes $K = 1, ..., L$. During the training phase, fuzzy ARTMAP dynamics is govern by four hyperparameters: the choice parameter $\alpha > 0$, the learning parameter $\beta \in [0,1]$, the baseline vigilance parameter $\bar{\rho} \in [0,1]$, and the matchtracking parameter ϵ. In term of incremental learning, the learning algorithm is able to adjusts previously-learned categories, in response to familiar inputs, and to creates new categories dynamically in response to inputs different enough from those already seen.

The following describes fuzzy ARTMAP during supervised learning of a finite data set. When an input pattern $\mathbf{a} = (a_1, ..., a_M)$ is presented to the network and the vigilance parameter $\rho \in [0,1]$ is set to its baseline value $\bar{\rho}$. The original M dimensions input pattern $\mathbf{a}$ is complement-coded to make a $2M$ dimensions network's input pattern: $\mathbf{A} = (\mathbf{a}, \mathbf{a}^c) = (a_1, a_2, ..., a_M; a_1^c, a_2^c, ..., a_M^c)$, where $a_i^c = (1 - a_i)$, and $a_i \in [0,1]$. Each F_2 node is activated according to the *Weber law choice function*: $T(\mathbf{A}) = |\mathbf{A} \wedge \mathbf{w}_j| / (\alpha + |\mathbf{w}_j|)$, and the node with the strongest activation $J = \text{argmax} \{T_j : j = 1, ..., N\}$ is chosen. The algorithm then verifies if $\mathbf{w}_J$ is similar enough to $\mathbf{A}$ using the vigilance test: $|\mathbf{A} \wedge \mathbf{w}_J| / 2M \geq \rho$. If node J fails the vigilance test, it is disactivated and the network searches for the next best node on the F_2 layer. If the vigilance test is passed, then the map field F^{ab} is activated through the category J and fuzzy ARTMAP makes a class prediction $K = k(J)$. In the case of an incorrect class prediction $K = k(J)$, a match tracking signal raises $\rho = (|\mathbf{A} \wedge \mathbf{w}_J| / 2M) + \epsilon$. Node J is disactivated, and the search among F_2 nodes begins anew. If node J passes the vigilance test, and makes the correct prediction, its category is updated by adjusting its prototype vector: $\mathbf{w}_J$ to $\mathbf{w}_J' = \beta(\mathbf{A} \wedge \mathbf{w}_J) + (1 - \beta)\mathbf{w}_J$. On the other hand, if none of the nodes can satisfy both conditions (vigilance test and correct prediction), then a new F_2 node is initialed. This new node is assigned to class K by setting w_{Jk}^{ab} to 1 if $k = K$ and 0 otherwise.

Once the weights $\mathbf{W}$ and $\mathbf{W}^{ab}$ have been found through this process, the fuzzy ARTMAP can predict a class label from a input pattern by activating the best F_2 node J, which activates a class $K = k(J)$ on the F_{ab} layer. Predictions are obtained without vigilance and match tests.

3   Experimental Methodology

3.1   Experimental Protocole

In order to observe the impact on performance of training a classifier with supervised incremental learning for different data structures, several data sets were selected for computer simulations. The synthetic data sets are representative of pattern recognition problems that involve either (1) simple decision boundaries with overlapping class distributions, or (2) complex decision boundaries without overlap on decision boundaries. The synthetic data sets correspond to 2 classes problems, with a 2 dimensional input feature space. Each data subset is composed of an equal number of 10,000 patterns per class, for a total of 20,000 (2 classes) randomly-generated patterns.

Prior to a simulation trial, each data set is normalized according to the min-max technique and partitioned into two equal parts – the learning and test subsets. The learning subset is divided into training and validation subsets. They respectively contain 2/3 and 1/3 of patterns from each class of the learning subset. In order to perform block-wise hold-out validation over several training *epochs*[2], the training and validation subsets are again divided into b blocks. Each block D_i ($i = 1, 2, ..., b$) contains an equal number of patterns per class. To observe the impact on performance of learning new blocks of training data incrementally for different data block sizes, two different cases are observed. The first case consists in training with $b = 10$, where $|D_i| = 1000$ patterns, while the second one consists in training with $b = 100$, where $|D_i| = 100$ patterns.

During each simulation trial, fuzzy ARTMAP is trained using a batch learning and incremental learning process. For *batch learning*, $|D_i|$ is set to the smaller block size, in our case $|D_i| = 100$, and the number of blocks D_n used for training is progressively increased from 1 to 100. For the nth trial, performance is assessed after initializing a fuzzy ARTMAP network and training it until convergence on $B_n = D_i \cup ... \cup D_n$. Since there is 100 blocks D_i, there will be 100 trials.

On the other hand, *incremental learning* consists in training the ARTMAP networks, until convergence, over one or more training epochs on successive blocks of data D_i. The training of each data block is done in isolation without reinitializing the networks. In the case of incremental learning two block sizes will be tested: $|D_i| = 100$ and when $|D_i| = 1000$. At first, performance is assessed after initializing an ARTMAP network and training on D_1. Then it is assessed after training the *same* ARTMAP network incrementally on D_2, and so on, until all b blocks are learned.

For each trial, learning is performed using a hold-out validation technique, with network training halted for validation after each epoch [8]. The performance of fuzzy

---

[2] An epoch is defined as one complete presentation of all the patterns of a finite training data set.

ARTMAP was measured when using standard parameter settings that yield minimum network resources (internal categories, epochs, etc.): $\beta = 1$, $\alpha = 0.001$, $\bar{\rho} = 0$ and $\epsilon = 0.001$ [4].

Since ARTMAP performance is sensitive to the presentation order of the training data, the pattern presentation orders were always randomized from one epoch to the next. In addition, each simulation trial was repeated 10 times with 10 different randomly generated data sets (learning and test). The average performance of fuzzy ARTMAP was assessed in terms of resources requirements and generalisation error. The amount of resources is measured by compression and convergence time. *Compression* refers to the average number of training patterns per category prototype created in the F_2 layer. *Convergence time* is the number of epochs required to complete learning for a training strategy. It does not include presentations of the validation subset used to perform hold-out validation. *Generalisation error* is estimated as the ratio of incorrectly classified test subset patterns over all test set patterns. The combination of compression and convergence time provides useful insight into the amount of processing required by fuzzy ARTMAP during training to produce its best asymptotic generalisation error. Average results, with corresponding standard error, are always obtained, as a result of the 10 independent simulation trials.

3.2   Data Sets

Of the five synthetic data sets selected for simulations, two have simple decision boundaries with overlapping class distributions ($D_{2N}(\xi_{tot})$ and $D_{XOR}(\xi_{tot})$) and three have complex decision boundaries without overlap (D_{XOR-U}, D_{CIS} and D_{P2}). The two classes in D_{2N} and D_{XOR} are randomly generated with normal distributions and the total theoretical probability of error associated with these problems is denoted by ξ_{tot}. Data from the classes in D_{XOR-U}, D_{CIS} and D_{P2} are uniform distributions, and since class distributions do not overlap on decision boundaries, the total theoretical probability of error for these data sets is 0.

The $D_{2N}(\xi_{tot})$ data (Fig. 2a) consists of two classes, each one defined by a normal distribution in a two dimensional input feature space [8]. Both sources are described by variables that are independent, have equal covariance Σ, and their distributions are hyperspherical. With the $D_{XOR}(\xi_{tot})$ problem, data is generated by 2 classes according to bi-modal distributions (Fig. 2b). The four normal distributions are centered in the 4 squares of a classical XOR problem. For those two problems, the degree of overlap

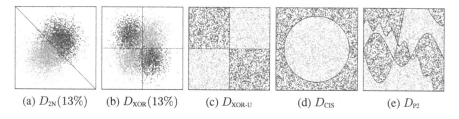

(a) $D_{2N}(13\%)$ (b) $D_{XOR}(13\%)$ (c) D_{XOR-U} (d) D_{CIS} (e) D_{P2}

Fig. 2. Example of the five data sets generated for a replication of each problem

is varied from a total probability of error ξ_{tot} = 1%, 13%, and 25%, by changing the covariance of each normal distribution.

With the D_{XOR-U} data, the 'on' and 'off' classes of the classical XOR problem are divided by a horizontal decision bound at $y = 0.5$, and a vertical decision bound at $x = 0.5$ (Fig. 2c). The Circle-in-Square problem D_{CIS} (Fig. 2d) requires a classifier to identify the points of a square that lie inside a circle, and those that lie outside a circle [2]. The circle's area equals half of the square. It consists of one non-linear decision boundary where classes do not overlap. Finally, with the D_{P2} problem (Fig. 2e), each decision region of its 2 classes is delimited by one or more of its four polynomial and trigonometric functions, and belongs to one of the two classes [14].

4   Simulation Results

4.1   Overlapping Class Distributions

Figure 3 presents the average performance achieved as a function of the training subset size, when fuzzy ARTMAP is trained using batch and incremental learning on the

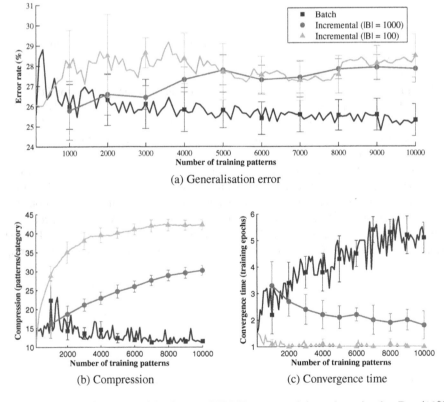

(a) Generalisation error

(b) Compression

(c) Convergence time

Fig. 3. Average performance of the fuzzy ARTMAP versus training subset size for $D_{XOR}(13\%)$ using batch and incremental learning. Each curve is shown along with 90% confidence interval.

$D_{\text{XOR}}(13\%)$ data set. For incremental learning, block sizes of 100 and 1000 are employed. Very similar tendencies are found in simulation results with other data sets with class distributions overlap ($D_{2N}(\xi\%)$ and $D_{\text{XOR}}(\xi\%)$).

As shown in Fig. 3a, the error rate obtained by training fuzzy ARTMAP through incremental learning is generally significantly higher than that obtained through batch learning. Using the smaller block size ($|D_i| = 100$) yields a higher error rate that with the larger block size ($|D_i| = 1000$), but this difference is not significant. In addition, error tends to grow with the number of blocks having been learned. For example, after the fuzzy ARTMAP network undergoes incremental learning of 100 blocks with $|D_i| = 100$, the average error is about 29.3%, yet after learning 10 blocks with $|D_i| = 1000$, the error is about 27,6%.

Although the error is greater, Fig. 3b indicates that the compression obtained when fuzzy ARTMAP is trained through incremental learning is significantly higher than if trained through batch learning, and it tends to grow as the block size is decreased. Incremental learning also tends to reduce the number of training epochs required for fuzzy ARTMAP to converge (see Fig. 3c). As the block size decreases, the convergence time tends towards 1. With incremental learning, the first blocks have a tendency to require a greater number of epochs. For example, after fuzzy ARTMAP undergoes incremental learning of 100 blocks with $|D_i| = 100$, the average compression and convergence time are about 40 patterns/category and 1.0 epoch, respectively. After learning 10 blocks with $|D_i| = 1000$, the compression and convergence time are about 30 patterns/category and 1.8 epochs. This compares favorably to fuzzy ARTMAP trained through batch learning, where the compression and convergence time are about 10 patterns/category and 5.0 epochs. In this case, the performance of fuzzy ARTMAP as the training set size grows is indicative of overtraining [8].

4.2 Complex Decision Boundaries

Fig. 4 presents the average performance achieved as a function of the training subset size, when the fuzzy ARTMAP is trained using batch and incremental learning on the D_{CIS} data set. Very similar tendencies are found in simulation results for other data set where complex boundaries and class distributions that do not overlap ($D_{\text{XOR-U}}$ and D_{P2}).

As shown in Fig. 4a, when the training set size increases, the average generalisation error of fuzzy ARTMAP trained with either batch or incremental learning decreases asymptotically towards its minimum. However, the generalisation error obtained by training fuzzy ARTMAP through incremental learning is generally significantly higher than that obtained through batch learning. As with the data that has overlapping class distributions, the error tends to grow as the block size decreases. However, after the fuzzy ARTMAP network performs incremental learning of 100 blocks with $|D_i| = 100$, the average error is comparable to after learning 10 blocks with $|D_i| = 1000$ (about 4.5%).

Again, training fuzzy ARTMAP through incremental learning yields a significantly higher compression than with batch learning (Fig. 4b). Furthermore, the convergence time associated with incremental learning is considerably lower than with batch learning (Fig. 4c). Results indicate that as the block size is decreased and the number of learned blocks increases, the convergence time with incremental learning tends towards

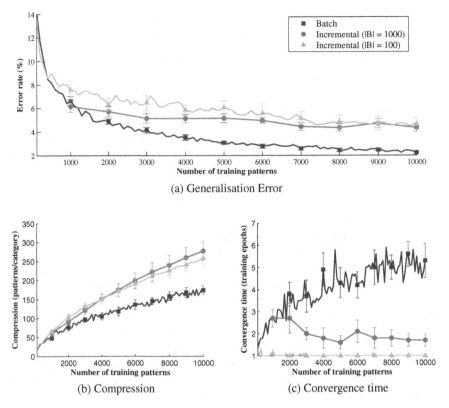

Fig. 4. Average performance of the fuzzy ARTMAP versus training subset size for D_{CIS} using batch and incremental learning. Each curve is shown along with 90% confidence interval.

1. For example, after fuzzy ARTMAP undergoes incremental learning of 100 blocks with $|D_i| = 100$, the average compression and convergence time are about 260 patterns/category and 1.0 epoch, respectively. After learning 10 blocks with $|D_i| = 1000$, the compression and convergence time are about 280 patterns/category and 1.8 epochs.

4.3 Discussion

Overall results indicate that when fuzzy ARTMAP undergoes incremental learning, the networks tend to become more compact, but the error rate tends to degrade. As shown in Fig. 5, this is reflected by decision boundaries among classes that become coarser as the block size decreases. Note that batch learning is equivalent to $|D_1| = 10000$. Since training on each block is performed in isolation, when fuzzy ARTMAP is trained on large data blocks, it has sufficient information to converge toward an optimal solution. Small data blocks represent the higher bound on the error rate. In our study, the smallest block size considered is $|D_i| = 100$ where each pattern must only be presented, on average, one time to the neural network for convergence (Figs. 3c and 4c). In this case, the network can create or update the model, but it appears to lack the necessary information to truly converge toward a solution over several training epochs.

Incremental learning ($|D_i| = 100$)

Incremental learning ($|D_i| = 1000$)

Batch learning

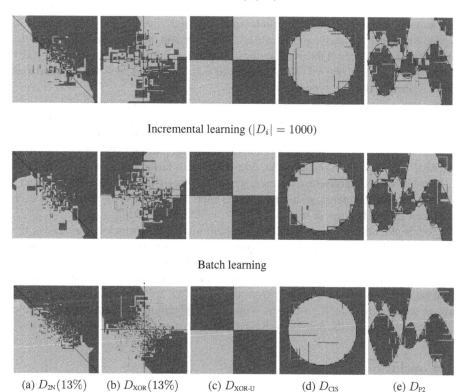

(a) $D_{2N}(13\%)$ (b) $D_{XOR}(13\%)$ (c) D_{XOR-U} (d) D_{CIS} (e) D_{P2}

Fig. 5. Decision boundaries for the best replication after learning all of the training patterns through batch and incremental learning, using the five data sets. The boundaries are shown for incremental learning with $|D_i| = 100$, incremental learning with $|D_i| = 1000$, and batch learning (i.e. $|D_1| = 10000$).

Table 1. Average generalisation error of fuzzy ARTMAP classifiers trained using batch and incremental learning with blocks of $|D_i| = 1000$ and $|D_i| = 100$ on all data from synthetic sets. Values are shown with the 90% confidence interval.

Data Set	Average generalisation error (%)		
	Batch	**Incremental ($|D_i| = 1000$)**	**Incremental ($|D_i| = 100$)**
$D_{2N}(1\%)$	2.1 ± 0.3	3.2 ± 0.5	2.8 ± 0.3
$D_{2N}(13\%)$	21.6 ± 0.6	23.8 ± 0.9	24.9 ± 1.1
$D_{2N}(25\%)$	35.8 ± 0.4	37.8 ± 0.4	38.4 ± 0.4
$D_{XOR}(1\%)$	1.2 ± 0.1	1.7 ± 0.5	1.7 ± 0.4
$D_{XOR}(13\%)$	25.3 ± 0.8	27.6 ± 0.6	29.3 ± 1.1
$D_{XOR}(25\%)$	43.0 ± 0.5	44.0 ± 0.5	44.0 ± 0.6
D_{XOR-U}	0.2 ± 0.1	0.6 ± 0.1	0.9 ± 1.1
D_{CIS}	2.2 ± 0.1	4.3 ± 0.3	4.7 ± 0.5
D_{P2}	5.1 ± 0.2	7.9 ± 0.3	9.4 ± 0.7

With overlapping data, even when the geometry of decision bounds matches the rectangular categories of fuzzy ARTMAP in $D_{\mathrm{XOR}}(\xi_{tot})$, the network still leads to the well known category proliferation problem. Compared with $D_{\mathrm{2N}}(\xi_{tot})$, the region of overlap isn't localized, so the proliferation is amplified and the error rate is higher. If the classes don't overlap, or if the overlapping is very low, results show that the complexity of boundaries with fuzzy ARTMAP is defined mainly by how well these boundaries can be represented using hyper-rectangles.

One key issue here is the internal mechanisms used by the network to learn new information. With the fuzzy ARTMAP network, only the internal vigilance parameter (ρ) is allow to grow dynamically during the learning process, over a range define by the baseline vigilance ($\bar{\rho}$). Since all network hyperparameter play an important role in fuzzy ARTAMP's ability to learn new data, one potential solution could imply optimizing the network's hyperparameters for incremental learning [8]. This way, all four fuzzy ARTMAP hyperparameters would be adapted such that the network would learn each data block $|D_i|$ to the best of its capabilities.

Another potential solution could be to exploit the learning block by organising the data in a specific way. Since the first blocks form the basis for future updates, results underline the importance of initiating incremental learning with blocks that contain enough representative data from the environment. With overlapping data, the first blocks could be organized to grow classes from the inside towards the overlapping regions, through some active learning strategy. With complex boundaries, the first blocks could be organized to define the non-linear bounds between classes.

5   Conclusion

In many practical applications, classifiers found inside pattern recognition systems may generalize poorly as they are designed prior to operations using limited training data. Techniques for on-line incremental learning would allow classifiers to efficiently adapt internal class models during operational phases, without having to retrain from the start using all the cumulative training data, and without corrupting the previously-learned knowledge structure. In this paper, fuzzy ARTMAP's potential for supervised incremental learning is assessed. An experimental protocole is proposed to characterize its performances for supervised incremental learning of new blocks of training data in an environment where class distributions are fixed. This protocole is based on a comprehensive set of synthetic data with overlapping class distributions and with complex decision boundaries, but no overlap.

Simulation results indicate that the average error rate obtained by training fuzzy ARTMAP through incremental learning is usually significantly higher than that obtained through batch learning, and that error tends to grow as the block size decreases. Results also indicate that training fuzzy ARTMAP through incremental learning often requires fewer training epochs to converge, and leads to more compact networks. As the block size decreases, the compression tends to increase and the convergence time tends towards one. The subject for futur work involve designing fuzzy ARTMAP networks that can approach the error rates of batch learning with incremental learning.

Some promising directions include organizing the blocks of training data through active learning and optimizing fuzzy ARTMAP hyperparameter values for incremental learning.

References

1. Carpenter, G.A., Grossberg, S.: A Massively Parallel Architecture for a Self-Organizing. Neural Pattern Recognition Machine, Computer, Vision, Graphics and Image Processing 37, 54–115 (1987)
2. Carpenter, G.A., Grossberg, S., Reynolds, J.H.: ARTMAP: Supervised Real-Time Learning and Classification of Nonstationary Data by a SONN. Neural Networks 4, 565–588 (1991)
3. Carpenter, G.A., Grossberg, S., Rosen, D.B.: Fuzzy ART: Fast Stable Learning and Categorization of Analog Patterns by an Adaptive Resonance System. Neural Networks 4(6), 759–771 (1991)
4. Carpenter, G.A., Grossberg, S., Markuzon, N., Reynolds, J.H., Reynolds, Rosen, D.B.: Fuzzy ARTMAP: A Neural Network Architecture for Incremental Supervised Learning of Analog Multidimensional Maps. IEEE Trans. on Neural Networks 3(5), 698–713 (1992)
5. Caragea, D., Silvescu, A., Honavar, V.: Towards a Theoretical Framework for Analysis and Synthesis of Agents That Learn from Distributed Dynamic Data Sources. In: Emerging Neural Architectures Based on Neuroscience. Springer, Berlin (2001)
6. Fritzke, B.: Growing Self-Organizing Networks - Why? In: Proc. European Symposium on Artificial Intelligence, pp. 61–72 (1996)
7. Granger, E., Rubin, M.A., Grossberg, S., Lavoie, P.: Classification of Incomplete Data Using the Fuzzy ARTMAP Neural Network. In: Proc. Int'l Joint Conference on Neural Networks, vol. iv, pp. 35–40 (2000)
8. Granger, E., Henniges, P., Sabourin, R., Oliveira, L.S.: Supervised Learning of Fuzzy ARTMAP Neural Networks Through Particle Swarm Optimization. J. of Pattern Recognition Research 2(1), 27–60 (2007)
9. Kasabov, N.: Evolving Fuzzy Neural Networks for Supervised/Unsupervised Online Knowledge-Based Learning. IEEE Trans. on Systems, Man, and Cybernetics 31(6), 902–918 (2001)
10. Maloof, M.: Incremental Rule Learning with Partial Instance Memory for Changing Concept. In: Proc. of the IEEE Int'l Joint Conf. on Neural Networks, vol. 14(1), pp. 1–14 (2003)
11. Okamoto, K., Ozawa, S., Abe, S.: A Fast Incremental Learning Algorithm with Long-Term Memory. In: Proc. Int'l Joint Conf. on Neural Network, Portland, USA, July 20-24, vol. 1(1), pp. 102–107 (2003)
12. Polikar, R., Udpa, L., Udpa, S., Honavar, V.: Learn++: An Incremental Learning Algorithm for MLP Networks. IEEE Trans. Systems, Man, and Cybernetics 31(4), 497–508 (2001)
13. Ruping, S.: Incremental Learning with Support Vector Machines. In: Proc. IEEE Int'l Conf. on Data Mining, San Jose, USA, November 29 - December 2, pp. 641–642 (2001)
14. Valentini, G.: An Experimental Bias-Variance Analysis of SVM Ensembles Based on Resampling Techniques. IEEE Trans. Systems, Man, and Cybernetics – Part B: Cybernetics 35(6), 1252–1271 (2005)

Discriminatory Data Mapping by Matrix-Based Supervised Learning Metrics

M. Strickert[1,*], P. Schneider[2], J. Keilwagen[1],
T. Villmann[3], M. Biehl[2], and B. Hammer[4]

[1] Leibniz Institute of Plant Genetics and Crop Plant Research Gatersleben
[2] Institute for Mathematics and Computing Science, University of Groningen
[3] Research group Computational Intelligence, University of Leipzig
[4] Institute of Computer Science, Technical University of Clausthal
stricker@ipk-gatersleben.de

Abstract. Supervised attribute relevance detection using cross-comparisons (SARDUX), a recently proposed method for data-driven metric learning, is extended from dimension-weighted Minkowski distances to metrics induced by a data transformation matrix Ω for modeling mutual attribute dependence. Given class labels, parameters of Ω are adapted in such a manner that the inter-class distances are maximized, while the intra-class distances get minimized. This results in an approach similar to Fisher's linear discriminant analysis (LDA), however, the involved distance matrix gets optimized, and it can be finally utilized for generating discriminatory data mappings that outperform projection pursuit methods with LDA index. The power of matrix-based metric optimization is demonstrated for spectrum data and for cancer gene expression data.

Keywords: Supervised feature characterization, adaptive matrix metrics, attribute dependence modeling, projection pursuit, LDA.

1 Introduction

Learning metrics constitute one of the most exciting topics in machine learning research [11,17,18]. The potential of metric adaptation needs exploration for facing challenges connected to the curse of dimensionality in high-throughput biomedical data sets. Mass spectra, gene expression arrays, or 2D electrophoretic gels, given as vectors of real-value measurements, are often characterized by only a low number of available experiments as compared to their huge dimensionality. Data-driven adaptation of a data metric can be used in many helpful ways. Applications of metric optimization range from attribute weighting via dimension reduction to data transformations into task-specific spaces.

The adaptive Euclidean distance $d_\lambda(\boldsymbol{x}, \boldsymbol{y}) = (\sum_{i=1}^{q} \lambda_i (x_i - y_i)^2)^{1/2}$, for example, relates attribute characterization to the choice of attribute scaling factors λ_i beneficial for the separation of labeled and unlabeled data in supervised and

---

** Corresponding author.

L. Prevost, S. Marinai, and F. Schwenker (Eds.): ANNPR 2008, LNAI 5064, pp. 78–89, 2008.
© Springer-Verlag Berlin Heidelberg 2008

unsupervised manners. This aim is shared with projection pursuit methods for which matrix parameters of a linear projection mapping are optimized with respect to criteria of data spreading and clusterability of the low-dimensional projections [4]. In this work, evaluation takes place in a space of original dimensionality where only the comparison criterion, the metric, is changed. If desired, attribute-related parameters with low impact, for example expressed as low scaling factors, can be pruned for dimension reduction, after adaptation. For the parametric Euclidean distance, small attribute scaling factors λ_i would indicate negligible attributes. Scaling factors can be also used for transforming the data to the non-adapted Euclidean space for further utilization of standard Euclidean methods. This kind of attribute characterization is different from many other methods for feature extraction [6], such as the recently suggested Iterative Relief algorithm [16] for which the attribute weights do not coincide with a canonic rescaling of the data space.

Matrix-based metrics help to extend the view of individual attribute processing to a model of dependence between pairs of attributes. Generally, matrix methods can be used for optimizing linear data transformations aiming at criteria related to the data spreading. In the unsupervised case, interesting transforms include sphering of the data covariance matrix to the unity matrix, or the projection of data to directions of maximum variance (PCA) or to directions along maximum non-Gaussianity (ICA) [10]. The projection pursuit method [4] is a very flexible approach to extract projections of interest by optimizing a target function, called the index of the projection. Such indices exist for unsupervised cases aiming at mappings to continuous or sharply clustered views. In addition, there are supervised indices like projection entropy and class separability according to linear discriminant analysis (LDA) criteria. A good environment for the study of projection pursuit and other matrix methods is, for example, provided by the free R statistical language with rGGobi and classPP packages [2,12], an application of classPP for the visualization of gene expression data, is provided in [3].

An alternative view on seeking optimum data transformations is data-driven adaptation of the data metric or, more generally, of a data similarity measure. Learning vector quantization (LVQ), for example, can implement metric adaptation for better data classification by boosting class-separating attributes between data prototype vectors. The generalized relevance LVQ method (GRLVQ) realizes such metric adaptation by using a misclassification cost function – minimized by gradient descent – making use of data labels for attribute rescaling [8]. For Euclidean distances, large-margin optimization is realized, but also non-Euclidean similarity-measures profit from parameter adaptation [7,14]. Recently, matrix learning has been integrated into the GRLVQ framework for modeling attribute-attribute dependencies by generalized Mahalanobis distance [13]. This allows to express scalings of the data space along arbitrary directions, and very good classification accuracies are obtained on difficult classification problems ranging from spectrum classification to image segmentation.

The success of matrix metric adaptation in GRLVQ classifiers initiated the present work. Here, no classifier will be build though, but the data space will be transformed: directions in the data space relevant to data label separation will be emphasized, while within-class variations will be damped. After all, relevant combinations of attributes, trained and expressed in form of a matrix, are identified for the discrimination task. Since only few data samples per class can be expected in costly and time-consuming biomedical studies, prototype-based data abstraction, like provided by GRLVQ, is avoided in order to keep maximum information. For the analyzed data sets it turned out that the transform matrix could be effectively compressed to only a few prominent eigenvectors, possibly only one, without significant loss of metric structure. After all, we are able to compute relatively compact discriminatory data models that allow hypotheses generation for supporting biomedical experts.

2 Method

Data. The q-dimensional row input vector $\mathbf{x} \in \mathbb{R}^{1 \times q}$ is taken from a data set containing n data vectors $\{\mathbf{x}^1, \mathbf{x}^2, \ldots, \mathbf{x}^n\}$. The proposed metric adaptation requires that each vector $\mathbf{x}^k$ is labeled with one class-specific index $\mathrm{c}(k)$, assuming at least two unique classes in the whole data set.

Metric. Most essential is the definition of the matrix-based metric $\mathrm{d}_{\boldsymbol{\Omega}}^{ij} \in [0; \infty)$ between data vectors $\mathbf{x}^i$ and $\mathbf{x}^j$:

$$\mathrm{d}_{\boldsymbol{\Omega}}^{ij} = \mathrm{d}_{\boldsymbol{\Omega}}(\mathbf{x}^i, \mathbf{x}^j) = (\mathbf{x}^i - \mathbf{x}^j) \cdot \boldsymbol{\Lambda} \cdot (\mathbf{x}^i - \mathbf{x}^j)^{\top}, \quad (\boldsymbol{\Lambda} = \boldsymbol{\Omega} \cdot \boldsymbol{\Omega}^{\top}) \in \mathbb{R}^{q \times q}. \quad (1)$$

Choosing the identity matrix $\boldsymbol{\Lambda} = \boldsymbol{\Omega} = \boldsymbol{I}$ induces the special case of the squared Euclidean distance; other diagonal matrices yield weighted squared Euclidean distances as discussed in [15]. Generally, metrics are obtained for arbitrary positive-definite matrices $\boldsymbol{\Lambda}$. Then the value expressed by $\boldsymbol{\Delta} \cdot \boldsymbol{\Lambda} \cdot \boldsymbol{\Delta}^{\top} \geq 0$, getting zero only for trivial difference vectors $\boldsymbol{\Delta} = \mathbf{0}$, is a metric. It is known that in context of metrics non-symmetric positive-definite matrices can be replaced by equivalent symmetric positive-definite matrices. Since any symmetric positive-definite matrix $\boldsymbol{\Lambda}$ can be decomposed by Cholesky decomposition into a product of a lower triangular matrix and its transposed, it is in principle sufficient to learn a lower triangular matrix $\boldsymbol{\Omega}$ for expressing $\boldsymbol{\Lambda}$. Alternatively, symmetric positive-definite $\boldsymbol{\Lambda}$ can be represented by the self-product $\boldsymbol{\Lambda} = \boldsymbol{\Omega} \cdot \boldsymbol{\Omega}$ of a symmetric $\boldsymbol{\Omega}$ [13]. Here, we consider products $\boldsymbol{\Lambda} = \boldsymbol{\Omega} \cdot \boldsymbol{\Omega}^{\top}$ with arbitrary $\boldsymbol{\Omega} \in \mathbb{R}^{q \times q}$. These full matrices $\boldsymbol{\Omega}$, possess more adaptive matrix elements than degrees of freedom needed for expressing the product solution space of $\boldsymbol{\Lambda}$. For the data sets discussed, the interaction of matrix element pairs Ω_{ij} and Ω_{ji} leads to a faster convergence of $\boldsymbol{\Lambda}$ during optimization, compared to the convergence properties obtained for symmetric or triangular matrices $\boldsymbol{\Omega}$.

Note that for some the data $\boldsymbol{\Lambda}$ might become positive-semidefinite during optimization, i.e. $\boldsymbol{\Delta} \cdot \boldsymbol{\Lambda} \cdot \boldsymbol{\Delta}^{\top} = 0$ with difference vectors $\boldsymbol{\Delta} \neq \mathbf{0}$. Then, the metric property gets relaxed to a mathematical distance with vanishing self-scalar product $(\boldsymbol{\Delta} \cdot \boldsymbol{\Omega}) \cdot (\boldsymbol{\Delta} \cdot \boldsymbol{\Omega})^{\mathsf{T}} = \langle \boldsymbol{\Delta} \cdot \boldsymbol{\Omega}, \boldsymbol{\Delta} \cdot \boldsymbol{\Omega} \rangle = 0$ becoming zero for certain configurations of $\boldsymbol{\Omega}$ with $\boldsymbol{\Delta} \cdot \boldsymbol{\Omega} = \mathbf{0}$, else positive.

Adaptation. Driven by the goal to minimize within-class differences while maximizing between class differences, the following cost function is minimized over pairs of all n data items:

$$s(\boldsymbol{\Omega}) := \frac{\sum_{i=1}^{n} \sum_{j=1}^{n} d_{\boldsymbol{\Omega}}(\mathbf{x}^i, \mathbf{x}^j) \cdot \delta_{ij}}{\sum_{i=1}^{n} \sum_{j=1}^{n} d_{\boldsymbol{\Omega}}(\mathbf{x}^i, \mathbf{x}^j) \cdot (1 - \delta_{ij})} = \frac{d_C}{d_D}, \quad \delta_{ij} = \begin{cases} 0 : c(i) \neq c(j) \\ 1 : c(i) = c(j) \end{cases} \quad (2)$$

Distances $d_{\boldsymbol{\Omega}}^{ij}$ between data vectors $\mathbf{x}^i$ and $\mathbf{x}^j$ depend on the adaptive matrix parameters $\boldsymbol{\Omega} = (\Omega_{kl})_{\substack{k=1\ldots q \\ l=1\ldots m}}$ of interest. The numerator represents within-class data scatter, which should be small; the denominator is related to inter-class distances, which should be large. Thus, optimization of $s(\boldsymbol{\Omega})$ handles both parts of the fraction simultaneously. Compromise solutions must be found in cases when within-class variation, potentially caused by outliers, needs compression, while inter-class separability would require inflation.

Using the chain rule, the cost function $s(\boldsymbol{\Omega})$ is iteratively optimized by gradient descent $\boldsymbol{\Omega} \leftarrow \boldsymbol{\Omega} - \gamma \cdot \frac{\partial s(\boldsymbol{\Omega})}{\partial \boldsymbol{\Omega}}$, which requires adaptation of the matrix $\boldsymbol{\Omega}$ in small steps γ into the direction of steepest gradient

$$\frac{\partial s(\boldsymbol{\Omega})}{\partial \boldsymbol{\Omega}} = \sum_{i=1}^{n} \sum_{j=1}^{n} \frac{\partial s(\boldsymbol{\Omega})}{\partial d_{\boldsymbol{\Omega}}^{ij}} \cdot \frac{\partial d_{\boldsymbol{\Omega}}^{ij}}{\partial \boldsymbol{\Omega}} . \quad (3)$$

The quotient rule applied to the fraction $s(\boldsymbol{\Omega}) = d_C / d_D$ in Eqn. 2 yields

$$\frac{\partial s(\boldsymbol{\Omega})}{\partial d_{\boldsymbol{\Omega}}^{ij}} = \frac{\delta_{ij} \cdot d_D}{d_D^2} + \frac{(\delta_{ij} - 1) \cdot d_C}{d_D^2} = \begin{cases} 1/d_D : c(i) = c(j) \\ -d_C/d_D^2 : c(i) \neq c(j) \end{cases} . \quad (4)$$

The right factor in Eqn. 3 is obtained by matrix derivative of Eqn. 1:

$$\frac{\partial d_{\boldsymbol{\Omega}}^{ij}}{\partial \boldsymbol{\Omega}} = 2 \cdot (\mathbf{x}^i - \mathbf{x}^j)^{\mathsf{T}} \cdot (\mathbf{x}^i - \mathbf{x}^j) \cdot \boldsymbol{\Omega} . \quad (5)$$

If desired, adaptation can be restricted to certain structures of $\boldsymbol{\Omega}$, such as to the lower triangular elements. In that case, undesired elements must be initially masked out by zeros in $\boldsymbol{\Omega}$. Additionally, the same zero masking pattern must be applied to the matrix resulting from Eqn. 5, because the equation calculates $\partial d_{\boldsymbol{\Omega}}^{ij}/\partial \boldsymbol{\Omega}$ correctly only for full adaptive matrices $\boldsymbol{\Omega}$. By consistent masking operations, though, the matrix of derivatives is mathematically correct. In practice, the gradient from Eqn. 3 is computed and reused as long the cost function decreases. Potential increase of $s(\boldsymbol{\Omega})$ triggers a recomputation of the gradient. The step size γ is dynamically determined as the initial size γ_0, being exponentially cooled down by rate η, divided by the maximum absolute element in the matrix $\partial s(\boldsymbol{\Omega})/\partial \boldsymbol{\Omega}$.

Initialization. Empirically, the initial step size γ_0 can be chosen from the interval $[0.05; 1)$, such as 0.75 in the conducted experiments. The number of iterations should be set to a value between 50 and 1000, depending on the saturation characteristics of the cost function. The exponential cooling rate should diminish the original step size by some orders of magnitude during training, for example, set to $\eta = 0.995$ for 1000 iterations.

The initialization of matrix Ω is of particular interest. If chosen as identity matrix $\Omega = I$, the algorithm starts from the usual squared Euclidean distance. For data sets with strong mutual attribute dependencies, i.e. prominent non-diagonal elements, the uniform structure of the identity matrix might lead to unnecessary iterations required for the symmetry breaking, as often encountered in neural network adaptations. Therefore, the alternatively proposed method is random matrix element sampling from uniform noise in the interval $[-0.5; 0.5]$. This noise matrix $A \in \mathbb{R}^{q \times q}$ is broken by QR-decomposition into $A = Q \cdot R$, of which the Q-part is known to form an orthonormal basis with $Q \cdot Q^\top = I$. This makes $\Omega = Q$ our preferred initial candidate.

Relation to LDA. At first glance, the proposed cost function looks quite similar to the inverse fraction of the LDA cost function for C classes that is maximized:

$$
S_{LDA} = \frac{v \cdot \left[\sum_{i=1}^{C} n_i \cdot (\mu_i - \mu)^\top \cdot (\mu_i - \mu) \right] \cdot v^\top}{v \cdot \left[\sum_{i=1}^{C} \Sigma_i \right] \cdot v^\top} , \quad n_i = |\{ \mathbf{x}^j : c(j) = i \}| . \quad (6)
$$

The numerator contains the between-class variation as the squared difference between class centers μ_i of all vectors $\mathbf{x}^j$ belonging to class i and the overall center $\mu = 1/n \cdot \sum_{k=1}^{n} \mathbf{x}^k$. The denominator describes the within-class variation over all classes i expressed by the sum of squared differences from class centers μ_i contained in the covariance matrices $\Sigma_i = \sum_{j:c(j)=i} (\mathbf{x}^j - \mu_i)^\top \cdot (\mathbf{x}^j - \mu_i)$.

LDA seeks an optimum direction vector v representing a good compromise of being collinear along the class centers (numerator, separating) and orthogonal to maximum within-class variation (denominator, compressing).

If multiple directions $V = (v_k)^\top$ are computed simultaneously, the products in the numerator and denominator of Eqn. 6, involving the matrices in square brackets, become matrices as well. In order to circumvent the problem of valid ratio calculation with matrices, determinants of the obtained matrices can be taken, as discussed in the LDA-based projection pursuit approach [12]. As a result, the LDA ratio optimizes low-dimensional projections onto discriminatory directions.

Our approach is structurally different, because the (inverse) LDA ratio in Eqn. 2 operates in the original data space, subject to the dynamically optimized metric. This explains the higher computational demands compared to LDA for which covariance matrices and class centers can be initially computed and then reused. As a benefit of the new approach, numerator and denominator of the new ratio in Eqn. 2 naturally contain sums of real-valued distances, which avoids problems of handling singular determinants in low-rank matrices.

3 Experiments

3.1 Tecator Spectral Data Set

The benchmark spectral data set, taken from the UCI repository of machine learning [1], contains 215 samples of 100-dimensional infrared absorbance spectra recorded on a Tecator Infratec Food and Feed Analyzer working in the wavelength range 850–$1050nm$ by the Near Infrared Transmission (NIT) principle. The original regression problem accompanying the data set is reformulated as attribute identification task for explaining the separation of 183 samples with low fat content and 77 high fat meat probes.

View 1. An exploratory data view is obtained from the left panel of Fig. 1 and from the PCA projection shown in the left scatter plot of Fig. 2. As expected, the strong spectrum overlap cannot be resolved by PCA projection. After application of the matrix learning all spectra were transformed according to $z = \mathbf{x} \cdot \Omega$, which realizes the left transformation part of the metric given in Eqn. 1; the right part is just z^{T}. The result of this data transformation leads to a good separation with

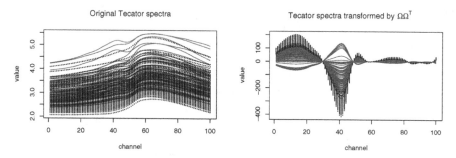

Fig. 1. Tecator spectra, raw (left) and transformed (right). Low fat content is reflected by dashed lines, high fat content by solid lines.

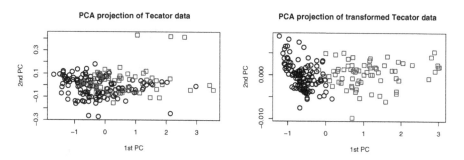

Fig. 2. Scatter plots of Tecator data. Bullets (o) denote low-fat samples, squares (□) high fat content. Left: PCA projection of original data. Right: PCA projection of data transformed by Ω.

almost no overlap in the PCA projection. This is shown in the right panel of Fig. 2.

View 2. By reformulating the metric definition in Eqn. 1 according to

$$d_\Omega^{ij} = (\mathbf{x}^i - \mathbf{x}^j) \cdot \Omega \cdot \Omega^{\mathsf{T}} \cdot (\mathbf{x}^i - \mathbf{x}^j)^{\mathsf{T}} = \langle \mathbf{x}^i \cdot \Omega \cdot \Omega^{\mathsf{T}} - \mathbf{x}^j \cdot \Omega \cdot \Omega^{\mathsf{T}}, \mathbf{x}^i - \mathbf{x}^j \rangle . \quad (7)$$

another interesting perspective on the data is obtained. This is a formal metric decomposition into a static part of difference vectors of the original data (right part of the scalar product) and a dynamically adapted transformation space of the data (left argument of the scalar product). A look into this space is obtained by the transformation to $\mathbf{x}^* = \mathbf{x} \cdot \Omega \cdot \Omega^{\mathsf{T}}$. The resulting transformed spectra with their amazingly separated attributes are shown in the right panel of Fig. 1.

The learned metric can be nicely presented by the matrices Ω and Λ shown in the left and right panel of Fig. 3, respectively. As displayed for Λ, attribute dependence is most prominent in the channel range 35–45. Strong emphasis of these channels around the diagonal is accompanied by simultaneous repression of the off-diagonal channels 5–30.

Matrix reduction. Since full matrices are quite big models, the study of their compressibility is important. Eigen decomposition of $\Lambda = S \cdot W \cdot W^{-1}$ into the diagonal eigenvalue matrix S and the eigenvectors matrix W helps to reach substantial compressions. In the current case, the highest eigenvalue contributes an amount of 95.3%, thus most variation in the learned matrix Ω can be explained by the corresponding eigenvector w, a column vector. Therefore, up to a scaling factor, a very good reconstruction of Λ by $w \cdot w^{\mathsf{T}}$ is obtained, as confirmed in the left matrix plot of Fig. 4. If the spectra are projected onto w, still a very good class separation is obtained, as demonstrated by the corresponding class-specific box plot in the right panel of Fig. 4.

The computational demands are quite high, though, requiring roughly one hour for 1000 updates of the matrix gradient. In contrast to that, the classPP [12] package is much faster, if only a class-separating projection is desired. In principle, classPP takes only several seconds or minutes, depending on the choice of

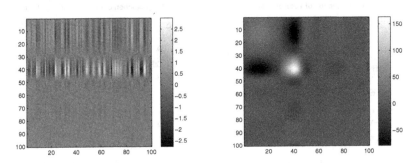

Fig. 3. Matrix representation of optimum metric for the 100-D Tecator data set. The learned matrix Ω is shown on the left, its squared counterpart $\Lambda = \Omega \cdot \Omega^{\mathsf{T}}$ on the right. Interesting dependencies are found around channel index 40.

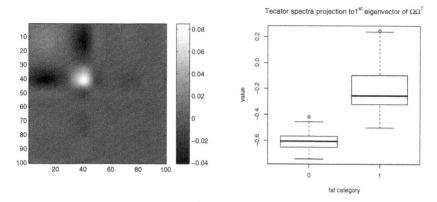

Fig. 4. Representation of Λ by its first eigenvector. Left: plot of reconstructed matrix. Right projection of Tecator spectra to the eigenvector.

annealing parameters. However, no stable solution could be obtained, because of the degeneration of the projection vectors, probably caused by near-singular matrix determinants during the computation. Training with our proposed method showed very stable results, converging to the presented solution for different random initializations of Ω. Projection displays are just a by-product of our method. It is important to remember that the original dimensionality of the data space is preserved by the transformation, enabling further utilization with any classification or projection method.

3.2 Gene Expression Analysis of AML/ALL Cancer

Many well-documented and deeply investigated data sets are freely available in cancer research. Since its publication in 1999 the leukemia gene expression dataset [5] used here has become a quasi-benchmark for testing feature selection methods. The original research aimed at the identification of the most informative genes for modeling and classification of two cancer types, acute lymphoblastic leukemia (ALL) and acute myeloid leukemia (AML). The training data covers 7129 genes by 27 cases of ALL and 11 cases of AML. The test data set contains 20 cases of ALL and 14 cases of AML.

The projection of the complete set of training and test data to the first two principal components yields the scatter plot shown in the left panel of Fig. 5. It is worth noticing that a systematic difference between AML training set and test set is indicated by the unbalanced distribution of closed and open bullets. Thus, a training set specific bias is induced during training. Matrix learning is computationally very expensive, because Ω is a 7129x7129 matrix. Thus, it takes roughly 40 hours on a 2.4 GHz system in order to achieve 500 gradient changes expressed by Eqn. 5. Yet, gradients are reused until first cost function degradation, creating several thousand updates, after all. Since experiment preparation in lab requires much more time, a two day calculation period is no principal problem.

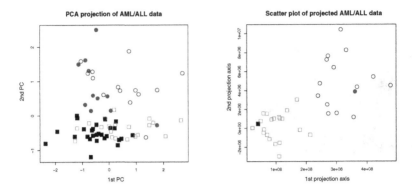

Fig. 5. Scatter plots of AML/ALL gene expression data. Bullets (○) denote expression samples of AML, squares (□) are related to ALL cancer. Closed symbols indicate training data, open symbols test data. Left panel: principal component projection without distinction between training and test data. Right panel: data projected to the first two eigenvectors of the trained interaction matrix $\Omega \cdot \Omega^T$. The training data is perfectly arranged, being very distinct and almost contracted to points.

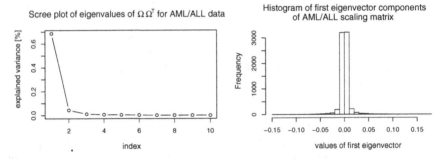

Fig. 6. Eigenvalues of the interaction matrix $\Omega \cdot \Omega^T$ (left) for the AML/ALL data, and the distribution of values in the most prominent eigenvector (right)

Several interesting results are obtained after training. The top ten eigenvalues of $\Omega \cdot \Omega^T$, displayed in the left panel of Fig. 6, point out a very strong explicative power of roughly 70% explained variance of the first eigenvector, dominating the all other eigenvectors. The right panel in Fig. 6 further indicates that only a small fraction of extreme values in that first eigenvector really contains interesting attribute magnifications. In case of data projection, many other attributes are transformed to near zero values, i.e. only few differentiating genes become emphasized.

The projection to the first two eigenvectors of the interaction matrix already yields a perfect separation of the training data into AML and ALL, as shown in the right panel of Fig. 5. Also a very strong compression of the within-class scatter almost to points is obtained. The test data, projected the same way and added to the plot, is still well-separated, but shows much larger variability. This is a clear indication of over-fitting. Sure, a huge 7129x7129 model has been trained; yet, the displayed projection only contains several hundred effective parameters, because the first two eigenvectors are much dominated by the few most prominent entries of the first eigenvector. A simple center-of-gravity model of the data, for example, would already be much larger.

For comparison, LDA indices of the projections shown in the right panel of Fig. 5 were calculated according to Eqn. 6 using the classPP package. An almost perfect near-one value of $S_{LDA} = 1 - 7.894 \cdot 10^{-12}$ was achieved for the projected training data and good value of $S_{LDA} = 0.8298$ for the test data. Using the built-in simulated annealing strategy, the classPP package itself reported a best seen index value of $S_{LDA} = 1 - 2.825 \cdot 10^{-6}$ for the optimized projections of the training data. Since the corresponding projection matrix is not returned correctly from the classPP package, an application to the test set was not possible. The authors of classPP have identified internal rounding errors in their package.

Individual gene variances correspond to gene-specific scaling factors, compensated by the cost function by adaptation of the related matrix entries. As a consequence, components in the most prominent eigenvector show low correlation with the variance in the original data set and, accordingly, systematically separating low-variance genes were able to gain high rankings in the eigenvector.

The real benefit of the proposed method is the possibility to infer putative gene-gene interactions responsible for cancer type separation. For that purpose the indices i, j corresponding to the most extreme (high and low) values in the matrix $\Omega \cdot \Omega^T$ are extracted and associated with the genes i and j. Because of symmetry, only the lower triangular matrix, including diagonal, is considered. The top 100 pairs extracted this way are compiled in Tab. 1. After all, 14 prominent self-dependent genes are detected as individual factors on the diagonal, three of them are coinciding with the list of Golub et al. of 50 genes. Three more genes of that study are detected on non-diagonal elements as dependent.

Table 1. Table of genes specific to separation of AML/ALL cancer in alphabetic reading order. The listed genes correspond to the 100 most extreme entries in the lower triangular part of the obtained symmetric matrix $\Omega \cdot \Omega^T$. As single genes participate multiple times in combination with others, only 30 different out of 200 possible genes appear in the table. Numbers indicate the frequencies of occurrence. Underlined genes appear also on the diagonal, stressing their individual importance. For illustration purposes, bold face genes are those acting in combination with M19507 which is the overall top-ranked gene. Asterisks mark genes coinciding with top-rated genes from the study of Golub et al.

D49824	HG3576-HT3779	L06797	L20688	L20941	M11147
6	13	3	1	1	1
M14328	M17733	M19507	M24485	M27891*	M28130_rna1*
1	1	26	1	3	1
M33600	M69043*	M77232_rna1	M91036_rna1	M91438	M96326_rna1*
4	1	1	13	1	14
S73591	U01317_cds4	U14968	V00594	X14046	X17042*
1	13	1	1	1	12
X78992	Y00433	Y00787*	Z19554	Z48501	Z70759
13	14	14	14	11	13

The most prominent gene found by the new method is M19507, which is not mentioned in the Golub study. Yet, the gene is confirmed as relevant in more recent publications, such as [9]. As this gene is connected to more than 20 other top-rated genes, its central role in the discriminatory transcriptome is clearly pointed out. Yet, the whole potential of the analysis, including proper interpretation of the findings, must be thoroughly worked out together with biological experts.

4   Conclusions and Outlook

A data-driven metric in flavor of a generalized Mahalanobis distance has been proposed that makes use of label information for emphasizing or repressing class-specific attribute combinations. Similar to LDA, metric optimization of $\Lambda = \Omega \cdot \Omega^{\top}$ seeks improved inter-class separation with simultaneous minimization of within-class variation. In contrast to LDA, it is not the low-dimensional projection to be optimized, but a transformation in the data space. The new method is not primarily designed for visual projection or classification, but it is a first step towards, because the resulting transformed data can be used as a preprocessing step for subsequent standard methods. As illustrated, visual data exploration is easily possible by projecting the data to the most prominent eigenvectors of Λ. No sophisticated optimization method is required, simple gradient descent works very reliably on the inverse LDA-like cost function. Both investigated data sets led to convergence to very useful label-specific metrics for different initializations of Ω. The main drawback of the new method is its long runtime for handling the potentially large matrices. Yet, as the discussed cases showed a strong dominance of only the first principal direction, future work will focus on the development of a sparse learning scheme for computing only the k most prominent eigenvectors instead of the whole matrix. This will help to reduce the model size and to speed up the optimization procedure. Finally, a better control of intra- and inter-class contributions to the cost function will be investigated.

Acknowledgment

The work is supported by grant XP3624HP/0606T, Ministry of Culture Saxony-Anhalt.

References

1. Blake, C., Merz, C.: UCI repository of machine learning databases (1998)
2. Cook, D., Swayne, D.: Interactive and Dynamic Graphics for Data Analysis with R and GGobi. Springer, Heidelberg (2007)
3. Faith, J., Mintram, R., Angelova, M.: Targeted projection pursuit for visualizing gene expression data classifications. Bioinformatics 22(21), 2667–2673 (2006)

4. Friedman, J.: Exploratory projection pursuit. Journal of the American Statistical Association 82, 249–266 (1987)
5. Golub, T., Slonim, D., Tamayo, P., Huard, C., Gaasenbeek, M., Mesirov, J., Coller, H., Loh, M., Downing, J., Caligiuri, M., Bloomfield, C., Lander, E.: Molecular classification of cancer: class discovery and class prediction by gene expression monitoring. Science 286(5439), 531–537 (1999)
6. Guyon, I., Gunn, S., Nikravesh, M., Zadeh, L.: Feature Extraction: Foundations and Applications. Springer, Berlin (2006)
7. Hammer, B., Strickert, M., Villmann, T.: Supervised neural gas with general similarity measure. Neural Processing Letters 21(1), 21–44 (2005)
8. Hammer, B., Villmann, T.: Generalized relevance learning vector quantization. Neural Networks 15, 1059–1068 (2002)
9. Hu, S., Rao, J.: Statistical redundancy testing for improved gene selection in cancer classification using microarray data. Cancer Informatics 2, 29–41 (2007)
10. Hyvärinen, A., Oja, E.: Independent component analysis: Algorithms and applications. Neural Networks 13(4–5), 411–430 (2000)
11. Kaski, S.: From learning metrics towards dependency exploration. In: Cottrell, M. (ed.) Proceedings of the 5th International Workshop on Self-Organizing Maps (WSOM), pp. 307–314 (2005)
12. Lee, E., Cook, D., Klinke, S., Lumley, T.: Projection pursuit for exploratory supervised classification. Journal of Computational and Graphical Statistics 14(4), 831–846 (2005)
13. Schneider, P., Biehl, M., Hammer, B.: Adaptive relevance matrices in learning vector quantization (Submitted to Machine Learning) (2008)
14. Strickert, M., Seiffert, U., Sreenivasulu, N., Weschke, W., Villmann, T., Hammer, B.: Generalized relevance LVQ (GRLVQ) with correlation measures for gene expression data. Neurocomputing 69, 651–659 (2006)
15. Strickert, M., Witzel, K., Mock, H.-P., Schleif, F.-M., Villmann, T.: Supervised attribute relevance determination for protein identification in stress experiments. In: Proc. of Machine Learning in Systems Biology (MLSB), pp. 81–86 (2007)
16. Sun, Y.: Iterative relief for feature weighting: Algorithms, theories, and applications. IEEE Transactions on Pattern Analysis and Machine Intelligence 29(6), 1035–1051 (2007)
17. Weinberger, K., Blitzer, J., Saul, L.: Distance metric learning for large margin nearest neighbor classification. In: Weiss, Y., Schölkopf, B., Platt, J. (eds.) Advances in Neural Information Processing Systems 18, pp. 1473–1480. MIT Press, Cambridge (2006)
18. Xing, E., Ng, A., Jordan, M., Russell, S.: Distance metric learning, with application to clustering with side-information. In: Becker, S., Thrun, S., Obermayer, K. (eds.) Advances in Neural Information Processing Systems 15 (NIPS), pp. 505–512. MIT Press, Cambridge (2003)

Neural Approximation of Monte Carlo Policy Evaluation Deployed in Connect Four

Stefan Faußer and Friedhelm Schwenker

Institute of Neural Information Processing, University of Ulm, 89069 Ulm, Germany
{stefan.fausser,friedhelm.Schwenker}@uni-ulm.de

Abstract. To win a board-game or more generally to gain something specific in a given Markov-environment, it is most important to have a policy in choosing and taking actions that leads to one of several qualitative good states. In this paper we describe a novel method to learn a game-winning strategy. The method predicts statistical probabilities to win in given game states using a state-value function that is approximated by a Multi-layer perceptron. Those predictions will improve according to rewards given in terminal states. We have deployed that method in the game Connect Four and have compared its game-performance with Velena [5].

1 Introduction

In the last 30 years, artificial intelligence methods like *Minimax* [8] have been used to build intelligent agents that have the task to choose a qualitative good move out of a set of all possible moves so they might win the game against another agent or even a human player. The basic approach in these methods are roughly the same: A game tree with a given depth will be calculated while exploring possible states and the move yielding to a state with minimum loss will be choosen using a heuristic evaluation function. While it is theoretical better to choose a large game tree depth, it is pratical impossible for games having an extensive set of possible states because of computing reasons. Furthermore such agents have not the ability to improve their computed strategy, although some agents might change it.

In contrast to the artificial intelligence methods above, the modern field of *Reinforcement learning* as defined in [1] were available since the late 1980s. Using those methods, the intelligent agent is learning by trial-and-error to estimate state values instead of exploring a game tree and taking the move with minimal maximum possible loss. In the specific case of Monte Carlo methods, this is achieved by first assigning each state an arbitrary initialized floating-point value that are then updated while playing in dependence of *Return values*, typically at intervall $[-1, +1]$, earned in terminal states. More generally, the intelligent agent is learning by experience to estimate state values and improves those estimations with each new generated episode. State values of all visited states in one episode will be updated by averaging their collected Returns. The simplest policy to take the best move is to choose the state out of all current possible states with the highest state-value.

While Monte Carlo methods require much computing time to estimate state values, because they theoretical need an infinite set of generated episodes for high quality estimations, it only takes linear time to use the derived policy to choose the best move. Overall Monte Carlo methods exhaust much less computing time than Minimax. Unfortunately in the classic way of assigning each state a state value, which is saving those values in tables, Monte Carlo methods cannot be used for environments with an extensive set of states because of the large requirement of computing space. This statement is also true for other reinforcement learning methods like *dynamic programming* and *temporal-tifference learning*.

Tesauro however has desribed a method of combining a Multi-Layer Perceptron with the Temporal-Difference learning to save a lot of computing space which he has applied in TD-Gammon [2] in 1992-1995. The basic idea was to approximate the state-value function $V(s)$ using a neural network trained by backpropagating TD errors. Although TD-Gammon performed well against human opponents as stated in [2], details like the used learning rate η and values of the initialized weights w in the Backpropagation algorithm are not given.

Encouraged by such scientific progress in the field of game playing, we have found a method to combine Monte Carlo Policy Evaluation with a Multi-Layer Perceptron that will be described in Section 3. We have applied this method to the well known game *Connect Four*, which has been choosen because of its simple rules and extensive set of possible states.

In Section 2, the basic rules of Connect Four as well as the complexity of the game will be presented. Afterwards the main ideas leading to our training algorithm, the learning process and the assembly of the resulting Multi-Layer perceptron will be discussed. Moreover, Section 4 provides the experimental results following by a conclusion.

2   Connect Four Rules and Game Complexity

Connect Four is a two-player game whereas one player has 21 red and the other player has 21 yellow discs. By turns, the players drop one disc into one of seven columns, which is sliding down and landing on the bottom or on another disc. The object of the game is to align four own discs horizontal, vertical or diagonal prior to the opponent. As no fortune is involved in winning the game, it is 100% a strategy game. Speaking of strategies, one of the better one is to arrange the own discs so that there are multiple opportunities to set the forth disc and win whilst preventing to loose.

Examining the game field, it is 42 fields, consisting of 7 columns and 6 rows, large and has 3 possible states per field. These states are red disc, yellow disc and empty. There are 3^{42} different possibilities to place up to 42 tokens on 42 fields. Plainly comparing this to english draughts, which is another popular board game, it has about $3^{42} * \frac{1}{5^{32}-5^{(32-24)}} \approx 0.0047 \approx \frac{1}{200}$ of its complexity.

3   Derivation of the Training Algorithm

Suppose we want to estimate the state-value function $V^\pi(s)$ for a given policy π using a standard Monte Carlo policy evaluation method as listed in [1]. Iteratively two main steps are repeated, whereas in step one a game episode is created using π and in step two the Return values following each visited s are added to a set $Return(s)$. The updated state value-function is then $V^\pi(s) = average(Return(s))$. Now assume only one Return $\in \{0, 1\}$, depending on the outcome of the game, i.e. the terminal state, is given per episode for all visited states s. This allows us to remove the neccessity of the set $Return(s)$:

$$V^\pi(s) = (n(s)V^\pi(s) + Return)\frac{1}{n(s) + 1}, \text{ for each } s \text{ visited in episode} \quad (1)$$

Therefore $V^\pi(s)$ is the mean of $n(s) + 1$ Return values that were received in observed terminal states, starting from s under policy π. Now consider policy π is a function that chooses one successor state s' out of all possible states $S_{successor}$ with the highest state-value:

$$\pi(s) = \underset{s'}{\operatorname{argmax}}(V^\pi(s')|s' \in S_{successor}) \quad (2)$$

This equation implies that an improvement of the estimation of $V^\pi(s)$ results in a more accurate choice of the successor state s'. Thus, learning to estimate $V^\pi(s)$ results in learning policy π. In general, generating a game episode under policy π is an interaction between our learning agent that does his move decision under policy π and a more or less intelligent opponent agent or human player. Note that in following notations, the state-value function is shortened to $V(s)$ because only one policy is targeted.

To evaluate and improve the policy, it is required to save the state-values on a computer storage. Having the sizes of the Connect Four game field, as introduced in Section 2, we calculate how much computing space the function itself consumes, if saved in tables: $3^{42} * 4$ bytes per state $\approx 40,76 * 10^{10}$ TByte. As this is much more space than a state-of-the-art computer, at present time, can deliver, it cannot be done in a straight-forward manner. Following Cybenko's Theorem [9] which denotes, that a Multi-Layer Perceptron (MLP) with at least one hidden layer and a sigmoid transfer function is capable of approximating any continuous function to an arbitrary degree of accuracy, we use a MLP to neural approximate the state-value function $V(s)$.

3.1   Neural Approximation of the State-Value Function $V(s)$

Assume we have generated an episode $\{s_1, s_2, ..., s_{TC}\}$ and have received one Return value $\in \{0, 1\}$ in terminal state s_{TC}. Let us now train a Multi-Layer Perceptron (MLP) as shown in Figure 1, so that the assigned output values $\{V(s_1), V(s_2), ..., V(s_{TC})\}$, which shall represent the statistical probabilities to win in given game states $\{s_1, s_2, ..., s_{TC}\}$, will be approximately updated like it

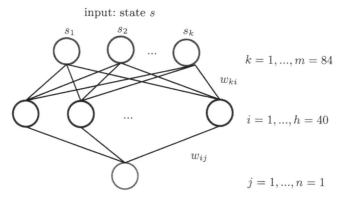

input: state s

s_1 s_2 s_k

$k = 1, ..., m = 84$

w_{ki}

$i = 1, ..., h = 40$

w_{ij}

$j = 1, ..., n = 1$

output: estimation of $V(s)$

Fig. 1. Multi-Layer Perceptron schemata

is done in Monte Carlo Policy evaluation described above. In general, we would declare the training signals exactly like in equation (1), but we don't have space for the number of received Returns $n(s)$, similar like we don't have space for the state-values $V(s)$ as discussed in Section 3. Instead, our proposal is to define a training signal that itself approximates equation (1) due to the nature of the MLP to only slightly reducing the error function E if learning rate η is rather small. Our training signals T_s for each state s are defined as follows:

$$T_s = Return\gamma^{TC-t(s)} \tag{3}$$

In this equation, T_s equals the discounted Return value, whereas the discounting factor is build up by TC which represents the amount of all visited states in this episode and $t(s)$ which returns the index number of state s between 1 and TC. Apparently there is no discounting in the last state s_{TC} and increasing discounting of the Return value towards the first state s_1 in the episode. The discounting strength can be manipulated by discounting parameter $0 < \gamma \leq 1$, whereas higher values cause smaller discounting steps. The Return value is given in terminal state s_{TC}:

$$Return = \begin{cases} 1, & \text{if agent has won} \\ 0, & \text{if agent has lost} \end{cases} \tag{4}$$

As $V(s)$ for a specific s should converge T_s to a certain degree, both are used to express the error function E_s:

$$E_s = ||T_s - V(s)||^2 \tag{5}$$

Using error E_s, which is the mean squared error (MSE) between T_s and $V(s)$, we can define the error of the whole episode:

$$E = \sum_s E_s \tag{6}$$

Having target function E defined and properly introduced, it can be minimized by updating the weights in the MLP. This can be achieved by calculating a gradient vector in the error surface, build-up by given weights, with a starting position equal to the current weight values and following the gradient vector in opposite direction. Approach to derivate the weights w_{i1} in the output layer:

$$\Delta w_{i1} = \eta - \nabla E_s = -\eta \frac{\partial E_s(w_{i1})}{\partial w_{i1}} = -\eta \frac{\partial E_s(u^{(2)})}{\partial u^{(2)}} \frac{\partial u^{(2)}(w_{i1})}{\partial w_{i1}} \quad (7)$$

Approach to derivate the weights w_{ki} in the input layer:

$$\Delta w_{ki} = \eta - \nabla E_s = -\eta \frac{\partial E_s(w_{ki})}{\partial w_{ki}} = -\eta \frac{\partial E_s(u_i^{(1)})}{\partial u_i^{(1)}} \frac{\partial u_i^{(1)}(w_{ki})}{\partial w_{ki}} \quad (8)$$

Analyzing the target function (6), it has the following effects on our learning agent:

- Due to the applied gradient descent method defined above, the weights are not updated immediately to match $E = 0$ but rather will be updated to minimize E slightly, taking a maximum step η of the gradient descent vector. For small values of η, this has the impact, that $V(s)$ for all s visited in episode are increasing slightly if Return is 1 or are decreasing slightly if Return is 0. Another view is that a new state value $V(s)$ is calculated based on weighted older state values of the same state s. State values occuring more often have a stronger weight than state values occuring seldom. Summed up, the behavior is similar to averaging Returns like it is done in the standard Monte Carlo Policy Evaluation method
- Discounting factor $\gamma^{TC-t(s)}$ causes states, that are more closely to the terminal state, to get higher state values. This forces the intelligent agent to maximize the received Return values in the long run
- As the error of the whole episode is minimized, the performance is not affected by state ordering in the episode (Offline / Batch learning)

3.2   Implementation and Assembly Details of the MLP

Before actually feeding a game state s into the input layer of the MLP, it has to be encoded. As stated in Section 2, Connect Four has 42 fields with 3 states per field which means that we are in need of ≥ 2 neurons per field if one neuron equates one bit. To reach a balanced distribution, we have choosen bit sequence 01 for red disc, 10 for yellow disc and 00 for empty field. Counting from left to right, the first two input neurons define the upper left and the last two input neurons define the lower right portion of the game field as seen in Figure 2. The coding of the output value is simple, as it represents one state-value $V(s)$ at intervall $[0, 1]$, for which only one single neuron is needed. Considering the prior defined coding scheme, the MLP needs $m = 84$ input neurons in the input layer and $n = 1$ output neuron in the output layer. The accurate count of hidden neurons h in the hidden layer does not result immediate out of the count of neurons $m+n$

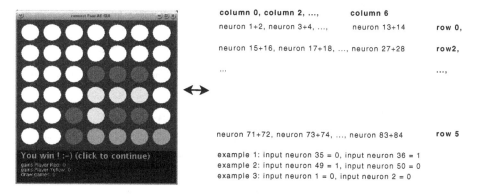

column 0, column 2, ...,	column 6	
neuron 1+2, neuron 3+4, ...,	neuron 13+14	row 0,
neuron 15+16, neuron 17+18, ..., neuron 27+28		row2,
...		...,
neuron 71+72, neuron 73+74, ..., neuron 83+84		row 5

example 1: input neuron 35 = 0, input neuron 36 = 1
example 2: input neuron 49 = 1, input neuron 50 = 0
example 3: input neuron 1 = 0, input neuron 2 = 0

Fig. 2. Relation of the input neurons and the game field demonstrated by our Connect Four software

but depends on it and on the complexity of the problem. Typical for Multi-Layer Perceptrons, each neuron in the input layer is weighted connected to each neuron in the hidden layer and each neuron in the hidden layer is weighted connected to each neuron in the output layer. Positive weights are supporting and negative weights inhibitoring to the dendritic potential of the target neuron. All weights are initialized at intervall $[-a, +a]$, whereas the exact value of a as well as hidden neurons h is available in Section 4. For the transfer function in the hidden layer and the output layer, we have choosen a nonlinear and sigmoid logistic function which seems to be natural because we want to assign probabilities:

$$f(x) = \frac{1}{1 + \exp(-x)} \tag{9}$$
$$f'(x) = f(x)(1 - f(x)) \tag{10}$$

As usual, the transfer function in the input layer is the identity-function.

3.3 Overview of the Training Algorithm

Exploitation and Exploration. Following the interaction cycle between the intelligent agent and an opponent as shown in Figure 3, the intelligent agent is spawning a pseudo-random value at intervall $[0, 1]$ after he received state s_t of the opponent and has builded possible successor states $s_{ta}, s_{tb}, ..., s_{tg}$ which result through an own move starting in s_t. If this value is $> \epsilon$ he is exploring, else he is exploiting the following game state s_{tx}. Exploiting denotes, that he calculates the state-values $V(s_{ta}), V(s_{tb}), ..., V(s_{tg})$ by feeding the game states one by one into the neural network to get the state values. Then he chooses the move resulting in state $s_{tx} \in \{s_{ta}, s_{tb}, ..., s_{tg}\}$ with the highest assigned state value $V(s_{tx})$. Exploration is carried out at rate $1 - \epsilon$ which means, that the intelligent agent randomly sets s_{tx} to one of successor states $s_{ta}, s_{tb}, ..., s_{tg}$ that he will visit but not include in the training set. This is not only important to speed up learning,

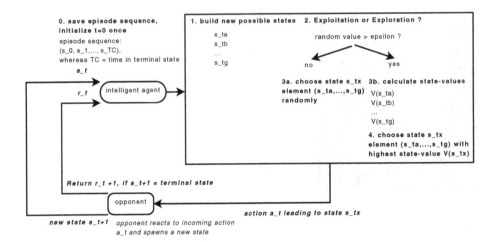

Fig. 3. Modell showing the interaction between the intelligent agent and an opponent (environment). Note that this cycle ends if the agent enters a terminal state or receives a terminal state of the opponent. In that case he would get a Return value and train the MLP with visited episode sequence and Return value to better estimate state-values in general.

but rather crucial to make good learning possible. The explanation is, that the once arbitrarily initialized weights in the MLP may have values so that certain state values $V(s)$ are fairly small. That in turn causes the intelligent agent to never visit those states s using Exploitation because of their low state values $V(s)$ and his policy to choose the highest state value. Exploration however enables to visit one of those states and to learn based on this state if the following state will be exploited.

4   Experiments and Results

4.1   Setting

To train the intelligent agent, we use an opponent that shares the same knowledge base, which is the weights of the MLP, to plan his moves. Additional, the opponent does random moves at rate $1 - \epsilon$ to avoid generating same episodes and to reach a balanced spawn of Return values. Having trained the intelligent agent with a certain amount of episodes, we are using another different opponent that does 100% random moves to plainly measure the quality of the learned policy. The quantity of successes of the intelligent agent is given in a winning quota after $500,000$ test-episodes:

$$\text{winning-quota} \ = \ \frac{\text{number of games won intelligent agent}}{\text{number of games won opponent}} \tag{11}$$

Note that the upper defined winning-quota is disregarding the number of draw games, which were in all test cases pretty low. As the behaviour of the intelligent

Table 1. Algorithm 1: Offline (Batch) training of a MLP to estimate state values using Monte Carlo Policy Evaluation

---

- **Given**: Episode sequence $\{s_1, s_2, ..., s_{TC}\}$ consisting of TC game states s, Return value $Return$ and MLP weights and parameters
- **for** $index = 1, ..., TC$
    - Choose pattern: $s = s_{index}$
    - Calculate $V(s)$:
        - * **for** $i = 1, ..., h$
            - · $u_i^{(1)} = \sum_{k=1}^{m} s_k w_{ki}^{(1)} - \theta_i^{(1)}$
            - · $y_i^{(1)} = f(u_i^{(1)})$
        - * **end for**
        - * $u^{(2)} = \sum_{i=1}^{h} y_i^{(1)} w_{i1}^{(2)} - \theta^{(2)}$
        - * $V(s) = f(u^{(2)})$
    - Calculate error on output neuron for pattern s:
        - * $T_s = Return * \gamma^{TC-t(s)}$
        - * $\delta_{index}^{(2)} = 2(T_s - V(s)) * f'(u^{(2)})$
    - Calculate error on hidden neurons for pattern s:
        - * **for** $i = 1, ..., h$: $\delta_{index,i}^{(1)} = \delta_{index}^{(2)} * w_{i1}^{(2)} * f'(u_i^{(1)})$
- Update weights:
    - **for** $k = 1, ..., m$
        - * **for** $i = 1, ..., h$
            - · $w_{i1}^{(2)} = w_{i1}^{(2)} + \eta \sum_{index=1}^{TC} \delta_{index}^{(2)} * y_{s,i}^{(1)}$
            - · $w_{ki}^{(1)} = w_{ki}^{(1)} + \eta \sum_{index=1}^{TC} \delta_{index,i}^{(1)} * s_{index,k}$
            - · $\theta^{(2)} = \theta^{(2)} - \eta \sum_{index=1}^{TC} \delta_{index}^{(2)}$
            - · $\theta_i^{(1)} = \theta_i^{(1)} - \eta \sum_{index=1}^{TC} \delta_{index,i}^{(1)}$
        - * **end for**
    - end for

---

agent is sensitive to six parameters, i.e. the number of hidden neurons h, gradient step η, also referred as learning rate, the number of generated training-episodes, weights interval a, discounting parameter γ and exploitation parameter ϵ, we have trained more than one agent to compare their performances and have assigned them a number. Prior each training, the weights of the MLPs have been initialized once at weight interval a. The results are listed in Table 2, whereas parameters $a = 0.77$, $\gamma = 0.97$ and $\epsilon = 0.7$ are identical for each MLP.

Further on we have developed a software interface, which allows the intelligent agent to play against Velena [5]. Velena is a shannon C-type program that is based on the theory of Victor Allis master thesis [6]. It combines eight rules and a PN-search engine and claims to play Connect Four perfectly in difficulty 'C'. Further difficulty levels 'A' and 'B' are available, which limit Velena's ability to look ahead. In various test runs we have observed that Velena is superior in all difficulty levels to an opponent that does 100% random moves. Table 3 lists the results, whereas we have to note that Velena is not acting deterministic,

Table 2. Winning-quota measure of multiple intelligent agents, performing versus opponent that does 100% random moves

MLP No.	winning-quota	h	η	episodes
1	146.00	40	0.1	$2,000,000$
	246.63		0.05	$+4,000,000$
	317.66		0.025	$+8,000,000$
2	722.56	60	0.1 to 0.025	$14,000,000$
3	1245, 86	100	0.1 to 0.025	$14,000,000$
4	1514.14	120	0.5 to 0.03125	$6,000,000$
5	**5049.42**	250	0.5 to 0.0625	$14,000,000$

Table 3. Performance measure of multiple intelligent agents versus Velena. *1st* notes that our intelligent agent had the first move, *ep diff* is the observed count of different episode sequences out of 500 possible.

MLP No.	diffic.	ep diff 1st	ep diff 2nd	won 1st	won 2nd	draw 1st	draw 2nd
1	A	92	161	18.2%	11.4%	6.8%	3%
	B	2	176	0%	16.8%	0%	1%
2	A	2	193	0%	25.4%	0%	10.8%
	B	2	194	0%	18%	0%	10.4%
3	A	44	265	1.6%	21.4%	3%	17.2%
	B	40	284	2%	25%	2.2%	18.4%
4	A	182	267	31%	40%	0.2%	14.6%
	B	176	258	36%	35%	0%	13.8%
5	A	57	323	89.6%	46.6%	0.004%	38.6%
	B	31	323	81%	47%	0%	36%
	C	11	35	0%	**0.002%**	0%	**0.004%**

i.e. is using a pseudo-random number generator in his policy. Different observed episode sequences are counted in column *ep diff*.

4.2   Discussing the Results

Analyzing the results in Table 2 it is apparent, that the success of gaining a good game-winning strategy is in dependence of the right parameter values. As expected, the winning rate improves with an increasing learning rate η as well as with an increasing number of training episodes. However, both values have to be limited, because if η or the number of training episodes get too large, then the success measured by the winning-rate is flattening. Further concentrating on the number of hidden neurons h, we have observed, that a MLP with a higher amount of h and more training-episodes performs a cut above which shows a well scaling behavior.

Recalling that our training signal T_s is just directing to the true value but is not equal to it, it is intuitive clear that the learning rate has to be decreased with an increasing amount of training episodes. Promising results have been reached by starting with a higher gradient step $\eta = 0.1$ that is step-wise decreasing to about $\eta = 0.025$. We have set the weights intervalls $a = 0.77$ as sugggested in [3] where they have produced the best neural approximators with the highest generalization rate:

An fixed weight intervall of 0.2, which corresponds to a weight range of $[-0.77, 0.77]$, gave the best mean performance for all the applications tested in this study.

Overall, the best result was obtained by using a step-wise decreasing learning rate $\eta = 0.5$ to $\eta = 0.0625$, hidden neurons $h = 250$, $a = 0.77$ and $14,000,000$ generated training episodes, which rewarded us with a winning-rate of about $5,000$.

Observing the results of our intelligent agents performing against Velena in Table 3, it is striking, that the intelligent agents with more hidden neurons h are better than those with less ones. Analyzing agents 1-3, they won average 20% of all games or ended about 10% draw, if they had the second move in difficulty 'A' and 'B'. Having the first move, it is unlikely for them to win a game. Observing this behavior, it is clear that they have not found their perfect opening position. Furthermore, intelligent agent 5 performed best: He lost only about 10% of all games in 'A' and 'B' while he even won once in difficulty 'C' and reached two draws.

5   Conclusion

We have described a method to approximate the Monte Carlo Policy Evaluation using a Multi-Layer Perceptron in this paper. In the experiments, we have statistically determined parameters for that algorithm to apply well in gaining a game-winning strategy in Connect Four. After the training had been carried out, we have used that strategy to perform against Velena. Although Velena claims to be unbeatable in difficulty 'C' and slightly weaker in difficulty 'A' and 'B', our intelligent agent 5 has experimental proven it's strength in all difficulties, where he has lost only about 10% of all games in 'A' and 'B' while he even *won once* in difficulty 'C' and reached two draws. Overall, we've achieved a good game-winning policy in Connect Four that can compete against experienced human players.

References

1. Sutton, R.S., Barto, A.G.: Reinforcement Learning: An Introduction. MIT Press, Cambridge (1998)
2. Tesauro, G.: Temporal Difference Learning and TD-Gammon. Communications of the ACM 38(3) (1995)
3. Thimm, G., Fiesler, E.: High order and multilayer perceptron initialization. IEEE Transactions on Neural Networks 8(2), 249–259 (1997)

4. Thimm, G., Fiesler, E.: Optimal Setting of Weights, Learning Rate and Gain. IDIAP Research Report, Dalle Molle Institute for Perceptive Artificial Intelligence, Switzerland (April 2007)
5. Bertoletti, G.: Velena: A Shannon C-type program which plays connect four perfectly (1997), http://www.ce.unipr.it/~gbe/velena.html
6. Allis, V.: A Knowledge-based Approach of Connect-Four, Department of Mathematics and Computer Science, Vrije Universiteit, Amsterdam (1998)
7. Lenze, B.: Einführung in die Mathematik neuronaler Netze. Logos Verlag, Berlin (2003)
8. Russel, S.J., Norvig, P.: Artificial Intelligence: A Modern Approach, 2nd edn. Prentice Hall, Englewood Cliffs (2002)
9. Cybenko, G.V.: Approximation by Superpositions of a Sigmoidal function. Mathematics of Control, Signals and Systems 2, 303–314 (electronic version) (1989)

Cyclostationary Neural Networks for Air Pollutant Concentration Prediction

Monica Bianchini, Ernesto Di Iorio, Marco Maggini, and Augusto Pucci

Dipartimento di Ingegneria dell'Informazione
Via Roma 56, I-53100 Siena (Italy)
{monica,diiorio,maggini,augusto}@dii.unisi.it

Abstract. There are many substances in the air which may impair the health of plants and animals, including humans, that arise both from natural processes and human activity. Nitrogen dioxide NO_2 and particulate matter (PM_{10}, $PM_{2.5}$) emissions constitute a major concern in urban areas pollution. The state of the air is, in fact, an important factor in the quality of life in the cities, since it affects the health of the community and directly influences the sustainability of our lifestyles and production methods. In this paper we propose a cyclostationary neural network (CNN) model for the prediction of the NO_2 and PM_{10} concentrations. The cyclostationary nature of the problem guides the construction of the CNN architecture, which is composed by a number of MLP blocks equal to the cyclostationary period in the analyzed phenomenon, and is independent from exogenous inputs. Some experiments are also reported in order to show how the CNN model significantly outperforms standard statistical tools and linear regressors usually employed in these tasks.

1 Introduction

There are many substances in the air which may impair the health of plants and animals, including humans, that arise both from natural processes and human activity. Substances not naturally found in the air, or at greater concentrations, or in different locations from usual, are referred to as *pollutants*. Pollutants can be classified as either primary or secondary. Primary pollutants are substances directly produced by a process, such as ash from a volcanic eruption or the carbon monoxide gas from a motor vehicle exhaust. Instead, secondary pollutants are not emitted. Rather, they form in the air when primary pollutants react or interact. An important example of a secondary pollutant is ozone – one of the many secondary pollutants that constitute the photochemical smog. Note that some pollutants may be both primary and secondary: that is, they are both emitted directly and formed as combinations of other primary pollutants. Primary pollutants produced by human activity include:

- Oxides of sulfur, nitrogen and carbon;
- Organic compounds, such as hydrocarbons (fuel vapour and solvents);
- Particulate matter, such as smoke and dust;

L. Prevost, S. Marinai, and F. Schwenker (Eds.): ANNPR 2008, LNAI 5064, pp. 101–112, 2008.

- Metal oxides, especially those of lead, cadmium, copper and iron;
- Toxic substances.

Secondary pollutants include some particles formed from gaseous primary pollutants and compounds in the photochemical smog, such as nitrogen dioxide, ozone and peroxyacetyl nitrate (PAN).

The main oxides of nitrogen present in the atmosphere are nitric oxide (NO), nitrogen dioxide (NO_2) and nitrous oxide (N_2O). Nitrous oxide occurs in much smaller quantities than the other two, but it is of interest as it represents a powerful greenhouse gas and thus contributes to global warming. The major human activity which generates oxides of nitrogen is fuel combustion, especially in motor vehicles. Oxides of nitrogen form in the air when fuel is burnt at high temperatures. This is mostly in the form of nitric oxide with usually less than 10% as nitrogen dioxide. Once emitted, nitric oxide combines with ozone (O_3) to form nitrogen dioxide, especially in warm sunny conditions. These oxides of nitrogen may remain in the atmosphere for several days and, during this time, chemical processes may generate nitric acid, and nitrates and nitrites as particles. The oxides of nitrogen play a major role in the chemical reactions which generate the photochemical smog. Nitrogen dioxide is also a respiratory irritant which may worsen the symptoms of an existing respiratory illness.

On the other hand, particulate matter (PM_{10}) pollution consists of very small liquid and solid particles floating in the air. Sources of PM_{10} in both urban and rural areas are motor vehicles, wood burning stoves and fireplaces, dust from construction, landfills, and agriculture, wildfires and waste burning, industrial sources, windblown dust from open lands. In particular, of greatest concern to public health, are the PM_{10} particles small enough to be inhaled into the deepest parts of the lung [16,19,20]. These particles are less than 10 microns in diameter, and result from a mixture of materials that can include smoke, soot, dust, salt, acids, and metals.

Many modelling efforts have been recently spent for controlling the NO_2 and PM_{10} concentrations in order to enable the development of tools for pollution management and reduction. One approach to predict future pollutant concentrations is to use detailed atmospheric diffusion models (see [2], for a review). Such models aim at solving the underlying physical and chemical equations that control pollutant concentrations and, therefore, require clean emission data and meteorological fields. An alternative approach is to devise statistical models which attempt to determine the underlying relationship between a set of input data and targets. Regression modelling is an example of such a statistical approach and has been applied to air quality modelling and prediction in a number of studies [7,21,22]. One of the limitations imposed by linear regression tools is that they will underperform when used to model non–linear systems. Instead, artificial neural networks can model non–linear systems and have been succesfully used for predicting air pollution concentrations (see, f.i., [5,6,9,11,12,13,14,15,18]).

In this paper, we propose a cyclostationary neural network (CNN) architecture to model and estimate hourly the NO_2 concetrations, and to obtain a 1–day ahead prediction for the PM_{10}. The cyclostationary nature of the problem guides the construction of the CNN, which is composed by a number of MLP blocks equal to the duration of the cyclostationary period in the analyzed phenomenon, specifically 24 hours for the prediction of the NO_2 concentration and a week (7 days) for the PM_{10}. The novelty of our approach particularly lies on its independence from exogenous data, in that it uses only the time series of NO, NO_2, and PM_{10}, respectively, to predict their future values. In fact, meteorological data are not taken (explicitly) into account, suggesting that the network is able to detect the necessary exogenous information directly from the pollution data. Therefore, the proposed CNN architecture is robust w.r.t. geographical and seasonal changes. Some experimentation was carried out on the data gathered by ARPA (Agenzia Regionale per la Protezione dell'Ambiente — Regional Environmental Protection Agency) of Lombardia (northern Italy). ARPA supplies a real–time air quality monitoring system to defend the people health and the region ecosystem quality. Experimental results are very promising and show that the CNN model significantly outperforms standard statistical tools — like AutoRegressive eXogenous (ARX) models — and linear regressors, usually employed for this task [4].

The paper is organized as follows. In the next section, the CNN architecture is introduced and the data preprocessing, aimed at creating a learning set tailored to the CNN model, is reported. Section 3 describes the experimental setup, respectively for the nitrogen dioxide and the particulate matter concentrations, comparing the performance of the proposed method with AR models and linear regression tools. Finally, Section 4 reports the conclusions.

2   Cyclostationary Neural Networks

A discrete time random process $X(t)$ is a rule for assigning to every outcome of an experiment ζ a function $X(t, \zeta)$. The domain of ζ is the set of all the experimental outcomes and the domain of t is a set of natural numbers [17]. Thus, a random process is a set of random variables, one for each time instant t. If the statistics of a random process change over time, then the process is called *nonstationary*. The subclass of nonstationary processes whose statistics vary periodically with time are called *cyclostationary*.

Whenever the cyclostationarity period T is known, a set of T stationary processes can be derived from the original one [10], on which different neural networks can be trained independently to predict the outcomes of the related random variables. Therefore, the CNN consists of a set of T independent — but with an identical architecture — MLPs, each modelling a random variable of the original cyclostationary process. Formally speaking, for a cyclostationary process X with period T, the set of all the outcomes $A^* = \{a_j \,|\, j \in [0, \infty)\}$ can be partitioned into T subsets, one for each random variable, that is $A^* = \{A_0, A_2, \ldots, A_{T-1}\}$, where $A_i = \{a_j \,|\, i = j \mod T\}$. The i–th MLP will be trained on the subset of the outcomes concerning the i–th random variable of the process.

2.1   Prediction of the NO₂ Concentration

This prediction task consists in modeling the NO_2 time series, based on the past concentrations of NO and NO_2. In this case, it is easily verifiable that a strong correlation exists between the past NO data and the current value of the NO_2 concentration, with a periodicity of 24 hours. This means that the NO_2 pollution at time t depends on the NO sampled at $t - 24$, $t - 48$, etc. Therefore, we claim that the process we are analyzing has a cyclostationary period $T = 24$, i.e. a daily periodicity, and, consequently, a CNN model will be composed by 24 MLP blocks. In particular, each MLP — one for each random variable of the cyclostationary process — will exploit $NO(t - T)$ and $NO_2(t - 1)$ to predict $NO_2(t)$. Formally:

$$NO_2(t)= f_k(NO_2(t - 1), NO(t - T)), \quad k = t \bmod T,$$

where $T = 24$ and f_k, $k = 0, \ldots, T - 1$, represents the k–th approximation function realized by the k–th MLP block.

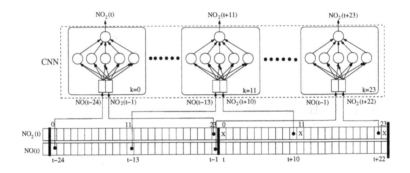

Fig. 1. The CNN architecture and the data sampling procedure

It is worth mentioning that the CNN model relies just on NO and NO_2 time series values. In fact, it is completely independent of exogenous data, such as weather condition (i.e. pressure, wind, humidity, etc.) and geographic conformation. This is just an interesting feature, since we can avoid to predict such weather conditions and focus only on the NO_2 concentration prediction. The resulting model will obviously be much more robust against noise and prediction error.

2.2   Prediction of the PM₁₀ Concentration

In order to predict the particulate matter pollution, and since PM_{10} time series underlies clear periodicities at the yearly and weekly level, we consider a CNN neural network composed by 7 MLP blocks to forecast the PM_{10} daily average concentration. To be used as an input for the predictor, each monitored parameter has to be grouped from the original hourly series to a daily time series; this has been accomplished calculating the average over the whole day. Therefore, the

k–th MLP block calculates the function f_k, for the 1–day prediction of PM_{10}, based on the today–value and on the concentration measured a week before:

$$PM_{10}(t)= f_k(PM_{10}(t-1), PM_{10}(t-W)), \quad k = t \bmod W,$$

where $W = 7$ and f_k, $k = 0, \ldots, W-1$, represents the k–th approximation function realized by the k–th MLP block.

As observed for the nitrogen dioxide, even for the case of the particulate matter, the CNN network is able to make the 1–day prediction without taking into account other environmental conditions except for the past concentrations of the pollutant to be predicted.

3   Experimental Results

In order to assess the capability of the proposed neural network model to predict cyclostationary phenomena and, in particular, air pollutant concentrations, several experiments have been performed to compare the CNN model to other models both for the prediction of the NO_2 hourly concentrations and for the prediction of the daily concentration of the PM_{10}, which is known to be a harder task (see, f.i. [8,9,11]) to be faced with connectionist models.

In this work, we used data gathered by the ARPA of Lombardia (northern Italy). ARPA supplies a real–time air quality monitoring system to defend the people health and the region ecosystem quality. The ARPA air quality monitoring system is composed by mobile and fixed stations.

The first dataset is made up by the nitric oxide and dioxide concentrations detected hourly by several stations in Bergamo and Brescia (two important cities in Lombardia) and by the unique station in Breno (a small city close to Brescia)[1]. Instead, for the particulate matter prediction task, the data are gathered from a monitoring station located in a residential area of Milan. The dataset is constituted by a hourly time series, with a missing value rate ranging between 5% and 10%.

3.1   Experiments on the Prediction of the NO_2 Concentration

To test the CNN model for the prediction of the nitrogen dioxide concentration, we exploited 21 different sets of data, gathered from seven monitoring stations, three in Bergamo (Via Garibaldi, Via Goisis, Via Meucci), three in Brescia (Broletto, via Trimplina, Via Turati), and the unique one in Breno. Three different datasets were defined corresponding to the measurements collected during different (or partially overlapped) time periods, as shown in Table 1.

For each dataset, the performance of the CNN architecture based on 24 MLPs, as described in Section 2, was compared with a similar architecture based on a set of 24 AutoRegressive eXogenous input (ARX) models. By a trial and error

---

[1] The dataset and some related information are available at the web site http://www.arpalombardia.it/qaria/doc_RichiestaDati.asp

Table 1. Datasets used in the experiments

Label	Learning set	Test set
2003-2004/2005	1-1-2003 1:00 a.m. to 1-1-2005 0:00 a.m.	1-1-2005 1:00 a.m. to 1-1-2006 0:00 a.m.
2003/2004	1-1-2003 1:00 a.m. to 1-1-2004 0:00 a.m.	1-1-2004 1:00 a.m. to 1-1-2005 0:00 a.m.
2004/2005	1-1-2004 1:00 a.m. to 1-1-2005 0:00 a.m.	1-1-2005 1:00 a.m. to 1-1-2006 at 0:00 a.m.

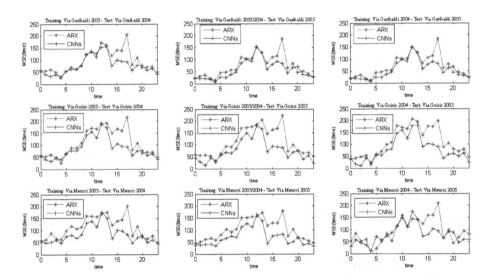

Fig. 2. Average absolute error for the prediction of the NO_2 hourly concentration at the three station in Bergamo. Each column corresponds to one of the three datasets of Table 1.

procedure, a two–layer neural network architecture with five hidden neurons was chosen for the CNN model. Figure 2 shows the results for the three stations in Bergamo. In each plot, the x–axis corresponds to the time of the day, while the y–axis corresponds to the average value of the absolute prediction error computed for all the days in the test set. The absolute prediction error is defined as $e(t) = |y(t) - \hat{y}(t)|$, where $y(t)$ is the current NO_2 value at time t and $\hat{y}(t)$ is the model estimation. By comparing the two error curves, it can be noted that the CNN absolute error is quite often significantly smaller than the error for the ARX model, both with respect to the station and the time of the day. Similar results were obtained for the stations in Brescia and Breno.

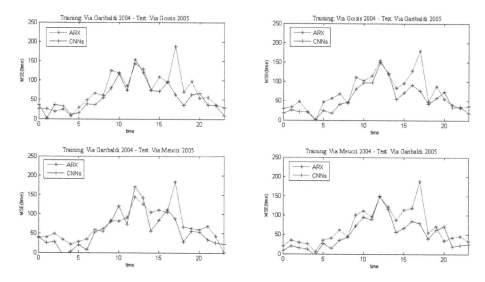

Fig. 3. Experimental results obtained by using the NO–NO$_2$ hourly concentration of a Bergamo's station to train the CNN, whose performance is then tested based on data gathered from another Bergamo's station.

A second set of experiments was aimed at evaluating the generalization performance of the CNN model with respect to the position of the monitoring station used to train the model and to the station considered for the prediction. Thus, the CNN was trained by using the NO–NO$_2$ time series gathered at a specific station and tested on the concentration collected, in the same period, at a different monitoring point. In particular, three different types of experiments were carried out:

- **Stations of the same city.** For each city, a CNN was trained for each NO–NO$_2$ time series and then tested on the data gathered at the other stations of the same city;
- **Stations of close cities.** A CNN was trained based on the NO–NO$_2$ time series measured at a certain station of a particular city, and then tested on the data gathered at some stations of a close city (for example, we use the concentration of the station at Brescia – Via Turati for the training and that of the station of Breno for the testing phase);
- **Stations of far cities.** A CNN was trained based on the NO–NO$_2$ time series measured at a certain station of a particular city, and then tested on the data gathered at some stations of a far city (e.g., using data from Brescia for the training and from Bergamo for the testing phase).

Figure 3 reports the results when considering four different combinations of the three measuring stations in Bergamo for the 2004/2005 dataset. The plots show that even in this case the CNN model performs better than the ARX

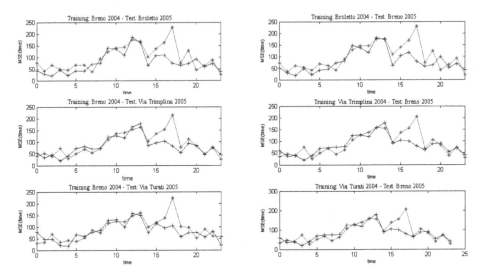

Fig. 4. Experimental results obtained on the 2004/2005 dataset by using the NO–NO$_2$ hourly concentration of the Breno's station for training and data gathered at some Brescia's stations for testing.

model. In particular, it is interesting to note that this experimental setup does not lead to a relevant CNN performance degradation.

In Figure 4, the results obtained by training the CNN on the hourly concentration of the Breno's station and testing it on the Brescia's stations are shown. This is the case of stations in nearby cities, and, as shown in the plots, there is not a significant performance degradation for the CNN model. We obtained similar results also by training the CNN model on a given station in Brescia and testing it on the data of the Breno's station.

In Figure 5, the plots show the average absolute error when the CNN is trained by using the hourly concentration of the Breno's station and then tested on the Bergamo's stations (and vice versa). In this case, the geographic distance among the stations is furthermore increased, nevertheless maintaining the CNN performance almost unchanged.

Finally, we investigated how the size of the learning set affects the prediction accuracy of the proposed model. We considered the 2004/2005 dataset that corresponds to two years of data and we adopted the following scheme. For each week, we considered the previous w weeks with $w = 1, \ldots, 25$. Starting from the 26–th week in the dataset, for each value of w, we predicted the hourly NO$_2$ concentration for each day, using a CNN model trained on the data collected for the days in the previous w weeks. For each learning set size w, we first computed the Mean Square Error for the hourly prediction in each week, and then we further averaged this value on all the weeks in the test set. The results are shown in the plots in Figure 6, where the x–axis is labelled by the parameter w — the number of weeks on which the model was trained — whereas the corresponding

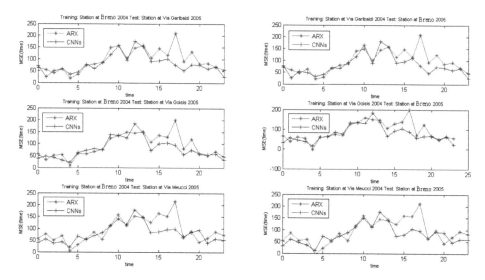

Fig. 5. Experimental results obtained by using the NO–NO$_2$ hourly concentration of the Breno's station for training and data gathered at some Bergamo's stations for testing and vice versa.

average MSE is reported on the y–axis. The different curves refer to the three stations in Brescia — Figure 6 (a) — and in Bergamo — Figure 6 (b). It is easy to note a common trend in each dataset, that corresponds to a performance improvement when the training window size increases up to 12–14 weeks, whereas there are not very significant changes when $w > 14$. Interestingly, this size for w corresponds exactly to an entire season. This result reveals a seasonal periodicity in the considered phenomenon.

3.2 Experiments on the Prediction of the PM$_{10}$ Concentration

The second experimental setup is aimed at presenting a comparison among CNNs, a neural network approach (also based on MLPs) proposed in [1], and a linear predictor described in [3], on the PM$_{10}$ 1–day ahead prediction. In this case the CNN architecture is based on a set of 7 two–layer MLPs, with 15 hidden neurons each. This architecture was determined by a trial and error procedure and assumes a weekly cyclostationarity period. The dataset is constituted by a hourly time series collected in the period 1999-2002 from a monitoring station located in a residential area of Milan. The data are grouped from the original hourly series into a daily time series and then splitted to form the training set (1999-2000), the validation set (2001), and the test set (2002), respectively.

In [1], many exogenous features are used as inputs for the neural network model to perform the PM$_{10}$ 1–day ahead prediction, including an autoregressive PM$_{10}$ term, the past concentrations of NO, NO$_2$, and SO$_2$, and a wide set of meteorological variables, such as temperature, humidity, wind velocity, solar

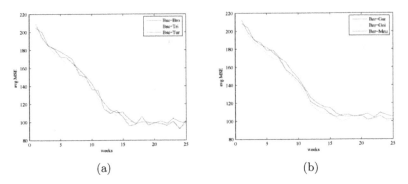

(a)   (b)

Fig. 6. Weekly Mean Square Error on the hourly NO$_2$ prediction task with respect to the number w of past weeks in the learning set. (a) Brescia stations. (b) Bergamo stations.

Table 2. Prediction Mean Absolute Error (MAE) on the PM$_{10}$ 1–day ahead prediction task

	CNNs	Cecchetti et al. [1]	Corani et al. [3]
MAE	8.34 mg/m^3	8.71 mg/m^3	11 mg/m^3

radiation, atmospheric pressure and the Pasquill stability class, a commonly used method of categorizing the amount of atmospheric turbulence[2]. Instead, in our approach, we just exploited the ad hoc past values of the PM$_{10}$ concetrations (those calculated the day before and one week before).

Table 2 reports the results obatined on the same data by CNNs, and the other approaches respectively proposed in [1] and [3].

4   Conclusions

In this paper, a connectionist model called Cyclostationary Neural Network (CNN) was introduced, particularly tailored for the prediction of cyclostationary phenomena. In particular, the CNN architecture was introduced to model and estimate hourly the nitrogen dioxide (NO$_2$) concetrations, and to obtain a 1–day ahead prediction for the particulate matter (PM$_{10}$), both pollutants playing a major role in the chemical reactions which generate the photochemical smog. One fundamental peculiarity of the proposed model is that of being independent of evironmental, tipically metereological, exogenous data, in that the NO$_2$ concentration prediction is based only on the previous level of NO$_2$

---

[2] The six stability classes are named A, B, C, D, E, and F, with class A being the most unstable or most turbulent class, and class F the most stable or least turbulent class. Each class is determined by the ranges of values for the surface windspeed, the daytime incoming solar radiation, and the nighttime cloud cover.

and of the nitric oxide (NO) — which combines with ozone (O_3) to form NO_2 —, whereas the PM_{10} prediction depends only on the previous values of the same pollutant. Some experimentation, carried out on the data gathered by ARPA Lombardia, looks very promising and shows that the CNN model significantly outperforms standard statistical tools, like ARX models, usually employed for the air pollutant prediction task.

References

1. Cecchetti, M., Corani, G., Guariso, G.: Artificial neural networks prediction of PM_{10} in the Milan area. In: 2nd International Environmental Modelling and Software Society Conference (2004)
2. Collet, R.S., Oduyemi, K.: Air quality modelling: A technical review of mathematical approaches. Metereological Applications 4(3), 235–246 (1997)
3. Corani, G., Barazzetta, S.: First results in the prediction of particulate matter in the Milan area. In: 9th Int. Conf. on Harmonisation within Atmospheric Dispersion Modelling for Regulatory Purposes (2004)
4. Finzi, G., Volta, M., Nucifora, A., Nunnari, G.: Real time ozone episode forecast: A comparison between neural network and grey box models. In: Proceedings of International ICSC/IFAC Symposium of Neural Computation, pp. 854–860. ICSC Academic Press, London (1998)
5. Gardner, M.W., Dorling, S.R.: Artificial neural networks (the multilayer perceptron) – A review of applications in the atmospheric sciences. Atmospheric Environment 32(14–15), 2627–2636 (1998)
6. Gardner, M.W., Dorling, S.R.: Neural network modelling and prediction of hourly NO_x and NO_2 concentration in urban air in London. Atmospheric Environment 33, 709–719 (1999)
7. Goyal, P., Chanb, A.T., Jaiswa, N.: Statistical models for the prediction of respirable suspended particulate matter in urban cities. Atmospheric Environment 40(11), 2068–2077 (2006)
8. Hooyberghs, J., Mensink, C., Dumont, G., Fierens, F., Brasseur, O.: A neural network forecast for daily average PM_{10} concentrations in Belgium. Atmospheric Environment 39(18), 3279–3289 (2005)
9. Kukkonen, J., Partanen, L., Karppinen, A., Ruuskanen, J., Junninen, H., Kolehmainen, M., Niska, H., Dorling, S.R., Chatterton, T., Foxall, R., Cawley, G.: Extensive evaluation of neural network models for the prediction of NO_2 and PM_{10} concentrations, compared with a deterministic modelling system and measurements in central Helsinki. Atmospheric Environment 37, 4539–4550 (2003)
10. Ljung, L.: System Identification — Theory for the User, 2nd edn. PTR Prentice Hall, Upple Saddle River (1999)
11. Lu, W.Z., Fan, H.Y., Lo, S.M.: Application of evolutionary neural network method in predicting pollutant levels in downtown area of Hong Kong. Neurocomputing 51, 387–400 (2003)
12. Lu, W.Z., Wang, W.J., Xu, Z.B., Leung, A.Y.: Using improved neural network model to analyze RSP, NO_x and NO_2 levels in urban air in Mong Kok, Hong Kong. Environmental Monitoring and Assessment 87(3), 235–254 (2003)
13. Morabito, F.C., Versaci, M.: Wavelet neural network processing of urban air pollution. In: Proceedings of IJCNN 2002, Honolulu (Hawaii), vol. 1, pp. 432–437. IEEE, Los Alamitos (2002)

14. Nunnari, G., Cannavò, F.: Modified cost functions for modelling air quality time series by using neural networks. In: Kaynak, O., Alpaydın, E., Oja, E., Xu, L. (eds.) ICANN 2003 and ICONIP 2003. LNCS, vol. 2714, pp. 723–728. Springer, Heidelberg (2003)

15. Ordieres, J.B., Vergara, E.P., Capuz, R.S., Salazar, R.E.: Neural network prediction model for fine particulate matter ($PM_{2.5}$) on the US–Mexico border in El Paso (Texas) and Ciudad Juárez (Chihuahua). Environmental Modelling and Software 20(5), 547–559 (2005)

16. Ostro, B., Chestnut, L., Vichit-Vadakan, N., Laixuthai, A.: The impact of particulate matter on daily mortality in Bangkok, Thailand. Journal of Air and Waste Management Association 49, 100–107 (1999)

17. Papoulis, A.: Probability, Random Variables, and Stochastic Processes, 3rd edn. McGraw–Hill, New York (1991)

18. Perez, P., Reyes, J.: Prediction of maximum of 24–h average of PM_{10} concentrations 30 h in advance in Santiago, Chile. Atmospheric Environment 36, 4555–4561 (2002)

19. Pope, C.A., Burnett, R., Thun, M.J., Calle, E.E., Krewskik, D., Ito, K., Thurston, G.D.: Lung cancer, cardiopulmonary mortality, and long term exposure to fine particulate air pollution. Journal of the American Medical Association 287, 1132–1141 (2002)

20. Pope, C.A., Thun, M.J., Namboodiri, M.M., Dockery, D.W., Evans, J.S., Speizer, F.E., Heath, C.W.: Particulate air pollution as predictor of mortality in a prospective study of US adults. American Journal of Respiratory and Critical Care Medicine 151, 669–674 (1995)

21. Wang, J.Y., Lord, E., Cannon, A., Walters, G.: Statistical models for spot air quality forecasts (O_3 and PM_{10}) in British Columbia. In: Proceedings of the 2005 Puget Sound Georgia Basin Research Conference, Seattle (2005)

22. Zolghadri, A., Henry, D.: Minimax statistical models for air pollution time series, Application to ozone time series data measured in Bordeaux. Environmental Monitoring and Assessment 98(1–3), 275–294 (2004)

Fuzzy Evolutionary Probabilistic Neural Networks

V.L. Georgiou[*], Ph.D. Alevizos, and M.N. Vrahatis

Computational Intelligence Laboratory (CI Lab), Department of Mathematics,
University of Patras Artificial Intelligence Research Center (UPAIRC),
University of Patras, GR-26110 Patras, Greece
{vlg,philipos,vrahatis}@math.upatras.gr

Abstract. One of the most frequently used models for classification tasks is the Probabilistic Neural Network. Several improvements of the Probabilistic Neural Network have been proposed such as the Evolutionary Probabilistic Neural Network that employs the Particle Swarm Optimization stochastic algorithm for the proper selection of its spread (smoothing) parameters and the prior probabilities. To further improve its performance, a fuzzy class membership function has been incorporated for the weighting of its pattern layer neurons. For each neuron of the pattern layer, a fuzzy class membership weight is computed and it is multiplied to its output in order to magnify or decrease the neuron's signal when applicable. Moreover, a novel scheme for multi–class problems is proposed since the fuzzy membership function can be incorporated only in binary classification tasks. The proposed model is entitled Fuzzy Evolutionary Probabilistic Neural Network and is applied to several real-world benchmark problem with promising results.

1 Introduction

A rapid development of Computational Intelligence methods has taken place recently. A simple but promising model which combines statistical methods and efficient evolutionary algorithms is the recently proposed *Evolutionary Probabilistic Neural Network* (EPNN) [1,2]. Specifically, EPNN is based on the Probabilistic Neural Network (PNN) introduced by Specht [3] that has been widely used in several areas of science with promising results [4,5,6,7]. PNN is based on discriminant analysis [8] and incorporates the Bayes decision rule for the final classification of an unknown feature vector. In order to estimate the Probability Density Function (PDF) of each class, the Parzen window estimator or in other words the kernel density estimator is used [9]. The recently proposed EPNN employs the Particle Swarm Optimization (PSO) algorithm [10,11] for the selection of the spread parameters of PNN's kernels. Several other variants of PNN have been proposed in the literature. A Fuzzy PNN is proposed in [12], where a modification of the typical misclassification proportion is minimized in the training procedure.

---

[*] Corresponding author.

L. Prevost, S. Marinai, and F. Schwenker (Eds.): ANNPR 2008, LNAI 5064, pp. 113–124, 2008.
© Springer-Verlag Berlin Heidelberg 2008

Several other remarkable efforts have taken place so that fuzzy logic [13,14] can be incorporated into well known and widely used classification models. Such an effort has been made in [15], where a Fuzzy Membership Function (FMF) has been introduced and incorporated into the Perceptron algorithm. Moreover, a Fuzzy Kernel Perceptron has been proposed in [16] in order to form a fuzzy decision boundary that separates two classes. The FMF that was employed in [16], is the one proposed by Keller and Hunt [15].

In this contribution an extension of the EPNN is proposed which incorporates the aforementioned Fuzzy Membership Function (FMF). This function describes the degree of certainty that a given datum belongs to each one of the predefined classes. The FMF provides a way of weighting all the training vectors so that an even better classification accuracy can be achieved.

2   Background Material

For completeness purposes, let us briefly present the necessary background material. As it has already been mentioned, PNN is used mainly for classification tasks. The training procedure of a PNN is quite simple and requires only a single pass of the patterns of the training data which results to a short training time. The architecture of a PNN always consists of four layers: the *input layer*, the *pattern layer*, the *summation layer* and the *output layer* [1,3].

Suppose that an input feature vector $\mathbf{X} \in \mathbb{R}^p$ has to be classified into one of K predefined classes. The vector $\mathbf{X}$ is applied to the p neurons of PNN's input layer and is then passed to the pattern layer. The neurons of the pattern layer are connected with all the input layers' neurons and are organized into K groups. Each group of neurons in the pattern layer consists of N_k neurons, where N_k is the number of training vectors that belong to the class k, $k = 1, 2, \ldots, K$. The ith neuron in the kth group of the pattern layer computes its output using a kernel function that is typically a Gaussian kernel function of the form:

$$f_{ik}(X) = \frac{1}{(2\pi)^{p/2}|\Sigma_k|^{1/2}} \exp\left(-\frac{1}{2}(X - X_{ik})^T \Sigma_k^{-1} (X - X_{ik})\right), \quad (1)$$

where $\mathbf{X}_{ik} \in \mathbb{R}^p$ is the center of the kernel and Σ_k is the matrix of spread (smoothing) parameters of the kernel. The vector $\mathbf{X}_{ik}$ corresponds to the ith feature vector of the kth group of the training data set.

The summation layer comprises K neurons and each one estimates the conditional probability of its class given an unknown vector $\mathbf{X}$:

$$G_k(X) = \sum_{i=1}^{N_k} \pi_k f_{ik}(X), \quad k \in \{1, 2, \ldots, K\}, \quad (2)$$

where π_k is the prior probability of class k, $\sum_{k=1}^{K} \pi_k = 1$. Thus, a vector $\mathbf{X}$ is classified to the class that has the maximum output of the summation neurons.

Instead of utilizing the whole training data set of size N and create a pattern layer that consists of N neurons, a smaller training set is created so that the new PNN will have less memory requirements and will be much faster. The well-known K-medoids clustering algorithm [17] is applied to the training data of each class and K is set equal to 5% of the size of each class. Following this strategy, the proportion of instances of a class to the whole training data set remains the same. The adjacent training vectors are grouped as a cluster that is represented by the corresponding medoid. Then, the obtained medoids are used as centers to the corresponding PNN's kernel functions of the pattern layer's neurons. Thus, the pattern layer's size of the proposed PNN is about twenty times smaller than the corresponding PNN which utilizes all the available training data resulting to a much faster model.

For the estimation of the spread matrix Σ_k as well as the prior probabilities π_k, PSO algorithm is used. PSO is a stochastic population–based optimization algorithm [10] and it is based on the idea that a population of particles are released into a search space and travel with adaptable velocity in order to find promising regions into it [11,18]. Moreover, they retain a memory of the best position they have ever visited and at each step they intercommunicate to inform each other about their position and the value of the objective function at that particular point. The velocity of each particle is updated according to the particle's best value and the swarm best value.

Let $g(X)$ be the objective function that has to be minimized. Given a d–dimensional search space $S \subset \mathbb{R}^d$ and a swarm consisting of NP particles, let $Z_i \in S$ be the position of the ith particle and V_i be the velocity of this particle. Moreover, let BP_i be the best previous position encountered by the ith particle in S. Assume gl to be the index of the particle that attained the best previous position among all particles, and t to be the iteration counter. Then, the swarm is manipulated by the equations

$$V_i(t+1) = \chi \left[V_i(t) + c_1 \, r_1 \big(BP_i(t) - Z_i(t) \big) + c_2 \, r_2 \big(BP_{gl}(t) - Z_i(t) \big) \right], \quad (3)$$
$$Z_i(t+1) = Z_i(t) + V_i(t+1), \quad (4)$$

where $i = 1, 2, \ldots, NP$; χ is a parameter called *constriction coefficient*; c_1 and c_2 are two positive constants called *cognitive* and *social* parameter, respectively; and r_1, r_2, are random vectors that are uniformly distributed within $[0, 1]^d$ [19]. All vector operations in Eqs. (3) and (4) are computed component-wise and the best positions are then updated according to the equation

$$BP_i(t+1) = \begin{cases} Z_i(t+1), & \text{if } g\big(Z_i(t+1)\big) < g\big(BP_i(t)\big), \\ BP_i(t), & \text{otherwise.} \end{cases}$$

The particles are always bounded in the search space S and the constriction coefficient is derived analytically through the formula

$$\chi = \frac{2\kappa}{\left| 2 - \varphi - \sqrt{\varphi^2 - 4\varphi} \right|}, \quad (5)$$

for $\varphi > 4$, where $\varphi = c_1 + c_2$, and $\kappa = 1$, based on the stability analysis of Clerc and Kennedy [19,20].

A different matrix of spread parameters $\Sigma_k = \text{diag}(\sigma_{1k}^2, \ldots, \sigma_{pk}^2), k = 1, 2, \ldots, K$ is assumed for each class and a swarm of Σ_k created by PSO. The objective function that PSO should minimize is the misclassification proportion on the whole training data set.

3   The Proposed Approach

One of the desirable properties that a supervised classification model should possess is the ability to adjust the impact of each training sample vector to the final decision of the model. In other words, vectors of high uncertainty about their class membership should have less influence on the final decision of the model, while vectors of low uncertainty should affect more the model's decision. One way of obtaining this desirable property is to incorporate a Fuzzy Membership Function (FMF) into the model. Among the large variety of classification models we chose the EPNN due to its simplicity, effectiveness and efficiency [1,2] and we have incorporated the FMF proposed in [15] for weighting the pattern neurons of the EPNN. By this way, fuzzy class membership values are assigned to each pattern neuron and for this reason the proposed model is named *Fuzzy Evolutionary Probabilistic Neural Network* (FEPNN). The efficiency of the EPNNs and their variants is clearly presented in [2] where the EPNNs are compared with the best ever classification models on several problems.

Next, let us further analyze the proposed model. As it has already been mentioned in Section 2, $\mathbf{X}_{ik}$, $i = 1, 2, \ldots, N_k$, $k = 1, 2, \ldots, K$ is the i–th training sample vector that belongs to class k. Since we are dealing with a two-class classification problem, we consider $K = 2$. Suppose further that $u(\mathbf{X}) \in [0, 1]$ is a fuzzy membership function, then we define:

$$u(\mathbf{X}_{ik}) \equiv u_{ik} = 0.5 + \frac{\exp\left((-1)^k \left[d_1(\mathbf{X}_{ik}) - d_2(\mathbf{X}_{ik})\right] \lambda/d\right) - \exp(-\lambda)}{2\left(\exp(\lambda) - \exp(-\lambda)\right)},$$

where for $k = 1, 2$, $\mathbf{M}_k$ is the mean vector of class k, $d_k(\mathbf{X}) = \|\mathbf{X} - \mathbf{M}_k\|$ is the distance between vector $\mathbf{X}$ and mean vector of class k, $d = \|\mathbf{M}_1 - \mathbf{M}_2\|$ is the distance between the two mean vectors and λ is a constant that controls the rate at which fuzzy membership values decrease towards 0.5 [15]. The fuzzy membership values were designed so that if the vector is equal to the mean of the class that it belongs, then it should be 1.0. Also, if the vector is equal to the mean of the other class, it should be 0.5, meaning that this pattern neuron should not consider the most to the final decision. Moreover, if the vector is equidistant from the two means, then it should be near 0.5, since it cannot really help us to the final classification. In other words, as the vector gets closer to its mean and goes further away the other mean, its value should approach 1.0 exponentially. Moreover, the pseudocode of the proposed approach is presented in Table 1.

As it was previously mentioned, the proposed approach can be applied only to binary classification problems. In order to make it applicable to a wider spectrum of tasks, we propose a way of applying it to multi-class classification problems by using the following multi-class decomposition scheme. Assuming that $K > 2$, let

Create the clustered training set $\mathcal{T}_{cl\,tr}$ of size $N_{cl\,tr}$ from the
training data set $\mathcal{T}_{tr}$.

Select initial random values for Σ_k and π_k, $k = 1, 2$.
Construct PNN using $\mathcal{T}_{cl\,tr}$, Σ_k and π_k.
Compute $\mathbf{M}_1$, $\mathbf{M}_2$ and d.
For $i = 1, N_{cl\,k}$ and $k = 1, 2$ do:
    Compute fuzzy membership values u_{ik} using Eq. (3).
EndFor

Compute Σ_k and π_k by PSO

For $l = 1, NP$ do:
    Initialize a swarm
$$Z_l(0) = [\sigma_{11\,l}, \sigma_{12\,l}, \ldots, \sigma_{1\,p\,l}, \sigma_{21\,l}, \sigma_{22\,l}, \ldots, \sigma_{2\,p\,l}, \pi_{1\,l}, \pi_{2\,l}].$$
    Initialize best positions $BP_l(0)$.
EndFor
For $t = 1, \text{MaxGeners}$ do:
    For $l = 1, NP$ do:
        Update velocities $V_l(t+1)$ using Eq. (3).
        Update particles $Z_l(t+1) = Z_l(t) + V_l(t+1)$.
        Constrain each particle $Z_l(t+1) \in (0, \gamma]^{2p} \times [0, 1]^2$.
        Set $MP_l = 0$. (Misclassification Proportion)
        For $m = 1, N_{tr}$ do:
            Compute $Out(m) = \arg\max_k \left(G_k(\mathbf{X}_m) = \pi_k \sum_{i=1}^{N_{cl\,k}} u_{ik}\, f_{ik}(\mathbf{X}_m) \right)$
            If $(Out(m) \neq Target(m))$ Then $MP_l = MP_l + 1$.
        EndFor
        Set $g(Z_l(t+1)) = MP_l / N_{tr}$
        Update best position $BP_l(t+1)$.
    EndFor
EndFor

Write the optimal Σ_k and Π and the classification accuracy of the PNN
on $\mathcal{T}_{tr}$ and $\mathcal{T}_{te}$.

Fig. 1. Pseudocode of the proposed approach

$\mathbf{OM}$ be the overall mean vector of the whole data set and $\mathbf{OM}_k$ be the overall mean vector of the data set excluding the vectors of class k, $k = 1, 2, \ldots, K$. We calculate the Euclidean distances $D_k = \|\mathbf{OM} - \mathbf{M}_k\|$ and $D'_k = \|\mathbf{OM}_k - \mathbf{M}_k\|$ for all the classes and we sort the K classes according to their total distance $OD_k = D_k + D'_k$.

So, a sequence of $K - 1$ FEPNNs will be created for the final classification. Let s_k, $k = 1, 2, \ldots, K$ be the indices of the sorted classes. For the first FEPNN, we will utilize a training set consisting of the vectors of class s_1 as class 1 and the rest of the training set as class 2. Since this is a binary classification training set, we can use the proposed FEPNN in order to classify the "unknown" vectors of the test set that belong to class s_1. By this manner, we can record the number of correctly classified vectors of s_1 as C_{s_1}. This procedure is continued for the rest of the s_k, $k = 2, 3, \ldots, K - 1$ and at every step we do not take into account the vectors of the classes s_i, $i < k$ and we compute the classification accuracies C_{s_k}. At $K - 1$ step, we have only the last two classes left, so only one FEPNN is needed from which we compute $C_{s_{K-1}}$ and C_{s_K}. So the final classification accuracy is the sum of C_{s_k}, $k = 1, 2, \ldots, K$.

Several other multi-class decomposition schemes can be used such as the one–vs–others (1–vs–r) or the one–vs–one (1–vs–1) scheme. In the (1–vs–r) scheme, the problem is decomposed into a set of K two-class problems where for each class $k = 1, 2, \ldots, K$ a classifier is constructed that distinguishes between class k and the composite class consisting of all other classes. By this way, the training set always consists of all the exemplars of all classes while in our proposed scheme, in every step one class is excluded and only $K - 1$ classifiers are constructed that results in a faster scheme. On the other hand, using the (1–vs–1) scheme also known as pairwise coupling where a classifier is constructed for each distinct pair of classes using only the training samples for those classes, $K(K-1)/2$ classifiers are constructed. This demands $K/2$ times more classifiers to be constructed compared to our proposed scheme although we should note that in our case the training data sets will be larger in the first steps.

4   Experimental Results

The proposed model has been applied to four binary and two multi–class benchmark problems from several fields of science from Proben1 database [21] that come from the UCI repository [22] in order to evaluate its efficiency and performance.

(1) The first data set is the *Wisconsin Breast Cancer Database* (WBCD) and the target of this problem is to predict whether a breast tumour is benign or malignant[23]. We have 699 instances and for each one of them we have 9 continuous attributes based on cell descriptions gathered by microscopic examination such as the uniformity of cell size and shape; bland chromatin; single epithelial cell size; and mitoses.

(2) In the second data set (*Card*), we want to predict the approval or non-approval of a credit card to a customer. There are 51 attributes which are unexplained for confidential reasons and 690 customers.

(3) The third data set is the *Pima Indians Diabetes* data set and the input features are the diastolic blood pressure; triceps skin fold thickness; plasma glucose concentration in a glucose tolerance test; and diabetes pedigree function. The 8 inputs are all continuous without missing values and there are 768 instances. The aim is to classify whether someone is infected by diabetes or not, therefore, there are two classes.

(4) In the last binary classification data set, namely *Heart Disease*, its aim is to predict whether at least one of the four major vessels of the heart is reduced in diameter by more than 50%. The 35 attributes of the 920 patients are age, sex, smoking habits, subjective patient pain descriptions and results of various medical examinations such as blood pressure and cardiogram.

(5) The fifth data set, *Glass*, consists of 214 instances and its aim is to classify a piece of glass into 6 different types, namely float processed or non float processed building windows, vehicle windows, containers, tableware and heat lamps. The classification is based on 9 inputs, which are the percentages of content on 8 different elements plus the refractive index and this task is motivated by forensic needs in criminal investigation.

(6) The last data set is the *Horse* data set and its task is to predict the fate of a horse that has a colic. The prediction whether the horse would survive, would die or would be euthanized is based on 58 inputs of a veterinary examination of the horse and there are 364 instances.

Moreover, the proposed model was applied to the aforementioned benchmark data sets using 10 times 10-fold cross-validation where the folds were randomly selected and the obtained results are presented in Tables 1 and 2. In order to

Table 1. Test set classification accuracy percentage of two-class data sets

Data set	Model	Mean	Median	SD	Min	Max
WBCD	PNN	95.79	95.85	0.25	95.27	96.14
	GGEE.PNN	96.39	96.42	0.20	95.99	96.71
	Hom.EPNN	95.82	95.85	0.28	95.28	96.28
	Het.EPNN	95.32	95.21	0.57	94.42	96.14
	Bag.EPNN	96.85	96.78	0.46	96.14	97.85
	Bag.P.EPNN	97.17	97.14	0.16	96.86	97.43
	FEPNN	**97.61**	**97.56**	0.19	97.42	97.85
Card	PNN	82.10	81.96	0.76	80.87	83.48
	GGEE.PNN	84.31	84.28	0.63	83.48	85.51
	Hom.EPNN	85.35	85.22	0.38	84.93	86.09
	Het.EPNN	**87.67**	**87.76**	0.51	86.96	88.55
	Bag.EPNN	86.64	86.67	0.51	85.80	87.39
	Bag.P.EPNN	86.83	86.81	0.34	86.38	87.39
	FEPNN	87.42	87.39	**0.28**	87.10	87.97
Diabetes	PNN	65.08	65.08	**0.05**	64.99	65.15
	GGEE.PNN	69.43	69.24	0.68	68.53	70.38
	Hom.EPNN	67.67	67.58	0.88	66.03	68.80
	Het.EPNN	69.37	69.46	0.80	67.73	70.54
	Bag.EPNN	71.00	71.16	1.02	68.90	72.09
	Bag.P.EPNN	71.22	71.39	1.00	69.75	72.54
	FEPNN	**75.09**	**75.39**	0.88	73.59	76.22
Heart	PNN	79.23	79.13	0.48	78.59	80.00
	GGEE.PNN	80.68	80.65	0.52	79.89	81.41
	Hom.EPNN	81.50	81.52	**0.27**	80.87	81.74
	Het.EPNN	82.60	82.45	0.40	82.07	83.26
	Bag.EPNN	82.28	82.34	0.62	81.20	83.15
	Bag.P.EPNN	82.35	82.50	1.05	80.43	84.13
	FEPNN	**83.01**	**82.94**	0.32	82.72	83.80

Table 2. Test set classification accuracy percentage of multi-class data sets

Data set	Model	Mean	Median	SD	Min	Max
Glass	PNN	33.25	32.61	3.40	27.96	39.01
	GGEE.PNN	50.07	50.08	**1.44**	47.74	51.94
	Hom.EPNN	68.52	68.15	1.55	66.80	70.78
	Het.EPNN	75.36	75.30	1.77	73.31	77.60
	Bag.EPNN	54.91	55.09	3.98	49.16	63.47
	Bag.P.EPNN	52.74	51.54	4.13	48.84	63.14
	Mult.EPNN	75.79	75.73	2.95	72.19	80.93
	Mult.FEPNN	**77.28**	**77.60**	2.74	71.09	81.73
Horse	PNN	64.63	64.74	**0.72**	63.05	65.42
	GGEE.PNN	61.97	62.39	1.23	59.83	63.75
	Hom.EPNN	66.54	66.74	0.79	65.33	67.55
	Het.EPNN	68.48	68.36	0.97	67.08	69.75
	Bag.EPNN	66.47	66.40	1.40	64.56	69.19
	Bag.P.EPNN	66.16	66.33	1.56	63.33	67.97
	Mult.EPNN	72.23	72.14	1.89	69.89	74.72
	Mult.FEPNN	**72.78**	**72.75**	1.78	70.19	75.19

eliminate the influence of PSO initialization phase, we conducted 5 runs on each cross-validated data set and selected the results (σ's and π's) that were obtained by the run on which the classification accuracy was the median of the classification accuracies on each training set. In particular, the mean, median, standard deviation, minimum and maximum classification accuracy on the test sets is presented in the aforementioned tables. Moreover, the CPU training times are also reported in Tables 3 and 4. In order to evaluate the performance of our model, we have applied these six benchmark problems to Homoscedastic and Heteroscedastic Evolutionary Probabilistic Neural Networks [1] as well as to original PNNs and Bagging EPNNs [2]. For the original PNN's implementation, an exhaustive search for the selection of the spread parameter σ has been conducted in the interval $[10^{-3}, 5]$ and the σ that resulted to the best classification accuracy on the training set has been used for the calculation of PNN's classification accuracy on the test set. The number of functional evaluations for PNN's exhaustive search is the same with the one of EPNNs and FEPNNs. Moreover, a variation of the PNN that is proposed by Gorunescu *et al.* [24] has also been used by the name GGEE.PNN. In multi–class problems, the proposed approach that constructs a sequence of EPNNs or FEPNNs has been applied with and without the incorporation of the fuzzy membership function and is named Mult.FEPNN and Mult.EPNN respectively.

Searching for the most promising spread matrix Σ_k in EPNNs and FEPNNs, a swarm of 5 particles has been evolved for 50 generations for EPNN's homoscedastic case and a swarm of 10 particles for 100 generations for the other cases. The space that PSO was allowed to search in, was the aforementioned interval $[10^{-3}, 5]$ for Hom.EPNN, $[10^{-3}, 5]^K$ for Het.EPNN, $[10^{-3}, 5]^{Kp}$ for Bagging EPNN and $[10^{-3}, 5]^{2p}$ for FEPNN. On the Bagging EPNN case, an

Table 3. CPU time for the training of the models (seconds)

Data set	Model	Mean	Median	SD	Min	Max
WBCD	PNN	42.09	42.42	0.66	40.66	42.69
	GGEE.PNN	1.52	1.61	0.17	1.22	1.65
	Hom.EPNN	89.12	88.82	1.07	88.12	91.73
	Het.EPNN	171.78	171.75	1.07	170.21	174.04
	Bag.EPNN	82.78	78.07	8.86	76.22	99.75
	Bag.P.EPNN	90.01	89.86	0.92	88.97	92.12
	FEPNN	8.59	8.56	0.14	8.39	8.82
Card	PNN	182.01	186.37	7.88	169.82	187.93
	GGEE.PNN	5.46	5.45	0.06	5.38	5.53
	Hom.EPNN	266.10	274.39	74.56	168.72	342.27
	Het.EPNN	521.60	510.24	142.74	327.08	671.83
	Bag.EPNN	309.85	309.36	1.88	307.58	314.33
	Bag.P.EPN	309.73	309.84	2.62	305.26	314.95
	FEPNN	28.94	28.80	0.47	28.37	29.94
Diabetes	PNN	49.58	49.64	0.38	49.06	50.09
	GGEE.PNN	1.87	1.87	0.03	1.83	1.90
	Hom.EPNN	101.17	101.13	0.48	100.40	102.01
	Het.EPNN	195.27	195.66	0.92	193.82	196.62
	Bag.EPNN	106.42	106.53	0.92	104.25	107.73
	Bag.P.EPNN	106.24	106.26	0.81	105.31	108.06
	FEPNN	10.03	10.08	0.14	9.79	10.19
Heart	PNN	207.99	223.48	45.27	125.62	241.32
	GGEE.PNN	6.47	6.94	0.92	4.95	7.18
	Hom.EPNN	223.28	224.35	4.28	215.15	228.97
	Het.EPNN	438.10	440.29	6.82	422.45	449.24
	Bag.EPNN	394.49	392.36	5.93	387.13	404.55
	Bag.P.EPNN	393.22	391.47	4.95	388.02	401.03
	FEPNN	38.00	38.03	0.86	36.81	39.18

ensemble of 11 EPNNs was constructed. The value of the parameter f in the FMF was set to 0.5 after a trial-and-error procedure. In order to decide whether parametric or non parametric statistical tests should be conducted for the statistical comparison of the models' performance, a Kolmogorov-Smirnov test has been conducted on each sample of runs [25]. In all the samples, the normality assumption was met so a corrected resampled t–test was employed for the comparisons [26,27]. The level of significance in all the statistical tests was set to 0.05 and if a model's mean performance is statistically significantly superior than the second best performance, then it is depicted in a box. On the cancer data set, the best mean performance was achieved by the FEPNN and there was a statistically significant difference between its performance and the Bag.P.EPNN's performance which achieved the second best performance. Moreover, FEPNN obtained the lowest standard deviation of the classification accuracies. The best mean performance on the Card data set was obtained by the Het. EPNN but it was quite similar to the one that FEPNN obtained. However, FEPNN's standard

Table 4. CPU time for the training of the models (seconds)

Data set	Model	Mean	Median	SD	Min	Max
Glass	PNN	3.82	3.80	0.02	3.79	3.85
	GGEE.PNN	3.66	3.64	0.08	3.57	3.79
	Hom.EPNN	9.16	9.26	0.65	7.99	9.95
	Het.EPNN	17.21	17.47	0.76	16.04	18.27
	Bag.EPNN	29.03	29.03	0.14	28.80	29.19
	Bag.P.EPNN	28.30	28.31	0.11	28.10	28.48
	Mult.EPNN	6.02	6.08	0.31	5.40	6.41
	Mult.FEPNN	6.17	6.25	0.31	5.68	6.67
Horse	PNN	29.29	29.53	0.88	26.92	30.00
	GGEE.PNN	5.46	5.56	0.14	5.26	5.58
	Hom.EPNN	76.10	77.98	7.97	66.17	87.37
	Het.EPNN	169.92	169.92	23.39	147.73	192.11
	Bag.EPNN	126.76	126.84	2.02	124.68	129.96
	Bag.P.EPNN	123.03	121.61	2.24	121.35	126.92
	Mult.EPNN	17.61	17.69	0.74	16.50	18.65
	Mult.FEPNN	17.65	17.74	0.67	16.53	18.48

deviation was almost half of the Het.EPNN's. Besides that, the number of pattern layer's neurons of Het.EPNN was about 690 while in the FEPNN there were 34 neurons, which has as a result a much faster model both in training and response time as it is confirmed in Table 3.

On the diabetes data set, the statistically significant superiority of FEPNN is clear compared with the rest of the models. FEPNN's standard deviation is moreover similar to the rest of models' standard deviation except of the one obtained by the PNN which is much smaller but since there is such a great difference between the mean classification accuracies, it is not worth noting. On the heart data set, FEPNN obtained the best mean accuracy and there is a statistically significant difference with Het.EPNN's mean accuracy.

On the two multi–class problems, the proposed approach achieved the best performance and especially in Horse there was a statistically significant superiority than Het.EPNN. Summarizing the above, in five out of six cases the FEPNN had a superior performance and in four of them, the superiority was statistically significant.

Moreover, the proposed approach needs much less CPU training time than Bagging EPNNs and Het.EPNNs in all the benchmark problems as we can observe from Tables 3 and 4.

5   Concluding Remarks

In this contribution, a novel classification model has been proposed, namely the Fuzzy Evolutionary Probabilistic Neural Network that incorporates a fuzzy membership function for binary classification. A novel way of handling multi–class problems using binary classification models is also proposed.

It has been shown that the FEPNN can achieve similar or superior performance compared to other PNN variations both in binary and multi–class problems. Nevertheless, it is much faster in training and response times since it utilizes only a small fraction of the training data and achieves similar or superior accuracy. It is clear that the incorporation of the fuzzy membership function into Evolutionary Probabilistic Neural Network, helped it to obtain even more promising results.

The proposed approach is a general purpose method since it achieves promising results in classification problems on several areas of science either binary classification or multi-class classification problems.

Acknowledgment

We thank the European Social Fund (ESF), the Operational Program for Educational and Vocational Training II (EPEAEK II) and particularly the IRAKLEITOS Program for funding the above work.

References

1. Georgiou, V.L., Pavlidis, N.G., Parsopoulos, K.E., Alevizos, Ph.D., Vrahatis, M.N.: New self–adaptive probabilistic neural networks in bioinformatic and medical tasks. International Journal on Artificial Intelligence Tools 15(3), 371–396 (2006)
2. Georgiou, V.L., Alevizos, Ph.D., Vrahatis, M.N.: Novel approaches to probabilistic neural networks through bagging and evolutionary estimating of prior probabilities. Neural Processing Letters 27, 153–162 (2008)
3. Specht, D.F.: Probabilistic neural networks. Neural Networks 1(3), 109–118 (1990)
4. Ganchev, T., Tasoulis, D.K., Vrahatis, M.N., Fakotakis, N.: Locally recurrent probabilistic neural networks with application to speaker verification. GESTS International Transaction on Speech Science and Engineering 1(2), 1–13 (2004)
5. Ganchev, T., Tasoulis, D.K., Vrahatis, M.N., Fakotakis, N.: Generalized locally recurrent probabilistic neural networks with application to text-independent speaker verification. Neurocomputing 70(7–9), 1424–1438 (2007)
6. Guo, J., Lin, Y., Sun, Z.: A novel method for protein subcellular localization based on boosting and probabilistic neural network. In: Proceedings of the 2nd Asia-Pacific Bioinformatics Conference (APBC 2004), Dunedin, New Zealand, pp. 20–27 (2004)
7. Huang, C.J.: A performance analysis of cancer classification using feature extraction and probabilistic neural networks. In: Proceedings of the 7th Conference on Artificial Intelligence and Applications, Wufon, Taiwan, pp. 374–378 (2002)
8. Hand, J.D.: Kernel Discriminant Analysis. Research Studies Press, Chichester (1982)
9. Parzen, E.: On the estimation of a probability density function and mode. Annals of Mathematical Statistics 3, 1065–1076 (1962)
10. Kennedy, J., Eberhart, R.: Particle swarm optimization. In: Proceedings IEEE International Conference on Neural Networks, Piscataway, NJ, vol. IV, pp. 1942–1948. IEEE Service Center, Los Alamitos (1995)

11. Parsopoulos, K.E., Vrahatis, M.N.: Recent approaches to global optimization problems through particle swarm optimization. Natural Computing 1(2–3), 235–306 (2002)
12. Delgosha, F., Menhaj, M.B.: Fuzzy probabilistic neural networks: A practical approach to the implementation of bayesian classifier. In: Reusch, B. (ed.) Fuzzy Days 2001. LNCS, vol. 2206, pp. 76–85. Springer, Heidelberg (2001)
13. Zadeh, L.A.: Fuzzy sets. Information and Control 8(3), 338–353 (1965)
14. Zadeh, L.A.: Fuzzy logic. IEEE Computer 21(4), 83–93 (1988)
15. Keller, J.M., Hunt, D.J.: Incorporating fuzzy membership functions into the perceptron algorithm. IEEE Trans. Pattern Anal. Machine Intell. 7(6), 693–699 (1985)
16. Chen, J., Chen, C.: Fuzzy kernel perceptron. IEEE Transactions on Neural Networks 13(6), 1364–1373 (2002)
17. Kaufman, L., Rousseeuw, P.J.: Finding Groups in Data: An Introduction to Cluster Analysis. John Wiley and Sons, New York (1990)
18. Parsopoulos, K.E., Vrahatis, M.N.: On the computation of all global minimizers through particle swarm optimization. IEEE Transactions on Evolutionary Computation 8(3), 211–224 (2004)
19. Clerc, M., Kennedy, J.: The particle swarm–explosion, stability, and convergence in a multidimensional complex space. IEEE Transactions on Evolutionary Computation 6(1), 58–73 (2002)
20. Trelea, I.C.: The particle swarm optimization algorithm: Convergence analysis and parameter selection. Information Processing Letters 85, 317–325 (2003)
21. Prechelt, L.: Proben1: A set of neural network benchmark problems and benchmarking rules. Technical Report 21/94, Fakultät für Informatik, Universität Karlsruhe (1994)
22. Newman, D.J., Hettich, S., Blake, C.L., Merz, C.J.: UCI repository of machine learning databases (1998)
23. Mangasarian, O.L., Wolberg, W.H.: Cancer diagnosis via linear programming. SIAM News 23, 1–18 (1990)
24. Gorunescu, F., Gorunescu, M., Revett, K., Ene, M.: A hybrid incremental/monte carlo searching technique for the smoothing parameter of probabilistic neural networks. In: Proceedings of the International Conference on Knowledge Engineering, Principles and Techniques, KEPT 2007, Cluj-Napoca, Romania, pp. 107–113 (2007)
25. Kanji, G.K.: 100 Statistical Tests. Sage Publications, Thousand Oaks (1999)
26. Bouckaert, R.R., Frank, E.: Evaluating the replicability of significance tests for comparing learning algorithms. In: Dai, H., Srikant, R., Zhang, C. (eds.) PAKDD 2004. LNCS (LNAI), vol. 3056, pp. 3–12. Springer, Heidelberg (2004)
27. Nadeau, C., Bengio, Y.: Inference for the generalization error. Machine Learning 52(3), 239–281 (2003)

Experiments with Supervised Fuzzy LVQ

Christian Thiel, Britta Sonntag, and Friedhelm Schwenker

Institute of Neural Information Processing, University of Ulm, 89069 Ulm, Germany
christian.thiel@uni-ulm.de

Abstract. Prototype based classifiers so far can only work with hard labels on the training data. In order to allow for soft labels as input label and answer, we enhanced the original LVQ algorithm. The key idea is adapting the prototypes depending on the similarity of their fuzzy labels to the ones of training samples. In experiments, the performance of the fuzzy LVQ was compared against the original approach. Of special interest was the behaviour of the two approaches, once noise was added to the training labels, and here a clear advantage of fuzzy versus hard training labels could be shown.

1 Introduction

Prototype based classification is popular because an expert familiar with the data can look at the representatives found and understand why they might be typical. Current approaches require the training data to be hard labeled, that is each sample is exclusively associated with a specific class. However, there are situations where not only hard labels are available, but soft ones. This means that each object is assigned to several classes with a different degree. For such cases, several classification algorithms are available, for example fuzzy Radial Basis Function (RBF) networks [1] or fuzzy-input fuzzy-output Support Vector Machines (SVMs) [2], but none of them allows for an easy interpretation of typical representatives. Thus, we decided to take the well-known Learning Vector Quantisation (LVQ) approach and enhance it with the ability to work with soft labels, both as training data and for the prototypes. The learning rules we propose are presented in section 2.

A related approach is the Soft Nearest Prototype Classification [3] with its extension in [4], where prototypes also are assigned fuzzy labels in the training process. However, the training data is still hard labeled. Note that also some variants of fuzzy LVQ exist, for example those of Karayiannis and Bezdek [5] or in [6], that have a hard labeling of the prototypes, and a soft neighbourhood function, exactly the opposite of our approach in this respect.

To assess the classification power of the suggested fuzzy LVQ, we compare its performance experimentally against LVQ1, and study the impact of adding various levels of noise to the training labels.

L. Prevost, S. Marinai, and F. Schwenker (Eds.): ANNPR 2008, LNAI 5064, pp. 125–132, 2008.

2   Supervised Fuzzy LVQ

Like the original Learning Vector Quantisation approach proposed by Kohonen [7], we employ r prototypes, and learn the locations of those. But, our fuzzy LVQ does not work with hard labels, but soft or fuzzy ones. It is provided with a training set M

$$
\begin{aligned}
& M = \{(x^\mu, l^\mu)\mid \mu = 1, \ldots, n\} \\
& x^\mu \in \mathbb{R}^z : \text{the samples} \\
& l^\mu \in [0;1]^c, \ \sum_{i=1}^{c} l_i^\mu = 1 : \text{their soft class labels}
\end{aligned}
\tag{1}
$$

where n is the number of training samples, z their dimension, and c the number of classes. Feeding a hitherto unseen sample x^{new} to the trained algorithm, we get a response l^{new} estimating a soft class label, of the form described just above.

As second major difference to the original LVQ network, each prototype $p^\eta \in \mathbb{R}^z$ is associated with a soft class label pl^η, which will influence the degree to which the prototypes (and their labels) are adapted to the data during training.

Note that the techniques presented can be applied to various prototype based learning procedures, for example LVQ3 or OLVQ.

The basic procedure to use the algorithm for classification purposes is split into three stages: initialisation, training and answer elicitation. The initialisation part will be explained in section 3, while obtaining an answer for the new sample x^{new} is quite simple: look for the prototype p^η closest to x^{new} (using the distance measure of your choice, we employed the Euclidean distance), and answer with its label pl^η. The training of the network is the most interesting part and will be described in the following.

2.1   Adapting the Prototypes

The basic training procedure of LVQ adapts the locations of the prototypes. For each training sample, the nearest prototype p^* is determined. If the label pl^* of the winner p^* matches the label l^μ of the presented sample x^μ, the prototype is shifted into the direction of the sample. When they do not match, the prototype is shifted away from the sample. This is done for multiple iterations.

Working with fuzzy labels, the similarity between two labels should no longer be measured in a binary manner. Hence, we no longer speak of a *match*, but of a similarity S between labels. Presenting a training sample x^μ, the update rule for our fuzzy LVQ is:

$$
p^* = p^* + \theta \cdot (\ S(l^\mu, pl^*) \ - e) \cdot (x^\mu - p^*)
\tag{2}
$$

Still, the prototype is shifted to or away from the sample. The direction now is determined by the similarity S of their labels: if S is higher than the preset threshold e, p^* is shifted towards x^μ, otherwise away from it. The learning rate θ controls the influence of each update step.

As similarity measure S between two fuzzy labels, we opted to use alternatively the simple scalar product (component wise multiplication, then summing up) and the S_1 measure introduced by Prade in 1980 [8] and made popular again by Kuncheva (for example in [9]):

$$S_1(l^a, l^b) = \frac{||l^a \cap l^b||}{||l^a \cup l^b||} = \frac{\sum_i min(l^a_i, l^b_i)}{\sum_i max(l^a_i, l^b_i)} \qquad (3)$$

2.2 Adapting the Labels

The label of the prototypes does not have to stay fixed. In the learning process, the winners' labels pl^* can be updated according to how close the training sample $s^\mu \in M$ is. This is accomplished with the following update rule:

$$pl^* = pl^* + \theta \cdot (l^\mu - pl^*) \cdot exp(-\frac{||x^\mu - p^*||^2}{\sigma^2}) \qquad (4)$$

The scaling parameter σ^2 of the exponential function is set to the mean Euclidean distance of all prototypes to all training samples. Again, θ is the learning rate.

3   Experimental Setup

The purpose of our experiments was twofold: Assessing the classification power of fuzzy LVQ, and its ability to stand up to noise added to the labels of its training data, all in comparison with standard LVQ1.

As test bed, we employed a fruits data set coming from a robotic environment, consisting of 840 pictures from 7 different classes (apples, oranges, plums, lemons,... [10]). Using results from Fay [11], we decided to use five features: *Colour Histograms* in the RGB space. Orientation histograms on the edges in a greyscale version of the picture (we used both the *Sobel* operator and the *Canny* algorithm to detect edges). As weakest features, colour histograms in the black-white opponent colour space $APQBW$ were calculated, and the mean color information in HSV space. Details on these features as well as references can be found in the dissertation mentioned just above.

The fruits data set originally only has hard labels (a banana is quite clearly a banana). As we need soft labels for our experiments, the original ones had to be fuzzified, which we accomplished using two different methods, K-Means and Keller. In the fuzzy K-Means [12] approach, we clustered the data, then assigned a label to each cluster centre according to the hard labels of the samples associated with it to varying degrees. Then, each sample was assigned a new soft label as a sum of products of its cluster memberships with the centres' labels. The fuzzifier parameter of the fuzzy K-Means algorithm, which controls the smoothness or entropy of the resulting labels, was chosen by hand so that the correspondence between hardened new labels and original ones was around 70%. The number of clusters was set to 35. In the second approach, based on work by Keller [13], the fuzzyness of a new soft label l^{new} can be controlled very straightforwardly, and using a mixture parameter $\alpha > 0.5$ it can now be ensured that the hardened l^{new} is the same as the original class C^{orig}:

$$l_i^{new} = \begin{cases} \alpha + (1 - \alpha) \cdot \frac{n_i}{k} \text{ , if } i = C^{orig} \\ (1 - \alpha) \cdot \frac{n_i}{k} \quad \text{ , otherwise} \end{cases}$$

The fraction counts what portion of the k nearest neighbours are of class i. In our experiments, $\alpha = 0.51$ and $k = 5$ were used.

All results were obtained using 5-fold cross validation. Hardening a soft label was simply accomplished by selecting the class i in the label with the highest value l_i, and our accuracy measure compares the hardened label with the original hard one.

Finding a suitable value for the threshold e in equation 2, which controls at what similarity levels winning prototypes are attracted to or driven away from samples, is not straightforward. We looked at intra- versus inter-class similarity values[1], whose distributions of course overlap, but in our case seemed to allow a good distinction at $e = 0.5$.

The 35 starting prototypes for the algorithms were selected randomly from the training data. With a learning rate of $\theta = 0.1$ we performed 20 iterations. One iteration means that all training samples were presented to the network and prototypes and labels adjusted. The experiments were run in online mode, adjusting after every single presentation of a sample.

As mentioned, we also wanted to study the impact of noise on the training labels on the classification accuracy. To be able to compare the noise level on hard and fuzzy labels, noise was simply added to the hard labels, imparting corresponding noise levels to the soft labels derived from the noised hard ones. The procedure for adding noise to the hard labels consisted simply of randomly choosing a fraction (0% to 100%, the noise level) of the training labels, and randomly flipping their individual label to a different class.

4   Results and Interesting Observations

The experiments on the five different features gave rather clear answers. As had to be expected, if no noise is present, the original LVQ1 approach performs best[2]. Then, depending on the feature, at a noise level between 10% and 30%, its accuracy drops rather sharply, and one of the fuzzy LVQ approaches wins (see figure 1). Right after LVQ1 drops, the fuzzy approach with the Keller-initialisation and scalar product as similarity measure remains most stable. Adding more noise, the approach initialised with K-Means and similarity measure S_1 now becomes the clear winner with the highest classification accuracy. This means that once there is a not insignificant level of noise on the (training) labels, a fuzzy LVQ approach is to be preferred over the basic hard LVQ one.

---

[1] Intra- and inter-class determined with respect to the original hard labels.
[2] Keep in mind that the correspondence between hardened new labels and original ones was only around 70%, giving the LVQ1 approach a headstart.

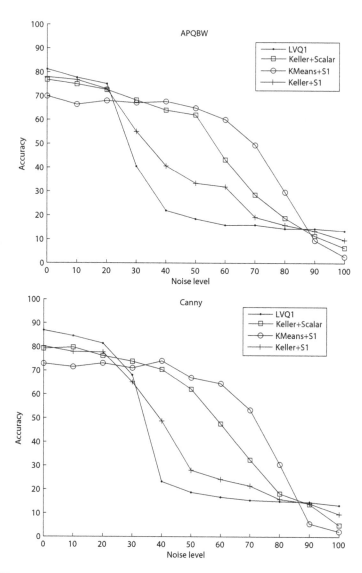

Fig. 1. Plotting the performance (accuracy given in %) of the original LVQ1 algorithm and our fuzzy LVQ approach when adding noise (noise level given in %). The fuzzy LVQ is plotted with different fuzzy label initialisation techniques (K-Means, Keller) and distance measures on the labels (S_1 and scalar product). Results given for two features, *APQBW* and *Canny*. The combination of K-Means and scalar product is omitted from this graph for clarity, as it is very similar to the one with Keller initialisation.

The poor performance of the fuzzy LVQ with Keller fuzzification and S_1 similarity measure has a simple explanation: setting the Keller mixing parameter α to 0.5 and $k := 5$, the possible values for intra- [0.34;1] and inter-class similarities [0;0.96] overlap largely. A similar explanation, experimentally obtained

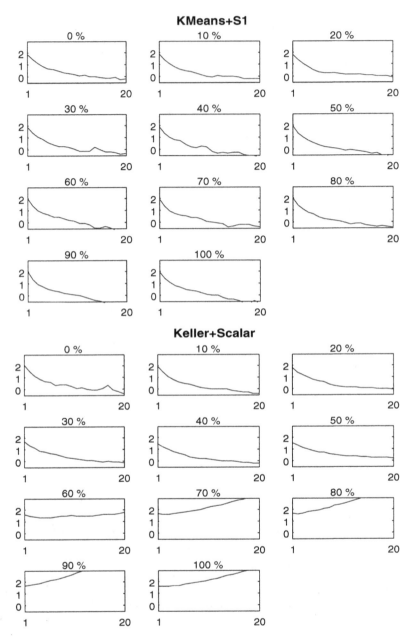

Fig. 2. Showing how much the location of the prototypes changes from iteration to iteration (x-axis, 20 training epochs), depending on how much noise is present (0 % to 100%). Location change is the sum of quadratic distances between the prototypes' position before and after each iteration, logarithmised to base 10. Plots given for Canny feature, and two fuzzy LVQ algorithm variants: K-Means coupled with S_1 and Keller coupled with scalar product.

this time, holds for K-Means fuzzification and the scalar product as distance measure. Initially, the same accuracy as with the S_1 measure is achieved, but this does not hold once noise is added (results not present in figure 1 for reasons of readability).

For higher noise levels, the winning approach is a fuzzy LVQ with K-Means label fuzzification, and S_1 similarity measure. This shows nicely the effect we were hoping to achieve with the process of fuzzifying the labels; the clustering of the training data, and usage of neighbouring labels for the soft labels, encodes knowledge about the label space into the labels itself. Knowledge, which can then be exploited by the fuzzy LVQ approach.

Examining how the labels of the prototypes changed from iteration to iteration (equation 4) of the fuzzy LVQ, we found that they remain rather stable after the first 10 rounds.

Looking closer at how the locations of the prototypes change (compare figure 2, and equation 2) across iterations, we could make an interesting observation. When no noise was added, the movement of the prototypes went down continuously with each iteration. But as soon as we added noise to the labels, the situation changed. The tendency of the volume of the movement was not so clear any more, for some algorithms it would even go up after some iterations before going down again. Reaching a noise level of 40 to 60 percent, the trend even reversed, and the movements got bigger with each iteration, not stabilising any more. The only exception here was the Fuzzy LVQ with K-Means initialisation and S_1 as similarity measure, which explains why it performs best of all variants on high noise levels.

The non-settling of the prototype locations also solves the question why, at a noise level of 100%, the original LVQ has an accuracy of 14%, which is exactly the random guess. It turned out that the cloud of prototypes shifts far away from the cloud of samples, in the end forming a circle around the samples. One random but fixed prototype is now the closest to all the samples, leading to the effect described.

5   Summary

We presented a prototype-based classification algorithm that can take soft labels as training data, and give soft answers. Being an extension of LVQ1, the prototypes are assigned soft labels, and a similarity function between those and a training sample's label controls where the prototype is shifted. Concerning the classification performance, in a noise free situation, the original approach yields the best results. This quickly changes once noise is added to the training labels, here our fuzzy approaches are the clear winners. This is due to information about the label-distribution inherently encoded in each of the soft labels. A finding which seems to hold for other hard-vs-soft scenarios, too, so we are currently investigating RBFs and SVMs in a multiple classifier systems scenario.

References

1. Powell, M.J.D.: Radial basis functions for multivariate interpolation: A review. In: Mason, J.C., Cox, M.G. (eds.) Algorithms for Approximation, pp. 143–168. Clarendon Press, Oxford (1987)
2. Thiel, C., Scherer, S., Schwenker, F.: Fuzzy-Input Fuzzy-Output One-Against-All Support Vector Machines. In: Apolloni, B., Howlett, R.J., Jain, L. (eds.) KES 2007, Part III. LNCS (LNAI), vol. 4694, pp. 156–165. Springer, Heidelberg (2007)
3. Seo, S.: Clustering and Prototype Based Classification. PhD thesis, Fakultät IV Elektrotechnik und Informatik, Technische Universität Berlin, Germany (November 2005)
4. Villmann, T., Schleif, F.M., Hammer, B.: Fuzzy Labeled Soft Nearest Neigbor Classification with Relevance Learning. In: Fourth International Conference on Machine Learning and Applications, pp. 11–15 (2005)
5. Karayiannis, N.B., Bezdek, J.C.: An integrated approach to fuzzy learning vector quantization andfuzzy c-means clustering. IEEE Transactions on Fuzzy Systems 5(4), 622–628 (1997)
6. Wu, K.L., Yang, M.S.: A fuzzy-soft learning vector quantization. Neurocomputing 55(3), 681–697 (2003)
7. Kohonen, T.: Self-organizing maps. Springer, Heidelberg (1995)
8. Dubois, D., Prade, H.: Fuzzy Sets and Systems: Theory and Applications. Academic Press, London (1980)
9. Kuncheva, L.I.: Using measures of similarity and inclusion for multiple classifier fusion by decision templates. Fuzzy Sets and Systems 122(3), 401–407 (2001)
10. Fay, R., Kaufmann, U., Schwenker, F., Palm, G.: Learning Object Recognition in a NeuroBotic System. In: Groß, H.M., Debes, K., Böhme, H.J. (eds.) 3rd Workshop on SelfOrganization of AdaptiVE Behavior SOAVE 2004. Fortschritt-Berichte VDI, Reihe 10, vol. 743, pp. 198–209. VDI (2004)
11. Fay, R.: Feature Selection and Information Fusion in Hierarchical Neural Networks for Iterative 3D-Object Recognition. PhD thesis, University of Ulm, Germany (2007)
12. MacQueen, J.B.: Some methods for classification and analysis of multivariate observations. In: 5th Berkeley Symposium on Mathematical Statistics and Probability, pp. 281–298. University of California Press (1967)
13. Keller, J., Gray, M., Givens, J.: A Fuzzy K Nearest Neighbor Algorithm. IEEE Transactions on Systems, Man and Cybernetics 15(4), 580–585 (1985)

A Neural Network Approach to Similarity Learning

Stefano Melacci, Lorenzo Sarti, Marco Maggini, and Monica Bianchini

DII, Università degli Studi di Siena
Via Roma, 56 — 53100 Siena (Italy)
{mela,sarti,maggini,monica}@dii.unisi.it

Abstract. This paper presents a novel neural network model, called similarity neural network (SNN), designed to learn similarity measures for pairs of patterns. The model guarantees to compute a non negative and symmetric measure, and shows good generalization capabilities even if a very small set of supervised examples is used for training. Preliminary experiments, carried out on some UCI datasets, are presented, showing promising results.

1 Introduction

In many pattern recognition tasks, appropriately defining the distance function over the input feature space plays a crucial role. Generally, in order to compare patterns, the input space is assumed to be a metric space, and Euclidean or Mahalanobis distances are used. In some situations, this assumption is too restrictive, and the similarity measure could be learnt from examples.

In the last few decades, the perception of similarity received a growing attention from psychological researchers [1], and, more recently, how to learn a similarity measure has attracted also the machine learning community. Some approaches are proposed to compute iteratively the similarity measure, solving a convex optimization problem, using a small set of pairs to define the problem constraints [2,3]. Other techniques exploit EM–like algorithms [4], Hidden Markov Random Fields [5], and constrained kernel mappings [6,7]. However, the existing approaches are generally strictly related to semi–supervised clustering, and the presence of some class labels or pairwise constraints on a subset of data is exploited to improve the clustering process.

In this paper a similarity learning approach based on SNNs is presented. The SNN architecture guarantees to learn a non negative and symmetric function, and preliminary results, carried out using some UCI datasets, show that the generalization performances are promising, even if a very small set is used for training.

The paper is organized as follows. In the next section, the network architecture and its properties are presented. In Section 3 some experimental results are reported, comparing them with commonly used distance functions. Finally, some conclusions are drawn in Section 4.

L. Prevost, S. Marinai, and F. Schwenker (Eds.): ANNPR 2008, LNAI 5064, pp. 133–136, 2008.
© Springer-Verlag Berlin Heidelberg 2008

2   Similarity Neural Networks

A SNN consists in a feed–forward multilayer perceptron (MLP) trained to learn a similarity measure between couples of patterns $x_i, x_j \in \mathbb{R}^n$.

Humans are generally able to provide a supervision on the similarity of object pairs in a dyadic form (similar/dissimilar), instead of to associate intermediate degrees of similarity, that cannot be easily defined in a coherent way. Hence, SNNs are trained using dyadic supervisions. In detail, a training set $T = \{(x_i, x_j, t_{i,j}); \ i = 1, ..., m; \ j = 1, ..., m\}$, collects a set of triples $(x_i, x_j, t_{i,j})$, being $t_{i,j}$ the similar/dissimilar label of $(x_i, x_j) \in \mathbb{R}^{2n}$, which represents the input vector to the SNN.

The SNN model is composed by a single hidden layer with an even number of units and by a unique output neuron with sigmoidal activation function, that encloses the output range in the interval $[0, 1]$. The hidden neurons are fully connected both with the inputs and the output. The training is performed using Backpropagation for the minimization of the squared error function. Learning a similarity measure is a regression task but, due to the dyadic supervision, it could also be considered as a two-class classification task by applying a threshold to the output of the network.

If $sim() : \mathbb{R}^{2n} \rightarrow [0, 1]$ is the function computed by a trained SNN, then the following properties hold for any pair x_i, x_j: $sim(x_i, x_j) \geq 0$, and $sim(x_i, x_j) = sim(x_j, x_i)$. The first property is guaranteed by the sigmoidal activation function of the output unit. The second one is forced by exploiting weight sharing along the structure of the network. In Fig. 1(a) the shared weights can be observed, while Fig. 1(b) shows the unfolding of the network over the shared weights. The SNN is essentially composed by a "duplicated" input layer, formed by the original and the exchanged pair, and by two networks that share the corresponding weights (see Fig. 1(b)).

The learnt function is a similarity measure but not necessarily a metric, since $sim(x_i, x_i) = 0$ and the triangle inequality are not guaranteed. Those properties could be learnt from data, but they are not forced in any way by the structure of the network. SNNs are an instance of the neural networks proposed in [8] to process Directed Acyclic Graphs, hence it can be shown that they are universal approximators.

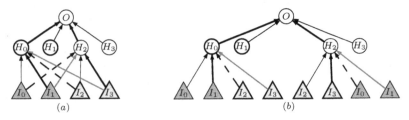

Fig. 1. The SNN architecture. Shared weights between two layers of neurons are drawn with the same gray level and mark. For the sake of simplicity, only some connections are depicted. (a) SNN for pairs of bidimensional vectors, $x_i = [I_0, I_1]'$, $x_j = [I_2, I_3]'$. (b) SNN unfolded over the shared weights.

3   Experimental Results

SNNs were trained over some datasets from the UCI repository, whose patterns are divided in distinct classes. In order to approach the described similarity measure learning problem, pairs of patterns that belong to the same class were considered similar, otherwise they were labeled dissimilar. The training set for SNNs contains both similarity and dissimilarity pairs, collected in the sets S and D respectively. Such sets have been iteratively created by adding a new randomly selected similarity or dissimilarity pair until a target number of connected components of $G_S = (Dataset, S)$ and $G_D = (Dataset, D)$ (the *similarity* and *dissimilarity graphs*) has been obtained, following the sampling criterion of similarity pairs defined in [2].

Moreover, the S set has been enriched by applying the transitive closure over the similarity graph, whereas other dissimilarity pairs were added to D exploiting both the similarity and dissimilarity relationships (if a is similar to b and b is dissimilar to c, then a is dissimilar to c), as suggested in [4]. The training set generation allows also to define the test set; as a matter of fact, given a training set T, the test set collects all the pattern pairs except those belonging to T.

The accuracies of many SNNs were evaluated varying both the network architecture (the number of hidden units) and the amount of supervision; the obtained results are reported in Table 1. Accuracy has been computed by rejecting all outputs $o_{i,j}$ such that $|o_{i,j} - t_{i,j}| > \epsilon$, where $t_{i,j}$ is the target label, in order to evaluate the capability of the network to correctly separate examples belonging to different classes. Constraints size is expressed by $Kc * |Dataset|$, that represents the number of connected components of the similarity and dissimilarity graphs generated with the given supervisions.

The quality of the learned similarity measure has been compared against common distance functions (Euclidean and Mahalanobis distances), over the *cumulative neighbor purity* index. Cumulative neighbor purity measures the percentage of correct neighbors up to the K-th neighbor, averaged over all the data points.

Table 1. SNNs accuracy on 3 UCI datasets. Results are averaged over 20 random generations of constraints for each Kc. Best results for each architecture/supervision are reported in bold.

Kc	ϵ	Iris				Balance				Wine			
		Hidden											
		4	6	10	16	10	12	16	30	18	22	28	36
0.9	0.3	**90.9**	90.7	90.4	89.1	81.1	**81.9**	80.4	79.8	78.6	79.6	79.5	**80.1**
	0.2	**90.3**	90.1	89.6	88.2	80.4	**81.1**	79.3	78.2	75.9	**76.9**	76.5	76.9
	0.1	**89.7**	89.1	88.5	86.9	79.3	**79.9**	77.4	75.7	71.7	**72.4**	71.6	71.6
0.8	0.3	92.3	**93.1**	92.3	92.4	85.4	84.7	84.9	**85.0**	**90.4**	89.7	90.4	90.2
	0.2	92.1	**92.7**	91.9	91.9	**84.6**	84.0	84.3	84.4	**88.8**	87.9	88.5	88.4
	0.1	91.6	**92.1**	91.2	91.3	83.3	82.9	83.4	**83.5**	**86.0**	84.9	85.2	84.8
0.7	0.3	93.4	**93.4**	93.4	92.9	87.1	86.8	86.7	**87.9**	94.9	**95.1**	95.0	95.1
	0.2	93.1	**93.2**	93.1	92.5	85.9	85.9	85.9	**87.4**	94.1	**94.2**	94.1	94.1
	0.1	92.7	92.8	**92.8**	91.9	84.1	84.3	84.5	**86.5**	**92.6**	92.5	92.4	92.5

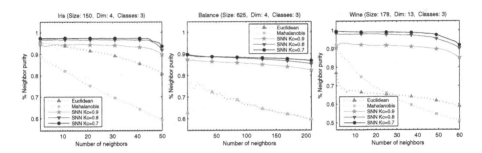

Fig. 2. Cumulative neighbor purity calculated over 3 datasets from the UCI repository. For each dataset, three results obtained by the SNNs trained with three differently sized sets of constraints, are shown. Each result is averaged over 10 random realization of constraints.

The maximum number of neighbors has been chosen such that $K \approx \frac{|Dataset|}{3}$. Results are reported in Fig. 2, showing that SNNs outperform common distance functions even if a small supervision is used.

4   Conclusions and Future Work

In this paper a neural network approach to similarity learning has been presented, showing encouraging results compared to common distance functions even with a small supervision. The proposed architecture assures to learn symmetric and non negative similarity relationship, and can also be trained to incorporate other properties of the data. Future work includes the application of the proposed similarity measure to clustering tasks.

References

1. Tversky, A.: Features of Similarity. Psychological Review 84(4), 327–352 (1977)
2. Xing, E., Ng, A., Jordan, M., Russell, S.: Distance metric learning, with application to clustering with side-information. Advances in Neural Information Processing Systems 15, 505–512 (2003)
3. De Bie, T., Momma, M., Cristianini, N.: Efficiently learning the metric using side-information. In: Proc. Int. Conf. on Algorithmic Learning Theory, pp. 175–189 (2003)
4. Bilenko, M., Basu, S., Mooney, R.: Integrating constraints and metric learning in semi-supervised clustering. In: Proc. Int. Conf. on Machine Learning, pp. 81–88 (2004)
5. Basu, S., Bilenko, M., Mooney, R.: A probabilistic framework for semi-supervised clustering. In: Proc. Int. Conf. on Knowledge Discovery and Data Mining, pp. 59–68 (2004)
6. Bar-Hillel, A., Hertz, T., Shental, N., Weinshall, D.: Learning a Mahalanobis Metric from Equivalence Constraints. J. Machine Learning Research 6, 937–965 (2005)
7. Tsang, I., Kwok, J.: Distance metric learning with kernels. In: Proc. Int. Conf. on Artificial Neural Networks, pp. 126–129 (2003)
8. Bianchini, M., Gori, M., Scarselli, F.: Processing directed acyclic graphs with recursive neural networks. IEEE Trans. on Neural Networks 12(6), 1464–1470 (2001)

Partial Discriminative Training of Neural Networks for Classification of Overlapping Classes

Cheng-Lin Liu

National Laboratory of Pattern Recognition (NLPR)
Institute of Automation, Chinese Academy of Sciences
95 Zongguancun East Road, Beijing 100190, P.R. China
`liucl@nlpr.ia.ac.cn`

Abstract. In applications such as character recognition, some classes are heavily overlapped but are not necessarily to be separated. For classification of such overlapping classes, either discriminating between them or merging them into a metaclass does not satisfy. Merging the overlapping classes into a metaclass implies that within-metaclass substitution is considered as correct classification. For such classification problems, I propose a partial discriminative training (PDT) scheme for neural networks, in which, a training pattern of an overlapping class is used as a positive sample of its labeled class, and neither positive nor negative sample for its allied classes (classes overlapping with the labeled class). In experiments of handwritten letter recognition using neural networks and support vector machines, the PDT scheme mostly outperforms cross-training (a scheme for multi-labeled classification), ordinary discriminative training and metaclass classification.

1 Introduction

In some pattern recognition applications, some patterns from different classes have very similar characteristics. In the feature space, such patterns of different classes correspond to co-incident or very close points, residing in an overlapping region. We call such classes as overlapping classes. A typical application is handwritten character recognition, where some classes such as letters 'O', 'o' and numeral '0' have identical shape, and it is neither possible nor necessary to separate them. Some other classes, such as upper-case letters 'A', 'M' and 'N', have many samples written in lower-case shapes (see Fig. 1). Thus, the upper-case letter and its corresponding lower-case have partially overlapping regions in the feature space. For such overlapping classes and all the pairs of upper-case/lower-case letters, it is not necessary to separate them at character level because it is easy to disambiguate according to the context.

Generally, there are two ways to deal with the overlapping classification problem. One way is to simply merge the overlapping classes into a metaclass and ignore the boundary between the classes. In metaclass classification, the substitution between overlapping classes (within a metaclass) is considered as correct.

L. Prevost, S. Marinai, and F. Schwenker (Eds.): ANNPR 2008, LNAI 5064, pp. 137–146, 2008.
© Springer-Verlag Berlin Heidelberg 2008

Fig. 1. Many samples of "AMN" are written in lower-case shape

In English letter recognition, the 52 letters can be merged into 26 case-insensitive classes. The 52-class letter classifier can also be evaluated at metaclass level by ignoring the substitution between upper and lower cases. Another way is to separate the overlapping classes by refining the classification boundary in the feature space, using multi-stage classifier or combining multiple classifiers [1,2]. Nevertheless, such attempt of discrimination is not necessary in the context of character recognition. The accuracy of overlapping classes separation is also limited by the inherent feature space overlap.

The problem of overlapping classification is similar to multi-labeled classification [3,4], where a pattern may belong to multiple classes. If we enhance the class label of a pattern from an overlapping class such that it belongs to the labeled class as well as the allied classes (those overlapping with the labeled class), the overlapping classification problem becomes a multi-labeled one. In evaluation, the classification of a pattern to any of its allied classes is considered correct.

For overlapping classification ignoring within-metaclass substitution, I propose a new scheme for training neural networks and support vector machines (SVMs). In my training scheme, called partial discriminative training (PDT), the pattern of an overlapping class is used as a positive sample of its labeled class, and neither positive nor negative sample of the allied classes. By contrast, in ordinary discriminative training of neural networks and SVMs, the pattern of a class is used as negative sample of all the other classes, and in multi-labeled classification (cross-training [3]), the pattern is used as positive sample of its allied classes.

To evaluate the performance of the proposed PDT method, I experimented on the C-Cube handwritten letter database [5,6] using neural networks and SVMs for classification. The results show that PDT mostly outperforms cross-training, ordinary discriminative training, and metaclass classification when evaluated at metaclass level.

In the rest of this paper, Section 2 briefly reviews the related works; Section 3 describes the proposed PDT scheme for neural networks and SVMs; Section 4 presents the experimental results, and Section 5 offers concluding remarks.

2   Related Works

Statistical classifiers [7] and artificial neural networks [8] have been popularly applied to pattern recognition. A parametric statistical classifier, which estimates the probability density function of each class without considering the boundary between classes, is ready for classification of overlapping classes. The overlap between classes will not affect the estimation of parameters of parametric statistical classifiers. In training neural networks, the connecting weights are iteratively adjusted by optimizing an objective of minimum squared error or cross-entropy between class outputs and desired targets [8]. The overlap between classes will affect the complexity of decision boundary. I will show in Section 3 that the training objective of neural networks can be decomposed into multiple binary (two-class) classification problems.

The support vector machine (SVM) [9] is an emerging classifier for solving difficult classification problems. Multi-class classification is usually accomplished by combining multiple binary SVMs encoded as one-versus-all, pairwise, or other ways. The binary SVM is trained (coefficients of kernel functions estimated) by maximizing the margin between two classes. The overlap between two classes also affects the boundary of the trained SVM.

Both neural networks and SVMs, as discriminative classifiers, attempt to separate different classes in the feature space. For overlapping classes, the decision boundary tends to be complicated. If we ignore the substitution between overlapping classes, as for handwritten letter recognition, the overlapping classes can be merged into a metaclass and then the ordinary discriminative classifiers can be applied to this reduced class set problem. Koerich [10] designed several neural network classifiers for recognizing 52 letters, 26 upper-case letters, 26 lower-case letters and 26 metaclasses, respectively, and showed that the metaclass classifier outperforms the 52-class classifier (evaluated at metaclass level) and the combination of upper-case and lower-case classifiers.

Blumenstein et al. [11] merged 52 letters into 36 metaclasses: all upper-case letters except "ABDEGHNQRT" are merged with their lower-case letters, and use a neural network for 36-class classification. Camastra et al. [6] use one-versus-all SVM classifiers for classifying handwritten letters in 52 classes, 26 classes and adaptively merged classes according to the overlap degree between upper and lower cases. Using classifiers of 52 classes, 38 classes and 26 classes, they obtained test accuracies (evaluated at 26-metaclass level) of 89.20%, 90.05% and 89.61%, respectively.

Multi-labeled classification methods have not been applied to overlapping classes problems, but I will test it in this case. Multi-labeled classification is generally transformed to multiple binary classification tasks, and different methods differ in the way of attaching binary labels to the training samples [4]. An effective method, called cross-training [3], uses each multi-labeled sample as the positive sample of each class it belongs to and not as negative sample for any of the labeled classes. For example, if a sample belongs to classes 'A' and 'a', it is used as positive sample when training the binary classifiers for 'A' and 'a', and as negative samples for the binary classifiers of other classes.

3   Partial Discriminative Training

Before describing the proposed partial discriminative training (PDT) scheme, I briefly review the training objectives of neural network classifiers.

3.1   Training of Neural Networks

Assume to classify a pattern (represented by a feature vector $\mathbf{x}$) to one of M classes $\{\omega_1, \ldots, \omega_M\}$. There are N training samples $(\mathbf{x}^n, c^n)$ (c^n is the class label of sample $\mathbf{x}^n$), $n = 1, \ldots, N$, for training a multi-class classifier. On an input pattern $\mathbf{x}$, the classifier outputs (sigmoidal) confidence values $y_k(\mathbf{x}, W)$ (W denotes the set of parameters) for classes $k = 1, \ldots, M$. The objective of neural network training is to minimize the squared error (SE):

$$\min_W SE = \min_W \sum_{n=1}^{N} \sum_{k=1}^{M} [y_k(\mathbf{x}^n, W) - t_k^n]^2, \tag{1}$$

where t_k^n denotes the target output:

$$t_k^n = \delta(c^n, k) = \begin{cases} 1, & k = c^n, \\ 0, & \text{otherwise.} \end{cases} \tag{2}$$

The SE in Eq. (1) can be re-written as

$$SE = \sum_{k=1}^{M} \sum_{n=1}^{N} [y_k(\mathbf{x}^n, W) - t_k^n]^2 = \sum_{k=1}^{M} E_k, \tag{3}$$

where $E_k = \sum_{n=1}^{N} [y_k(\mathbf{x}^n, W) - t_k^n]^2$ is the squared error of a binary classifier for class ω_k versus the others. Thus, the training of the multi-class neural network is equivalent to the training of multiple binary one-versus-all classifiers. Accordingly, the class output $y_k(\mathbf{x}, W)$ functions as the discriminant for separating class ω_k from the others.

The cross-entropy (CE) objective for neural networks can be similarly decomposed into multiple binary classifiers:

$$\begin{aligned} CE &= -\sum_{n=1}^{N} \sum_{k=1}^{M} [t_k^n \log y_k + (1 - t_k^n) \log(1 - y_k)]^2 \\ &= -\sum_{k=1}^{M} \sum_{n=1}^{N} [t_k^n \log y_k + (1 - t_k^n) \log(1 - y_k)]^2 \\ &= \sum_{k=1}^{M} CE_k. \end{aligned} \tag{4}$$

For multi-labeled classification, each sample $\mathbf{x}^n$ is labeled to belong to a subset of classes C^n. For training neural networks in such case, the objective is the same as Eq. (1), (3) or (4) except that the target output is changed to

$$t_k^n = \begin{cases} 1, \ k \in C^n, \\ 0, \ \text{otherwise.} \end{cases} \tag{5}$$

Due to the class modularity of objective functions SE and CE, for either single-labeled or multi-labeled classification, we can either train a multi-class classifier or multiple binary one-versus-all classifiers.

The overlapping classification problem is different from multi-labeled classification in that the training samples have single class labels, but I enhance the label of each sample with the allied classes (those overlapping with the labeled class) to convert the problem to be multi-labeled.

3.2   Partial Discriminative Training (PDT)

For overlapping classification with classifiers trained with single-labeled samples, the boundary between overlapping classes will be complicated by discriminative training to maximize the separation between overlapping classes. The over-complicated boundary will deteriorate the generalized classification performance and also affect the boundary between metaclasses. On the other hand, simply merging the overlapping classes into a metaclass will complicate the distribution of the metaclass.

If the substitution between overlapping classes is to be ignored, the training objective of neural networks, squared error (SE) or cross-entropy (CE), can be modified to ignore the error of the allied classes of each training sample. Denote the allied classes of ω_k as a set $\Lambda(k)$ (e.g., in alphanumeric recognition, the allied classes of 'O' are "o0"), the squared error of Eq. (3) is modified as

$$SE = \sum_{k=1}^{M} \sum_{n=1, k \notin \Lambda(c^n)}^{N} [y_k(\mathbf{x}^n, W) - t_k^n]^2. \tag{6}$$

This implies, the training pattern $\mathbf{x}^n$ is not used as negative sample for the allied classes of c^n. Note that the relation of alliance is symmetric, i.e., $k \in \Lambda(c) \Leftrightarrow c \in \Lambda(k)$.

Excluding a training pattern from the negative samples of the allied classes of the label prevents the classifier from over-fitting the boundary between the labeled class and its allied classes (which are overlapping with the labeled class). Still, the number of classes remains unchanged (the structure of the multi-class classifier does not change), unlike in metaclass merging, the number of classes is reduced. Remaining the number of classes has the benefit that the classifier outputs confidence scores to each of the overlapping classes. If two allied classes are partially overlapped, a sample from the un-overlapped region can be classified to its class unambiguously. By class merging, however, the boundary between allied classes are totally ignored.

The PDT scheme can be applied to all types of binary classifiers, with multiple binary classifiers combined to perform multi-class classification. For multi-class classification using one-versus-all SVMs, when training an SVM for a class ω_k, if ω_k is an allied class of a sample from a different class, this sample is excluded from the negative samples of ω_k.

3.3   Specific Classifiers

I have applied the PDT scheme to five types of neural networks and SVMs with two types of kernel functions. The neural classifiers are single-layer neural network (SLNN), multi-layer perceptron (MLP), radial basis function (RBF) network [8], polynomial network classifier (PNC) [12,13], and class-specific feature polynomial classifier (CFPC) [14]. Two one-versus-all SVM classifiers use a polynomial kernel and an RBF kernel, respectively.

The neural classifiers have a common nature that each class output is the sigmoidal (logistic) function of the weighted sum of values of the previous layer. In SLNN, the input feature values are directly linked to the output layer. The MLP that I use has one layer of hidden units and all the connecting weights are trained by back-propagation. The RBF network has one hidden layer of Gaussian kernel units, and in training, the Gaussian centers and variance values are initialized by clustering and are optimized together with the weights by error minimization. The PNC is a single-layer network with the polynomials of feature values as inputs. For reducing the number of polynomial terms, I use a PNC with the binomial terms of the principal components [13]. Unlike the PNC that uses a class-independent principal subspace, the CFPC uses class-specific subspaces as well as the residuals of subspace projection [14].

For saving the computation of projection onto class-specific subspaces, the CFPC is trained class by class [14], i.e., the binary one-versus-all classifiers are trained separately. The other four neural networks are trained for all classes simultaneously. The weights of the neural networks are trained by minimizing the squared error criterion by stochastic gradient descent.

The one-versus-all SVM classifier has multiple binary SVMs each separating one class from the others. In my implementation of SVMs using two types of kernel functions, the pattern vectors are appropriately scaled for the polynomial kernel, with the scaling factor estimated from the lengths of the sample vectors. For the Gaussian (RBF) kernel, the kernel width σ^2 is estimated from the variance of the sample vectors. I call the SVM classifier using polynomial kernel SVM-poly and the one using Gaussian kernel SVM-rbf. In partial discriminative training (PDT) of a binary SVM for class ω_k, the only change is to remove from negative samples the ones of the allied classes of ω_k.

4   Experimental Results

I evaluated the partial discriminative training (PDT) scheme and related methods with different classifiers on a public database of handwritten letters, C-Cube

[5,6][1]. This database contains 57,293 samples of 52 English letters, partitioned into 38,160 training samples and 19,133 test samples. The samples were segmented from handwritten words, so the character shapes are very cursive and the number of samples per class is seriously imbalanced. In addition to confusion between upper-case and lower-case letters, the confusion between different case-insensitive letters is also considerable. By k-NN classification based on vector quantization, the authors ordered the overlap degree of upper/lower cases of each letter for merging the cases of selected letters. The database provides binary images as well as extracted feature values (34D) of the samples. Since my intention is to evaluate classifiers, I do not improve the features, but use the given features in the database.

I consider three numbers of classes as those in [5,6]: 52 case-sensitive letters, 38 classes by merging the upper/lower cases of 14 letters ("CXOWYZMKJUN-FVA"), and 26 case-insensitive letters. In the cases of 38 classes and 26 letters, each merged upper-case letter is allied with its lower-case and vice versa. In all cases, I set the number of hidden units of MLP as 100, the number of hidden units of RBF network as 150. The PNC uses linear and binomial terms of the original features without dimensionality reduction. The CFPC uses 25D class-specific subspaces. The SVM-poly uses 4-th order polynomial kernel, and the SVM-rbf uses an RBF kernel with kernel width fixed at 0.5 times the average within-class variance.

First, I trained four multi-class neural networks (SLNN, MLP, RBF, and PNC) with three training schemes optimizing the squared error criterion: ordinary discriminative training, PDT, and cross-training (enhancing the label of each sample with its allied classes). The accuracies on test samples are shown in Table 1, where each row gives the accuracies evaluated at a number of metaclasses (52, 38 or 26, within-metaclass substitution is ignored), and each column corresponds to a number of metaclasses in training. By ordinary discriminative training, the number of classes is reduced by class merging, whereas by PDT and cross-training, the number of classes remains unchanged but the samples are attached allied classes or multi-labeled. Each classifier can be evaluated at a reduced number of classes by ignoring within-metaclass substitution. At each row (evaluated at a number of metaclasses), the highest accuracy is highlighted in bold face, and the accuracies of merged metaclass training and PDT are boxed.

Apparently, the ordinary all-class discriminative training (3rd column of Table 1) gives the highest accuracy for 52-class classification. This is reasonable because all the classes are aimed to be separated in this case, while PDT ignores the separation between allied classes. When evaluated at reduced number of classes, however, merged metaclass training (4th and 5th columns) and PDT (6th and 7th columns) may give higher accuracies than all-class training. In seven of eight cases (two class numbers 38 and 26 combined with four classifiers), PDT gives higher accuracies than all-class training and merged metaclass training. The inferior performance of cross-training can be explained that the

---

[1] Downloadable at http://ccc.idiap.ch/

Table 1. Test accuracies (%) of multi-class neural networks on the C-Cube Letter database. Each row gives the accuracies evaluated at a number of metaclasses, and each column corresponds to a number of metaclasses in training. 4th and 5th columns correspond to merged metaclass training.

Classifier	#Class	Discriminative training 52	38	26	Partial training 38	26	Cross-training 38	26
SLNN	52	**66.15**			65.49	65.23	52.66	34.18
	38	71.93	70.40		**72.75**	72.47	70.38	52.66
	26	72.53	70.94	67.94	73.36	**73.44**	70.92	67.98
MLP	52	**78.97**			78.21	78.20	60.83	45.46
	38	84.64	84.98		**85.42**	85.06	84.37	68.08
	26	85.00	84.34	84.42	**85.79**	85.57	84.81	83.62
RBF	52	**78.06**			77.71	77.75	61.20	44.84
	38	83.76	**84.37**		84.25	84.31	84.00	66.68
	26	84.16	84.72	84.28	84.61	**84.81**	84.36	83.70
PNC	52	**81.09**			80.67	80.64	63.34	43.34
	38	86.87	87.11		**87.62**	87.61	86.80	65.41
	26	87.18	87.42	86.29	87.93	**88.03**	87.10	85.65

framework of multi-labeled classification does not match the problem of overlapping classification.

On three one-versus-all classifiers (CFPC, SVM-poly and SVM-rbf), I used two training schemes: ordinary discriminative training and PDT. The test accuracies are shown in Table 2. Again, all-class discriminative training (3rd column of Table 2) gives the highest accuracies for 52-class classification. When evaluated at reduced number of metaclasses, both merged metaclass training (4th and 5th columns) and PDT (6th and 7th columns) gives higher accuracies than all-class training. For the CFPC, PDT outperforms merged metaclass training. For the SVM classifiers, merged metaclass training gives the highest accuracies of metaclass classification, but the accuracies of PDT are closely competitive.

Overall, when evaluated at metaclass level, PDT gives higher accuracies than ordinary all-class discriminative training on all the seven classifiers, outperforms merged metaclass training on five neural classifiers, and performs comparably with merged metaclass training on two SVM classifiers. On the C-Cube database of handwritten letters, the remaining classification error rate of 26 metaclasses is still appreciable (over 10%) because of the inherent confusion of handwritten shapes between different letters. This can be alleviated by extracting more discriminant features which provide better between-class separation.

Compared to merged metaclass training, PDT has an advantage that it still outputs confidence scores for all classes. Thus, if a pattern of partially overlapping classes resides in the non-overlape region, it can still be classified unambiguously. By merged metaclass training, however, the boundary between partially overlapping classes is totally ignored.

Table 2. Test accuracies (%) of one-versus-all classifiers on the C-Cube Letter database. Each row gives the accuracies evaluated at a number of metaclasses, and each column corresponds to a number of metaclasses in training. 4th and 5th columns correspond to merged metaclass training.

Classifier	#Class	Discriminative training			Partial training	
		52	38	26	38	26
CFPC	52	**81.07**			80.73	80.71
	38	86.76	86.65		**87.22**	87.21
	26	87.09	86.99	85.86	87.56	**87.70**
SVM-poly	52	**82.19**			81.97	81.88
	38	88.13	**88.66**		88.61	88.51
	26	88.41	88.94	**89.03**	88.88	88.99
SVM-rbf	52	**82.65**			82.42	82.35
	38	88.73	**89.12**		89.10	89.02
	26	89.00	89.33	**89.43**	89.39	89.40

5   Conclusion

This paper proposed a partial discriminative training (PDT) scheme for classification of overlapping classes. It is applicable to all types of binary one-versus-all classifiers, including neural networks and SVM classifiers. The rationale of PDT is to ignore the difference between overlapping classes in training so as to improve the separation between metaclasses. Experiments in handwritten letter recognition show that when evaluated at metaclass level, the PDT scheme mostly outperforms ordinary all-class discriminative training. Compared to merged metaclass training, the PDT gives higher or comparable accuracies at metaclass level and provides more informative confidence scores. The PDT scheme is especially useful for such applications where overlapping classes are not necessarily discriminated before contextual information is exploited. This work will be extended by experimenting with different datasets.

Acknowledgements

This work is supported by the Hundred Talents Program of Chinese Academy of Sciences and the National Natural Science Foundation of China (NSFC) under grant no. 60775004 and grant no.60723005.

References

1. Lu, B.-L., Ito, M.: Task decomposition and modular combination based on class relations: a modular neural network for pattern classification. IEEE Trans. Neural Networks 10(5), 1244–1256 (1999)
2. Podolak, I.T.: Hierarchical classifier with overlapping class groups. Expert Systems with Applications 34(1), 673–682 (2008)

3. Boutell, M.R., Luo, J., Shen, X., Browm, C.M.: Learning multi-label scene classification. Pattern Recognition 37(9), 1757–1771 (2004)
4. Tsoumakas, G., Katakis, I.: Multi-label classification: an overview. Int. J. Data Warehousing and Mining 3(3), 1–13 (2007)
5. Camastra, F., Spinetti, M., Vinciarelli, A.: Offline cursive character challenge: a new benchmark for machine learning and pattern recognition algorithms. In: Proc. 18th ICPR, Hong Kong, pp. 913–916 (2006)
6. Camastra, F.: A SVM-based cursive character recognizer. Pattern Recognition 40(12), 3721–3727 (2007)
7. Fukunaga, K.: Introduction to Statistical Pattern Recognition, 2nd edn. Academic Press, London (1990)
8. Bishop, C.M.: Neural Networks for Pattern Recognition. Oxford University Press, Oxford (1995)
9. Burges, C.J.C.: A tutorial on support vector machines for pattern recognition. Knowledge Discovery and Data Mining 2(2), 1–43 (1998)
10. Koerich, A.L.: Unconstrained handwritten character recognition using different classification strategies. In: Gori, M., Marinai, S. (eds.) Proc. 1st IAPR Workshop on Artificial Neural Networks in Pattern Recognition, pp. 52–56 (2003)
11. Blumenstein, M., Liu, X.Y., Verma, B.: An investigation of the modified direction feature for cursive character recognition. Pattern Recognition 40(2), 376–388 (2007)
12. Shürmann, J.: Pattern Classification: A Unified View of Statistical and Neural Approaches. Wiley Interscience, Chichester (1996)
13. Kreßel, U., Schürmann, J.: Pattern classification techniques based on function approximation. In: Bunke, H., Wang, P.S.P. (eds.) Handbook of Character Recognition and Document Image Analysis, pp. 49–78. World Scientific, Singapore (1997)
14. Liu, C.-L., Sako, H.: Class-specific feature polynomial classifier for pattern classification and its application to handwritten numeral recognition. Pattern Recognition 39(4), 669–681 (2006)

Boosting Threshold Classifiers for High–Dimensional Data in Functional Genomics

Ludwig Lausser[1], Malte Buchholz[3], and Hans A. Kestler[1,2,⋆]

[1] Department of Internal Medicine I, University Hospital Ulm, Germany
[2] Institute of Neural Information Processing, University of Ulm, Germany
[3] Internal Medicine, SP Gastroenterology, University Hospital Marburg, Germany
`ludwig.lausser@uni-ulm.de, malte.buchholz@staff.uni-marburg.de,`
`hans.kestler@uni-ulm.de`

Abstract. Diagnosis of disease based on the classification of DNA microarray gene expression profiles of clinical samples is a promising novel approach to improve the performance and accuracy of current routine diagnostic procedures. In many applications ensembles outperform single classifiers. In a clinical setting a combination of simple classification rules, such as single threshold classifiers on individual gene expression values, may provide valuable insights and facilitate the diagnostic process. A boosting algorithm can be used for building such decision rules by utilizing single threshold classifiers as base classifiers. AdaBoost can be seen as the predecessor of many boosting algorithms developed, unfortunately its performance degrades on high-dimensional data. Here we compare extensions of AdaBoost namely MultiBoost, MadaBoost and AdaBoost-VC in cross-validation experiments on noisy high-dimensional artifical and real data sets. The artifical data sets are so constructed, that features, which are relevant for the class distinction, can easily be read out. Our special interest is in the features the ensembles select for classification and how many of them are effectively related to the original class distinction.

1 Introduction

The onset and progress of many human diseases, including most if not all human cancers, is associated with profound changes in the activity status of large numbers of genes. DNA Microarrays are high–throughput molecular biology devices capable of monitoring the expression levels of up to several thousand genes simultaneously. One important goal in biomedical research is to make use of this biological principle to develop novel approaches for the accurate differential diagnosis of diseases based on microarray analyses of clinical samples (e.g. tissue biopsy samples). Often single classifiers trained are not able to fulfill certain tasks satisfactorily. If this is the case, better results might be achieved by integrating the results of a whole ensemble of classifiers. Meta algorithms, which do so, are called ensemble methods. They use a basic learing algorithm, generate

---

⋆ Corresponding author.

L. Prevost, S. Marinai, and F. Schwenker (Eds.): ANNPR 2008, LNAI 5064, pp. 147–156, 2008.
© Springer-Verlag Berlin Heidelberg 2008

a set of base classifiers and combine them in order to get an improved classifier. Boosting algorithms are a subgroup of these methods. Boosting methodes have the characteristic to be able to combine classifiers with moderate accuracy (weak classifiers) to an ensemble with high accuracy [1]. One of the most popular Boosting algorithms is AdaBoost from Freund and Schapire [2]. The use of AdaBoost was evaluated for many different data sets, but it has also been shown that its performance did not match the expectations on high–dimensional data [3]. The concept of combining several, sometimes weak categorization rules, resembles in some aspects human medical decision making. This makes the representation more suitable for further investigation of functional dependencies. We here investigate its use in the context of expression profile classification. We apply several variants of AdaBoost to published gene expression profile data from different tumor types. Furthermore we investigate the performance on artificial high-dimensional data with different noise levels and a varying number of discriminating features among many irrelevant, which reflects the current belief (of biologists) of only a small number of genes (among all) being relevant for categorization.

2   AdaBoost

A pseudocode description of AdaBoost can be seen in Algorithm 2. AdaBoost iteratively creates an ensemble of T members. The algorithm receives an sample S of N training examples (x_i, y_i), where x_i is an element of the input space and $y_i \in \{-1, 1\}$ is its label. Before starting the iterations, an N-dimensional weight vector $D_1 = (1/N, \ldots, 1/N)^T$ is initialized. This vector influences the training of the weak classifier h_t in one of two ways. If AdaBoost is used as a Boosting by resampling algorithm, D_t will be used as a distribution for choosing the weak classifier's training examples. If AdaBoost is used as a Boosting by reweighting algorithm, the whole training set and the weight vector are used as input arguments for a weak learning algorithm which can deal itself with weighted training examples. This means that an example with an high weight influences the training of the weak classifier more than an example with an low weight (step 1). After h_t has been chosen, D_t is used to compute a weighted training error ϵ_t (step 2). With ϵ_t the parameter α_t is calculated (step 3), which determines the influence of h_t on the final ensemble h_f. According to α_t the weight vector D_t is updated as well (step 4). The weight of an example will be decreased if it was classified correctly and increased otherwise. In this way the training of new ensemble members will always concentrate such misclassified examples.

2.1   Base Classifier

The base classifier used in these experiments is chosen from the class $h_{c,d,e}(x)$:

$$h_{c,d,e}(x) = \begin{cases} sign(\mathbb{1}_{[e \leq x_d]} - 0.5), & \text{if } c = 1 \\ sign(\mathbb{1}_{[e \geq x_d]} - 0.5), & \text{otherwise} \end{cases} \tag{1}$$

Algorithm 1. AdaBoost(S,$WeakLearn$,T)

Input:

- sequence S of N labeled examples $\langle(\boldsymbol{x}_1, y_1), \ldots, (\boldsymbol{x}_N, y_N)\rangle$
  where $\boldsymbol{x}_i \in X$ and $y_i \in \{-1, 1\}$
- weak learning algorithm **WeakLearn**
- integer T specifying number of iterations

Init:

distribution $\boldsymbol{D}_1$ with $D_1^i = \frac{1}{N}$ for all $i \in \{1, \ldots, N\}$

Procedure:

Do for $t = 1, 2, \ldots, T$

1. Call **WeakLearn**, providing it with the distribution $\boldsymbol{D}_t$;
   get back a hypothesis $h_t : X \to \{-1, 1\}$.
2. Calculate the error of $h_t : \epsilon_t = \sum_{i=1}^{N} D_t(i) \mathbb{1}_{[h_t(\boldsymbol{x}_i) \neq y_i]}$.
3. Set $\alpha_t = \ln\left(\frac{1-\epsilon_t}{\epsilon_t}\right)$
4. Update weights vector

$$D_{t+1}^i = \frac{D_t^i \exp\left(-\alpha_t \mathbb{1}_{[h_t(\boldsymbol{x}_i) = y_i]}\right)}{Z_t}$$

where Z_t is a normalization factor

Output:

A hypothesis h_f

$$h_f(\boldsymbol{x}) = \begin{cases} 1, & \text{if } \sum_{t=1}^{T} \alpha_t h_t(\boldsymbol{x}) > 0 \\ -1, & \text{otherwise} \end{cases}$$

This class contains all simple threshold classifiers working on only one feature dimension. Here d is the chosen dimension and e is the chosen threshold. The parameter c determines the kind of inequation used by the classifier. In each iteration t the best classifier h_t is chosen:

$$h_t = \arg \min_{h_{c,d,e}} \sum_{i=1}^{N} D_t \mathbb{1}_{[h_{c,d,e}(\boldsymbol{x}) \neq y_i]} \tag{2}$$

2.2   Tested Algorithms

Most of the Boosting algorithms proposed after 1995 are more or less based on the AdaBoost algorithm. Differences appear most often in weighting schemes or in the used error formula. In this section the algorithms, which were used in these tests shall be described and their differences to AdaBoost shall be highlighted.

MultiBoost. In this methode the boosting idea is coupled with wagging [4]. Wagging is a derivate from Breiman's Bagging approach [5]. In the original Bagging methode the base classifiers are trained on bootstrap replicates from the original training data. For each example of the training set it is randomly chosen, if an example is placed into the replicate set or not. This process is continued until the replicate set is from the same size than the original training set. In this way some examples will be in the bootstrap replicate more than once and others won't be in there at all. The trained base classifiers are combined to an unweighted sum. Wagging is a variante of Bagging for base classifiers, which can deal with weighted examples. Here each base classifier is trained on a weighted version of the original data. The single weights are chosen after an distribution, which simulates the bootstrapping process. In the MultiBoost algorithm a continuous Poisson distribution is used for this simulation. The MultiBoost [6] algorithm itself builds a wagging ensemble of AdaBoost ensembles. This means that after a certain number of boostingsteps, all example weights are reset like in the Wagging approach. The next ensemble members will again be chosen as in the AdaBoost algorithm with the reseted weights as its initial weight vector. The final ensemble is a sum of all weighted base classifiers. Normally the MultiBoost algorithm receives a number of timesteps determining the sizes of the AdaBoost ensembles. In this work MultiBoost's default settings are used, which determine the size of a single AdaBoost ensemble to $\left\lfloor \sqrt{T} \right\rfloor$.

MadaBoost. The MadaBoost[1] algorithm was proposed by Carlos Domingo and Osamu Watanabe [7]. The algorithm differs from the original AdaBoost in its sample weighting scheme. If the weight of an single data point x_i $D_t^i \geq D_0^i$ this weight is reset to D_0^i. In this way MadaBoost counteracts the fact, that noisy data receives very high weights in the original AdaBoost weighting scheme.

AdaBoost-VC. The main difference between AdaBoost and AdaBoost-VC [9] is the new error formular ϵ_{VC} that is used by this algorithm:

$$\epsilon_t^{VC} = \epsilon_t + \frac{d}{N} \left(\log N + \sqrt{1 + \frac{\epsilon_t N}{d}} \right) \tag{3}$$

Here N denotes the number of training examples and ϵ_t the weighted training error as used in the original AdaBoost algorithm. The parameter d regulates the influence of the additional term. This term is inspired by theoretical foundings of Vapnik about an upper bound to the expectation error of classifiers [10]. Within these experiments d is set to $(1, 2, 3, 4)$. Another difference to the original AdaBoost is that each feature can only be used once in an ensemble. If a feature exists, which seperates the training data perfectly, a classifier using it will always minimize a weighted training error and will be selected in each iteration.

---

[1] Actually Carlos Domingo and Osamu Watanabe proposed different sub-versions of their algorithm [7] [8]. The version, which is talked about here is the batch learning version of MB1.

3   Data Sets

For the shown tests both artificial and real data sets are used. The main tests were done under the controlled conditions of the artificial data sets. How these sets are built is shown in 3.1. The real data sets are described in 3.2.

3.1   Artificial Data

An data set contains 200 data points $x \in [-1, 1]^{100}$. Each entry of x is drawn independently after a uniform distribution. The labels $y \in [-1, 1]$ are given by a function $f_j(x)$ which depends on the first j dimensions of x:

$$f_j(x) = \begin{cases} +1, & \text{if } \sqrt{\sum_{i=1}^{j}(x_i - 1)^2} < \sqrt{\sum_{i=1}^{j}(x_i - (-1))^2} \\ -1, & \text{else} \end{cases} \tag{4}$$

The function $f_j(x)$ signals if the distance from the subvector $(x_1, \cdots, x_j)^T$ to the j-dimensional unit vector is smaller than the distance from $(x_1, \cdots, x_j)^T$ the the j-dimensional negative unit vector. In this way the number of real features can easily be regulated by changing the parameter j. This methode will create approximately the same number of positiv and negativ examples. In the experiments perturbed labels y' are used. These labels are generated according to a noise rate ρ. For each x an random variable $\tau \in [0, 1]$ is drawn after a uniform distribution. The new label is built as follows:

$$y' = \begin{cases} +y, & \text{if } \tau >= \rho \\ -y, & \text{else} \end{cases} \tag{5}$$

Tests were made for $j \in \{2, 10, 20\}$ and $\rho \in \{0, 0.1, 0.2\}$.

3.2   Real Data

ALL-AML-Leukemia. The ALL-AML-Leukemia data set presented by Golub et al. [11] contains data from a microarray experiment concerning accute Leukemia. The data set contains examples for two different subtypes of the disease called ALL (acute lymphoblastic leukemia) and AML (acute myeloid leukemia). The 47 ALL and 25 AML examples contain 7129 probes for 6817 human genes.

Breast cancer. The breast cancer data set presented by van't Veer et al. [12] contains microarray data from patients who had developed distant metastases within 5 years (relapsed patients) and patients who remained healthy from the disease for at least 5 years (non-relapsed patients). The 34 relapse and the 44 non-relapse examples contain data from 24481 gene fragments. If the value of an attribute is not avaible in a single example the mean value is calculated over al values from the same class.

Colon cancer. The colon cancer data set presented by Alon et al. [13] contains 40 biopsis of tumor tissues (negative examples)and 22 of normal tissues (positive examples). Each expression profile consists of 2000 genes.

4 Results and Conclusion

We performed 10–fold cross–validation tests on the previously described gene expression profiles. The ensemble size was stepwise increased by 10 from 10 to 100. The results are given in Table 1.

Table 1. Error rates on gene expression data form different tumor types [11,12,13] (AdaBoost-VC: $d = 1 \ldots 4$). The number of ensemble members was increased from 10 to 100 by a stepsize of 10.

	breast cancer			colon cancer			leukemia		
	min	mean	var	min	mean	var	min	mean	var
AdaBoost	0.27	0.31	$6.0 \cdot 10^{-4}$	0.16	0.21	$6.0 \cdot 10^{-4}$	0.03	0.06	$3.0 \cdot 10^{-4}$
MadaBoost	0.27	0.31	0.001	0.20	0.23	$4.0 \cdot 10^{-4}$	0.03	0.04	$2.0 \cdot 10^{-4}$
MultiBoost	0.25	0.30	$8.0 \cdot 10^{-4}$	0.15	0.19	$5.0 \cdot 10^{-4}$	0.04	0.06	$1.0 \cdot 10^{-4}$
Adaboost-VC1	0.25	0.28	$6.0 \cdot 10^{-4}$	0.15	0.17	$4.0 \cdot 10^{-4}$	0.025	0.038	$1.9 \cdot 10^{-4}$
Adaboost-VC2	0.28	0.29	$1.7 \cdot 10^{-4}$	0.13	0.14	$1.3 \cdot 10^{-4}$	0.038	0.04	$8.5 \cdot 10^{-5}$
Adaboost-VC3	0.27	0.56	0.012	0.23	0.61	0.018	0.0375	0.0379	$1.7 \cdot 10^{-6}$
Adaboost-VC4	0.66	0.68	$3.2 \cdot 10^{-4}$	0.70	0.72	$8.8 \cdot 10^{-4}$	0.0625	0.2733	0.0272

Table 2. Minimal errors on the artificial training set. Here ρ and j determine the noise rate and the number of real features used for building the artificial data set. The labels *err* and *iter* determine the minimal test error and the iteration when it was achieved.

	$j = 2$						$j = 10$						$j = 20$					
	$\rho = 0$		$\rho = 0.1$		$\rho = 0.2$		$\rho = 0$		$\rho = 0.1$		$\rho = 0.2$		$\rho = 0$		$\rho = 0.1$		$\rho = 0.2$	
	err	iter	err	iter	err	iter	err	iter	err	iter	err	iter	err	iter	err	iter	err	iter
AdaBoost	0.09	14	0.14	3	0.26	3	0.15	72	0.21	130	0.31	30	0.21	118	0.28	13	0.34	9
MadaBoost	0.08	7	0.13	3	0.22	8	0.14	30	0.21	16	0.30	11	0.21	127	0.27	13	0.33	49
MultiBoost	0.07	134	0.13	86	0.21	16	0.13	150	0.18	139	0.29	150	0.25	138	0.27	125	0.31	116
AdaBoost-VC	0.20	3	0.22	100	0.24	12	0.17	9	0.21	13	0.30	26	0.27	48	0.26	16	0.34	24

On the artificially generated data we estimated the expected classification error by 10-fold cross-validation tests. Ensembles of different sizes from 1 to 150 were trained for the algorithms, AdaBoost, MultiBoost and MadaBoost. Because AdaBoost-VC discards on feature-dimension per iteration, AdaBoost-VC can only be evaluated up to an ensemble size of 100. The algorithms are tested on artificial data sets differing in their number of real features $j \in \{2, 10, 20\}$ and the used noise level $\rho \in \{0, 0.1, 0.2\}$, see Figure 1. The minimal test errors are shown in Table 2. Reference experiments with other standard classifiers are listed in Table 3.

Table 3. Reference experiments on the artificial data set. The classifiers, which were used for these experiments are 1-nearest neighbour (1NN), 5-nearest neighbour (5NN), support vector machine (SVM) with linear kernel and $C = 1$, nearest centroid ($NCen$) and nearest shrunken centroid classifier ($SCen$) with $\Delta = 0.1$ and 30 steps. The parameters ρ and j denote the noise rate and the number of real features respectively.

													ρ		
	1NN			5NN			SVM			$NCen$			$SCen$		
j	0	0.1	0.2	0	0.1	0.2	0	0.1	0.2	0	0.1	0.2	0	0.1	0.2
2	0.41	0.35	0.42	0.41	0.33	0.39	0.15	0.22	0.32	0.23	0.20	0.24	0.20	0.17	0.22
10	0.37	0.43	0.41	0.39	0.39	0.37	0.23	0.31	0.35	0.24	0.25	0.29	0.25	0.29	0.28
20	0.37	0.34	0.43	0.36	0.33	0.37	0.18	0.24	0.32	0.23	0.23	0.28	0.22	0.23	0.29

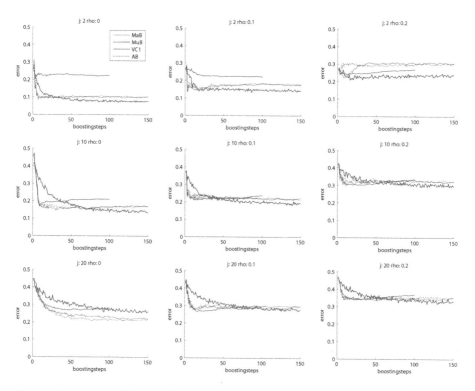

Fig. 1. Results of AdaBoost (AB, black line), MadaBoost (MaB, green), MultiBoost (MuB, red), AdaBoost-VC (VC1, pink) on the artifical data set. Please note that AdaBoost-VC can use at most 100 weak classifiers in this setting.

In a second round of experiments on the artificial data it was tested which features are selected by the different algorithms. Ensembles of 2 up to 50 members are trained as described for the cross-validation experiment. It was recorded how many relevant features were used. The training set contained all 200 points. The

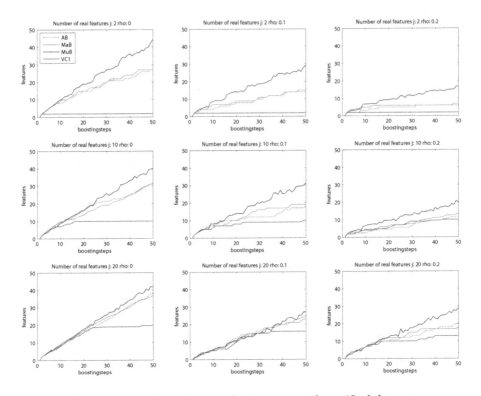

Fig. 2. Results of the feature selection test on the artifical data set

experiment was repeated 10 times with different permutations of the training set. The mean of these 10 experiments is given in Figures 2.

4.1   Results

The results for the experiments on the artificial data sets can be seen in Fig. 1. In the noise free case ($\rho = 0$), AdaBoost and MadaBoost achieve in the earlier iterations lower error rates than the other algorithms. This effect increases if the number of real features increases ($j \in \{10, 20\}$). If more noise is added, the effect is reduced. In the case of few real features MultiBoost achieves lowest error rates, if large ensembles are used. This effect decreases if j raises. The AdaBoost-VC experiments show, that the use of an higher value for d also increases the error rate. In the best tested case ($d = 1$), the error rate of AdaBoost-VC is the highest one in case $\rho = 0$ and $j = 2$. As the ensemble increases the error rate of AdaBoost-VC tends to increase.

The results on the real data sets can be seen in Table 1. For the ALL-AML-Leukemia data set and the colon cancer data set the accuracy of AdaBoost-VC outperformes the other algorithms. In all real data experiments, the error rates of MultiBoost outperform the error rates of AdaBoost if large ensembles are used. The results given for the ALL-AML-Leukemia and the colon cancer data set are comparable to those given in [9]. For the classifier, published with the breast cancer data set, a classification error of 17% is reported [12]. Note that for this classifier manually set cut-off parameters are used.

The results of the second round of experiment on the artificial data are given in Fig. 2. In this experiment MultiBoost choses more often real features than the other algorithms. This effect depends on the parameter f. As f raises, the effect is decreased. The maximal number of real features used by AdaBoost-VC is also determined by f. For higher values of f the number of used real features varies among the different varients of these algorithm. In thes case the varients with small d use more real features than the other.

4.2 Conclusion

The behaviour of MultiBoost can be attributed to its internal AdaBoost ensemble of size $\lceil \sqrt{T} \rceil$ which depends on the size T of the whole MultiBoost ensemble. If T is small, a single AdaBoost ensemble will not contain enough weak classifiers according to the number of effective features j. If the single AdaBoost ensembles reaches an appropriate size, MultiBoost outperforms AdaBoost. As no feature is presented twice, AdaBoost-VC can only choose j useful weak classifiers. So after $T \geq j$ iterations an AdaBoost-VC ensemble contains at least $T - j$ useless classifiers. The benefit of including these classifiers is mere chance. Note that this is also the reason why the ensembles outperform most of the mentioned distance based classifiers in this scenario. On the real data sets AdaBoost-VC seems to be more robust in the choise of parameter d. This phenomenon originates possibly from a greater number of informative features in the tumor gene expression profiles which may also be attributed to co-regulation or prior gene selection.

References

1. Schapire, R.E., Singer, Y.: Improved boosting algorithms using confidence-rated predictions. Machine Learning 37(3), 297–336 (1999)
2. Freund, Y., Schapire, R.E.: A decision-theoretic generalization of on-line learning and an application to boosting. In: Vitányi, P. (ed.) EuroCOLT 1995. LNCS, vol. 904, pp. 23–37. Springer, Heidelberg (1995)
3. Dudoit, S., Fridlyand, J., Speed, T.P.: Comparison of discrimination methods for the classification of tumors using gene expression data. Journal of the American Statistical Association 97(457), 77–87 (2002)
4. Bauer, E., Kohavi, R.: An empirical comparison of voting classification algorithms: Bagging, boosting, and variants. Machine Learning 36(1-2), 105–139 (1999)
5. Breiman, L.: Bagging predictors. Machine Learning 24(2), 123–140 (1996)
6. Webb, G.I.: Multiboosting: A technique for combining boosting and wagging. Machine Learning 40(2), 159–196 (2000)
7. Domingo, C., Watanabe, O.: Madaboost: A modification of adaboost. In: COLT 2000: Proceedings of the Thirteenth Annual Conference on Computational Learning Theory, pp. 180–189. Morgan Kaufmann Publishers Inc., San Francisco (2000)

8. Domingo, C., Watanabe, O.: Experimental evaluation of an adaptive boosting by filtering algorithm. Technical Report C-139, Tokyo Institut of Technology Department of Mathematical and Computing Sciences, Tokyo, Japan (December 1999)
9. Long, P.M., Vega, V.B.: Boosting and microarray data. Mach. Learn. 52(1-2), 31–44 (2003)
10. Vapnik, V.: Estimation of Dependences Based on Empirical Data: Springer Series in Statistics (Springer Series in Statistics). Springer-Verlag New York, Inc., Secaucus (1982)
11. Golub, T.R., Slonim, D.K., Tamayo, P., Huard, C., Gaasenbeek, M., Mesirov, J.P., Coller, H., Loh, M.L., Downing, J.R., Caligiuri, M.C., Bloomfield, C.D., Lander, E.S.: Molecular classification of cancer: class discovery and class prediction by gene expression monitoring. Science 286(5439), 531–537 (1999)
12. van 't Veer, L.J., Dai, H., van de Vijver, M.J., He, Y.D., Hart, A.A., Mao, M., Peterse, H.L., van der Kooy, K., Marton, M.J., Witteveen, A.T., Schreiber, G.J., Kerkhoven, R.M., Roberts, C., Linsley, P.S., Bernards, R., Friend, S.H.: Gene expression profiling predicts clinical outcome of breast cancer. Nature 415(6871), 530–536 (2002)
13. Alon, U., Barkai, N., Notterman, D.A., Gish, K., Ybarra, S., Mack, D., Levine, A.J.: Broad patterns of gene expression revealed by clustering analysis of tumor and normal colon tissues probed by oligonucleotide arrays. Proc. Natl. Acad. Sci. USA 96(12), 6745–6750 (1999)

Decision Fusion on Boosting Ensembles

Joaquín Torres-Sospedra, Carlos Hernández-Espinosa,
and Mercedes Fernández-Redondo

Departamento de Ingenieria y Ciencia de los Computadores, Universitat Jaume I,
Avda. Sos Baynat s/n, C.P. 12071, Castellon, Spain
{jtorres,espinosa,redondo}@icc.uji.es

Abstract. Training an ensemble of neural networks is an interesting way to build
a Multi-net System. One of the key factors to design an ensemble is how to combine the networks to give a single output. Although there are some important
methods to build ensembles, *Boosting* is one of the most important ones. Most of
methods based on *Boosting* use an specific combiner (*Boosting Combiner*). Although the *Boosting combiner* provides good results on boosting ensembles, the
results of previoues papers show that the simple combiner *Output Average* can
work better than the *Boosting combiner*. In this paper, we study the performance
of sixteen different combination methods for ensembles previously trained with
Adaptive Boosting and *Average Boosting*. The results show that the accuracy of
the ensembles trained with these original boosting methods can be improved by
using the appropriate alternative combiner.

1 Introduction

One technique often used to increase the generalization capability with respect to a
single neural network consists on training an ensemble of neural networks . This procedure consists on training a set of neural network with different weight initialization or
properties in the training process and combining them in a suitable way.

The two key factors to design an ensemble are how to train the individual networks
and how to combine the outputs provided by the networks to give a single output.
Among the methods of training the individual networks and combining them there are
an important number of alternatives. Our research group has performed some comparisons on methods to build and combine ensembles.

Reviewing the bibliography we can see that *Adaptive Boosting* (*Adaboost*) is one
of the best performing methods to create an ensemble [3]. *Adaboost* is a method that
construct a sequence of networks which overfits the training set used to train a neural
network with hard to learn patterns. A sampling distribution is used to select the patterns
we use to train the network.

In previoues papers, we successfully proposed three new boosting methods [13, 14,
15]. In those papers we noticed that in the majority of cases the *Output average* was
better than the specific *Boosting combiner*.

Some authors like Breiman [1], Kuncheva [9] or Oza [11] have deeply studied and
successfully improved *Adaboost* but any study on combining boosting methods has not
been done.

L. Prevost, S. Marinai, and F. Schwenker (Eds.): ANNPR 2008, LNAI 5064, pp. 157–167, 2008.

In this paper, we present a comparison of sixteen different combiners on ensembles previously trained with *Adaboost* and *Aveboost*, two of the most important boosting methods, in order to test if the *Boosting combiner* is the most appropriate way to combine boosting ensembles.

This paper is organized as follows. Firstly, some theoretical concepts are reviewed in section 2. Then, the ten databases used and the experimental setup are described in section 3. Finally, the experimental results and their discussion are in section 4.

2   Theory

2.1   Adaptive Boosting - Adaboost

In *Adaboost*, the successive networks are trained with a training data set T' selected at random from the original training data set T, the probability of selecting a pattern from T is given by a *sampling distribution* associated to the network $Dist_{net}$. The sampling distribution associated to a network is calculated when the previous network learning process has finished. *Adaboost* is described in algorithm 1.

Algorithm 1. AdaBoost $\{T, V, k\}$

Initialize Sampling Distribution: $Dist_x^1 = 1/m \ \forall x \in T$
for $net = 1$ to k **do**
    Create T' sampling from T using $Dist^{net}$
    MF Network Training T' , V
    Calculate missclassified vector:
$$miss_x^{net} = \begin{cases} 0 \text{ if } x \text{ is correctly classified} \\ 1 \text{ otherwise} \end{cases}$$
    Calculate error:
    $\epsilon_{net} = \sum_{x=1}^{m} Dist_x^{net} \cdot miss_x^{net}$
    Update sampling distribution:
$$Dist_x^{net+1} = Dist_x^{net} \cdot \begin{cases} \frac{1}{(2\epsilon_{net})} & \text{if } miss_x^{net} \\ \frac{1}{2(1-\epsilon_{net})} & \text{otherwise} \end{cases}$$
end for

Adaboost and *Aveboost* use an specific combination method, *Boosting combiner*, to combine the networks and get the final output or hypothesis eq.1.

$$h(x) = \underset{c=1,\dots,classes}{\arg\max} \sum_{net:h^{net}(x)=c}^{k} \log \frac{1 - \epsilon_{net}}{\epsilon_{net}} \qquad (1)$$

2.2   Averaged Boosting - Aveboost

Oza proposed in [11] *Averaged Boosting* (Algorithm 2). *Aveboost* is a method based on *Adaboost* in which the sampling distribution related to a neural network is also based on the number of networks previously trained. The whole description is detailed in algorithm 2.

Algorithm 2. Aveboost $\{T, V, k\}$

Initialize Sampling Distribution: $Dist_x^1 = 1/m \; \forall x \in T$
for $net = 1$ to k **do**
   Create T' sampling from T using $Dist^{net}$
   MF Network Training T', V
   Calculate missclassified vector:
$$miss_x^{net} = \begin{cases} 0 \text{ if } x \text{ is correctly classified} \\ 1 \text{ otherwise} \end{cases}$$
   Calculate error:
$$\epsilon_{net} = \sum_{x=1}^m Dist_x^{net} \cdot miss_x^{net}$$
   Update sampling distribution:
$$C_x^{net} = Dist_x^{net} \cdot \begin{cases} \frac{1}{(2\epsilon_{net})} & \text{if } miss_x^{net} \\ \frac{1}{2(1-\epsilon_{net})} & \text{otherwise} \end{cases}$$
$$Dist_x^{net+1} = \frac{net \cdot Dist_x^{net} + C_x^{net}}{net+1}$$
end for

2.3 Alternative Combiners

In this subsection, we briefly review the alternative combiners we have used to obtain the experimental results.

Average. This approach simply averages the individual classifier outputs across the different classifiers. The output yielding the maximum of the averaged values is chosen as the correct class.

Majority Vote. Each classifier provides a vote to a class, given by the highest output. The correct class is the one most often voted by the classifiers.

Winner Takes All (WTA). In this method, the class with overall maximum output across all classifier and outputs is selected as the correct class.

Borda Count. For any class c, the *Borda count* is the sum of the number of classes ranked below c by each classifier [5, 16]. The class with highest count is selected as correct class.

Bayesian Combination. This combination method was proposed in references [19]. According to this reference a belief value that the pattern x belongs to class c can be approximated by the following equation based on the values of the confusion matrix [16]

$$Bel(c) = \frac{\prod\limits_{net=1}^k P(x \in q_c | \lambda_{net}(x) = j_{net})}{\sum\limits_{i=1}^{classes} \prod\limits_{net=1}^k P(x \in q_i | \lambda_{net}(x) = j_{net})} \tag{2}$$

Weighted Average. This method introduces weights to the outputs of the different networks prior to averaging. The weights try to minimize the difference between the output of the ensemble and the *desired or true* output. The weights can be estimated from the error correlation matrix. The full description of the method can be found in [8, 16].

Choquet Integral. This method is based in the fuzzy integral [2, 4] and the Choquet integral. The method is complex, its full description can be found in [16].

Fuzzy Integral with Data Dependent Densities. It is another method based on the fuzzy integral and the Choquet integral. But in this case, prior to the application of the method it is performed a partition of the input space into n regions by frequency sensitive learning algorithm (*FSL*). The full description can be found in reference [16].

Weighted Average with Data Dependent weights. This method is the weighted average described above. But in this case, a partition of the space is performed by *FSL* algorithm and the weights are calculated for each partition. We have a different combination scheme for the different partitions of the space. The method is fully described in [16].

BADD Defuzzification Strategy. It is another combination method based on fuzzy logic concepts. The method is complex and the description can also be found in [16].

Zimmermann's Compensatory Operator. This combination method is based in the Zimmermann's compensatory operator described in [20]. The method is complex and can be found in [16].

Dynamically Averaged Networks. Two versions of *Dinamically Averaged Networks* were proposed by Jimenez [6, 7]. In these methods instead of choosing static weights derived from the network output on a sample of the input space, we allow the weights to adjust to be proportional to the certainties of the respective network output.

Nash Vote. In this method each voter assigns a number between zero and one for each candidate output. The product of the voter's values is compared for all candidates. The higher is the winner. The method is reviewed in reference [17].

Stacked Combiners (Stacked and Stacked+). The training in *Stacked Generalization* is divided into two steps. In the first one, the expert networks are trained. In the second one, the combination networks are trained with the outputs provided by the experts.

Stacked Generalization [18] can be adapted to combine ensembles of neural networks if the networks of the ensembles are used as expert networks. In [12], *Stacked* and *Stacked+*, two combiners based on *Stacked Generalization*, were successfully proposed.

3   Experimental Setup

In the experiments, the *Boosting combiner* and the alternative combiners have been applied on ensembles of 3, 9, 20 and 40 *MF* networks previously trained with *Adaptive*

Boosting and *Averaged Boosting* on the databases described in subsection 3.1 using the training parameters described in table 1. In the case of *Stacked combiners*, a single *MF* combination network has been applied.

Moreover, we have repeated the whole learning process 10 times using different training, validation and test sets. With this procedure we can obtain a mean performance of the ensemble for each database and an error in the performance calculated by standard error theory.

3.1   Datasets

We have used the following ten classification problems from the *UCI repository of machine learning databases* [10]:*Arrhythmia* (aritm), *Dermatology* (derma), *Ecoli* (ecoli), *Solar Flares* (flare), *Image segmentation* (img), *Ionosphere Database* (ionos), *Pima Indians Diabetes* (pima), *Haberman's Survival Data* (survi), *Vowel Database* (vowel) and *Wisconsin Breast Cancer* (wdbc).

The optimal parameters of the *Multilayer Feedforward* networks (*Hidden units, Adaptation step, Momentum rate* and *Number of iterations*) we have used to train the networks of the ensembles are shown in table 1.

Table 1. MF training parameters

database	hidden	step	mom	ite	accuracy
aritm	9	0.1	0.05	2500	75.6 ± 0.7
derma	4	0.1	0.05	1000	96.7 ± 0.4
ecoli	5	0.1	0.05	10000	84.4 ± 0.7
flare	11	0.6	0.05	10000	82.1 ± 0.3
img	14	0.4	0.05	1500	96.3 ± 0.2
ionos	8	0.1	0.05	5000	87.9 ± 0.7
pima	14	0.4	0.05	10000	76.7 ± 0.6
survi	9	0.1	0.2	20000	74.2 ± 0.8
vowel	15	0.2	0.2	4000	83.4 ± 0.6
wdbc	6	0.1	0.05	4000	97.4 ± 0.3

The optimal parameters of the *Multilayer Feedforward* networks we have used to train the combination networks of combiners *Stacked* and *Stacked+* on ensembles trained with *Adaboost* and *Aveboost* is shown in table 2.

Finally, we set to $n = 5$ the numbers of regions used in the combiners based on *data depend densities*. The parameters have been set after an exhaustive trial and error procedure using the training and validation sets.

4   Results and Discussion

Due to the lack of space, the general results on combining ensembles trained with *Adaboost* and *Aveboost* are shown in this section instead of showing the complete results. The general measurements used in this paper are described in subsection 4.1.

Table 2. Training parameters - Combiners Stacked and Stacked+

| | | Adaboost | | | | | | | Aveboost | | | | | | |
| | | Stacked | | | | Stacked+ | | | | Stacked | | | | Stacked+ | | | |
	nets	h.u	step	mom	ite	h.u	step	mom	ite	h.u	step	mom	ite	h.u	step	mom	ite
aritm	3	30	0.40	0.10	500	15	0.40	0.05	1750	30	0.40	0.01	500	24	0.40	0.20	4000
	9	24	0.05	0.05	500	14	0.40	0.20	500	19	0.40	0.20	500	21	0.40	0.20	1750
	20	25	0.05	0.01	500	5	0.40	0.20	500	28	0.05	0.20	500	30	0.10	0.20	500
	40	2	0.20	0.10	500	24	0.20	0.20	500	2	0.05	0.01	500	9	0.40	0.20	500
derma	3	4	0.40	0.20	2500	3	0.40	0.20	7500	24	0.10	0.10	1500	3	0.40	0.20	1500
	9	3	0.10	0.20	4000	3	0.10	0.05	7500	3	0.40	0.05	3000	3	0.20	0.10	3000
	20	3	0.20	0.20	6000	3	0.05	0.10	7500	3	0.20	0.05	4000	3	0.40	0.10	3000
	40	5	0.10	0.10	7500	3	0.10	0.01	2500	4	0.10	0.20	7500	3	0.40	0.05	1500
ecoli	3	3	0.10	0.20	1750	4	0.20	0.05	1750	3	0.40	0.05	1500	11	0.40	0.05	2500
	9	5	0.10	0.20	1500	3	0.40	0.10	6500	6	0.40	0.01	1500	7	0.40	0.10	4000
	20	25	0.40	0.20	500	26	0.40	0.20	500	29	0.40	0.20	500	7	0.40	0.05	5000
	40	19	0.40	0.10	1500	16	0.40	0.01	1500	25	0.10	0.01	2500	23	0.10	0.10	1750
flare	3	20	0.10	0.05	7500	2	0.10	0.01	7500	28	0.05	0.01	500	9	0.05	0.05	7500
	9	11	0.05	0.20	7500	30	0.05	0.10	500	30	0.05	0.10	500	21	0.40	0.01	4500
	20	16	0.10	0.05	2500	30	0.20	0.20	500	8	0.40	0.10	7500	27	0.05	0.05	500
	40	2	0.40	0.20	500	2	0.40	0.05	500	30	0.40	0.05	1500	5	0.10	0.10	6500
img	3	3	0.40	0.05	1500	5	0.40	0.20	500	13	0.40	0.20	500	8	0.40	0.10	500
	9	2	0.40	0.01	4000	3	0.10	0.05	1500	3	0.05	0.20	1500	2	0.20	0.10	7500
	20	2	0.40	0.05	4500	2	0.40	0.05	2500	2	0.20	0.01	3000	3	0.05	0.20	1500
	40	2	0.20	0.01	1750	3	0.05	0.05	7500	4	0.10	0.05	1500	3	0.05	0.05	1500
ionos	3	3	0.40	0.01	500	29	0.05	0.05	500	18	0.05	0.01	500	8	0.05	0.01	1500
	9	3	0.20	0.10	500	5	0.05	0.05	500	6	0.40	0.20	500	29	0.05	0.05	500
	20	2	0.40	0.20	500	2	0.40	0.05	500	16	0.05	0.20	500	21	0.05	0.05	500
	40	23	0.05	0.10	500	19	0.05	0.01	500	2	0.05	0.20	500	3	0.10	0.10	500
pima	3	28	0.05	0.10	500	30	0.05	0.20	500	2	0.20	0.20	1750	20	0.05	0.10	500
	9	7	0.20	0.10	500	13	0.20	0.10	500	16	0.10	0.05	500	21	0.40	0.05	500
	20	9	0.05	0.20	500	14	0.05	0.20	500	10	0.20	0.05	500	7	0.20	0.10	500
	40	10	0.05	0.01	500	4	0.05	0.05	500	7	0.20	0.20	500	20	0.40	0.20	1750
survi	3	2	0.40	0.20	1500	9	0.40	0.05	3000	5	0.40	0.20	500	10	0.20	0.20	500
	9	6	0.20	0.20	7500	2	0.40	0.20	1500	9	0.40	0.01	6500	8	0.20	0.20	4000
	20	2	0.40	0.20	1500	8	0.40	0.01	7500	25	0.20	0.20	3000	6	0.40	0.20	500
	40	3	0.40	0.05	1500	2	0.40	0.20	1500	24	0.40	0.10	6000	21	0.40	0.20	1500
vowel	3	14	0.10	0.05	1500	4	0.20	0.20	2500	11	0.05	0.01	5000	19	0.20	0.20	3000
	9	24	0.40	0.10	1500	11	0.20	0.01	7500	5	0.40	0.10	4000	6	0.10	0.01	7500
	20	6	0.10	0.05	6500	4	0.05	0.05	7500	6	0.10	0.05	7500	7	0.20	0.01	7500
	40	11	0.10	0.05	7500	12	0.05	0.20	7500	6	0.40	0.01	7500	4	0.20	0.01	4000
wdbc	3	29	0.40	0.20	500	29	0.40	0.20	500	30	0.40	0.20	500	28	0.40	0.20	500
	9	11	0.40	0.20	3000	14	0.40	0.20	7500	30	0.10	0.05	500	29	0.05	0.10	500
	20	15	0.05	0.20	4500	13	0.10	0.10	2500	30	0.20	0.01	500	30	0.10	0.05	500
	40	28	0.05	0.01	500	30	0.05	0.10	500	28	0.10	0.01	500	30	0.05	0.10	500

4.1   General Measurements

In our experiments, we have calculated the *Increase of Performance* (*IoP* eq.3) and the *Percentage of Error Reduction* (*PER* eq.4) of the results with respect to a single network in order to perform an exhaustive comparison. The *IoP* value is an absolute measurement whereas the *PER* value is a relative measurement. A negative value on these measurements mean that the ensemble performs worse than a single network.

$$IoP = Error_{SingleNet} - Error_{Ensemble} \tag{3}$$

$$PER = 100 \cdot \frac{Error_{SingleNet} - Error_{Ensemble}}{Error_{SingleNet}} \tag{4}$$

Finally, we have calculated the mean *IoP* and the mean *PER* across all databases to get a general measurement to compare the methods presented in the paper. The results on combining *Adaboost* are presented in subsection 4.2 whereas the results on combining *Aveboost* are in subsection 4.3.

4.2   Adaboost Results

In this subsection the results of the different combiners on ensembles trained with *Adaptive Boosting* are shown. Table 3 shows the mean *IoP* and the mean *PER* for the ensembles trained and combined with the *Boosting combiner* as in the original method *Adaboost* and for the same ensembles combined with the alternative combiners described in subsection 2.3.

Table 3. Adaptive Boosting - Mean IoP and PER among all databases

Method	Mean IoP				Mean PER			
	3 Nets	9 Nets	20 Nets	40 Nets	3 Nets	9 Nets	20 Nets	40 Nets
adaboost	0.3	0.86	1.15	1.26	1.33	4.26	9.38	12.21
average	−0.18	−0.15	−0.36	−0.35	−20.49	−19.71	−18.81	−22.05
voting	−0.97	−1.48	−1.74	−1.26	−27.18	−27.2	−25.53	−26.21
wta	−1.07	−4.78	−8.22	−11.07	−16.66	−78.22	−132.97	−184.84
borda	−2.74	−2.76	−2.84	−2.07	−50.55	−35.57	−32.71	−32.13
bayesian	−0.36	−1.28	−3.22	−5.46	−6.28	−16.17	−38.15	−65.05
wave	0.59	1.19	0.41	0.44	1.15	6.5	3.7	7.3
choquet	−1.26	−6.23	−	−	−23.68	−107.65	−	−
fidd	−1.37	−6.91	−	−	−27.85	−123.9	−	−
wave dd	0.68	0.72	−	−	4.18	0.67	−	−
badd	−0.18	−0.15	−0.36	−0.35	−20.49	−19.71	−18.81	−22.05
zimm	0.35	−1.16	−13.37	−13.02	−0.28	−28.25	−150.69	−212.35
dan	−12.54	−17.08	−20.23	−19.13	−123.17	−199.31	−244.94	−278.47
dan2	−12.59	−17.45	−20.46	−19.47	−123.5	−202.67	−248.34	−282.06
nash	−1.27	−1.76	−1.71	−1.57	−16.55	−30.14	−28.14	−29.13
stacked	0.69	0.83	0.7	0.51	3.39	2.87	4.7	4.27
stacked+	0.71	0.95	0.64	−0.14	5.43	6.25	4.39	1.81

Table 4. Averaged Boosting - Mean IoP and PER among all databases

	Mean IoP				Mean PER			
Method	3 Nets	9 Nets	20 Nets	40 Nets	3 Nets	9 Nets	20 Nets	40 Nets
aveboost	0.5	1.49	1.83	1.82	1.13	10.46	11.7	10.79
average	0.87	1.61	2	1.8	4.26	11.64	12.93	12.99
voting	0.37	1.54	1.76	1.91	0.28	11.15	12.67	13.01
wta	0.49	0.36	−0.38	−0.88	−0.1	−2.31	−9.2	−10.88
borda	−0.34	1.15	1.57	1.73	−6.73	8.13	11.05	12.12
bayesian	0.02	−0.13	−1.3	−2.87	−4.14	−7.94	−23.39	−40.81
wave	0.85	1.17	1.19	0.32	4.29	8.36	7.65	3.78
choquet	0.21	−0.19	−	−	−3.69	−10.02	−	−
fidd	0.14	−0.35	−	−	−3.76	−11.14	−	−
wave dd	0.92	1.62	−	−	5.62	11.88	−	−
badd	0.87	1.61	2	1.8	4.26	11.64	12.93	12.99
zimm	0.74	0.59	−2.75	−7.53	4.17	5.17	−18.5	−63.01
dan	−2.65	−3.04	−5.06	−5.13	−30.32	−25.63	−34.57	−34.21
dan2	−2.64	−3.13	−5.1	−5.27	−30.37	−26.56	−34.89	−35.38
nash	−0.09	1.03	1.63	1.4	−2.56	7.33	11.34	8.86
stacked	0.9	0.99	0.95	0.96	6.67	7.79	7.15	8.43
stacked+	0.95	1.02	0.83	0.8	6.42	6.22	7.64	6.84

Table 5. Adaboost - Best performance

	Boosting Combiner		Alternative Combiners		
Database	Performance	Nets	Performance	Method	Nets
aritm	73.8 ± 1.1	40	75.3 ± 0.9	bayes	9
derma	98 ± 0.5	3	98.1 ± 0.7	stacked+	3
ecoli	86 ± 1.3	20	87.2 ± 1.0	w.ave	3
flare	81.7 ± 0.6	3	82.2 ± 0.6	w.ave	3
img	97.3 ± 0.2	20	97.4 ± 0.3	average	20
ionos	91.6 ± 0.7	40	92 ± 0.9	average	20
pima	75.7 ± 1.0	3	76.6 ± 1.1	average	3
survi	75.4 ± 1.6	3	74.8 ± 1.0	zimm	3
vowel	97 ± 0.6	40	97.1 ± 0.6	average	40
wdbc	96.7 ± 0.9	40	96.6 ± 0.6	bayes	20

Moreover, table 5 shows the best performance for each database on ensembles trained with the original *Adaboost* (applying the *Boosting combiner*). The table also shows the best performance of these ensembles combined with the sixteen alternative combiners.

4.3 Aveboost Results

In this subsection the results of the different combiners on ensembles trained with *Averaged Boosting* are shown. Table 4 shows the mean IoP and the mean PER for the ensembles trained and combined with the *Boosting combiner* as in the original method *Aveboost* and for the same ensembles combined with the alternative combiners.

Table 6. Aveboost - Best performance

Database	Boosting Combiner		Alternative Combiners		
	Performance	Nets	Performance	Method	Nets
aritm	76.3 ± 1.0	40	77.0 ± 1.1	average	20
derma	97.9 ± 0.5	20	97.8 ± 0.6	w.avedd	3
ecoli	86.5 ± 1.2	9	87.6 ± 0.9	w.ave	3
flare	82.4 ± 0.7	20	82.5 ± 0.6	stacked+	3
img	97.5 ± 0.2	40	97.6 ± 0.2	stacked	40
ionos	91.6 ± 0.9	40	92.4 ± 1	zimm	3
pima	76.6 ± 1.0	9	77.1 ± 1	w.ave	3
survi	75.1 ± 1.2	3	75.1 ± 1.2	voting	3
vowel	96.4 ± 0.6	40	96.7 ± 0.4	w.ave	40
wdbc	96.6 ± 0.4	9	96.7 ± 0.3	zimm	9

Moreover, table 6 shows the best performance for each database on ensembles trained with the original *Aveboost* (applying the *Boosting combiner*). The table also shows the best performance of these ensembles combined with the sixteen alternative combiners.

4.4 Discussion

We see that the *Boosting combiner* is not the best alternative in *Adaboost* in all cases, *Stacked combiners* and *Weighted Average with Data-Depend Densities* are better (table 3) when the number of networks is reduced. If we analyse table 5, we can see that the *boosting combiner* only provides the best result on databases *survi* and *wdbc*.

We can also see in *Aveboost* that, in the majority of cases, the mean IoP and PER of the boosting combiner is always lower than for the *Output average* (table 4). Moreover, the *boosting combiner* only provides the best result for database *derma* (table 6).

Moreover, we can notice that in some cases the accuracy is highly improved by applying an alternative combiner while the number of networks required to get the best performance is reduced. We got better results by combining a smaller set of networks.

5 Conclusions

In this paper, we have performed a comparison among sixteen combiners on ensembles trained with *Adaptive Boosting* and *Averaged Boosting*. To carry out our comparison we have used ensembles of 3, 9, 20 and 40 networks previously trained with *Adaptive boosting* and *Averaged boosting* and the accuracy of the ensemble using the *Boosting Combiner*. In our experiments we have selected ten databases whose results are not easy to improve with an ensemble of neural networks. Alternatively, we have applied sixteen different combiners on these ensembles to test if the *boosting combiner* is the best method to combine the networks of a *boosting ensemble*. Moreover, we also want to know which is the most appropriate combiner in each case. Finally, we have calculated the mean *Increase of Performance* and the mean *Percentage of Error Reduction* with respect to a single network to compare the combiners. Furthermore, the best accuracy for each database with the original methods, *Boosting combiner* on *Adaboost* and *Aveboost*, and applying the sixteen alternative combiners on these ensembles have been shown.

According the general results, the *Boosting combiner* is not the most effective way to combine an ensemble trained with *Boosting* in a wide number of cases. The original results have been improved with the use of the appropriate alternative combiner. In general, the *Output average*, the *Weighted average* and the *Stacked combiners* are the best combiners on ensembles trained with *Adaboost*. In a similar way, the *Output average*, the *Voting* and the *Borda Count* are the best combiners on ensembles trained with *Aveboost*.

According the best performance for each database (tables 5 and 6), we can see that the *Output average* and both versions of the *Weighted average* should be seriously considered for combining *boosting ensembles* because the *Boosting combiner* provides the best results only in 16.6% of the cases. In addition, in a some cases not only have the accuracy of the ensembles been improved with the use of an alternative combiner, the numbers of networks to get the best result is also reduced. For instance, the best accuracy for database *ecoli* using the original *Adaboost* (applying the *Boosting combiner*) was got with the 20-network ensembles (86 ± 1.3). The best overall accuracy for this database using *Adaboost* was got by applying the *Weighted average* to the 3-networks ensembles (87.2 ± 1.0). The accuracy was improved in 1.2%, the error rate was reduced in 0.3% and the number of networks required were reduced from 20 to 3.

Nowadays, we are extending the comparison we have performed adding more methods and databases. The results we are getting also show that the *Boosting combiner* does not provide either the best general results (best mean IoP or PER) or the best performance for each database. Furthermore, we are working on an advanced combination method based on the *boosting combiner* we think could increase the accuracy of the *Boosting ensembles*.

We can conclude by remarking that the accuracy of a *boosting ensemble* can be improved and its size can be reduced by applying the *Output average* or advanced combiners like the *Weighted average* or the *Stacked combiners* on ensembles previously trained with *Adaboost* or *Aveboost*.

References

1. Breiman, L.: Arcing classifiers. The Annals of Statistics 26(3), 801–849 (1998)
2. Cho, S.-B., Kim, J.H.: Combining multiple neural networks by fuzzy integral for robust classification. IEEE Transactions on System, Man, and Cybernetics 25(2), 380–384 (1995)
3. Freund, Y., Schapire, R.E.: Experiments with a new boosting algorithm. In: International Conference on Machine Learning, pp. 148–156 (1996)
4. Gader, P.D., Mohamed, M.A., Keller, J.M.: Fusion of handwritten word classifiers. Pattern Recogn. Lett. 17(6), 577–584 (1996)
5. Ho, T.K., Hull, J.J., Srihari, S.N.: Decision combination in multiple classifier systems. IEEE Transactions on Pattern Analysis and Machine Intelligence 16(1), 66–75 (1994)
6. Jimenez, D.: Dynamically weighted ensemble neural networks for classification. In: Proceedings of the 1998 International Joint Conference on Neural Networks, IJCNN 1998, pp. 753–756 (1998)

7. Jimenez, D., Darm, T., Rogers, B., Walsh, N.: Locating anatomical landmarks for prosthetics design using ensemble neural networks. In: Proceedings of the 1997 International Conference on Neural Networks, IJCNN 1997 (1997)
8. Krogh, A., Vedelsby, J.: Neural network ensembles, cross validation, and active learning. In: Tesauro, G., Touretzky, D., Leen, T. (eds.) Advances in Neural Information Processing Systems, vol. 7, pp. 231–238. The MIT Press, Cambridge (1995)
9. Kuncheva, L., Whitaker, C.J.: Using diversity with three variants of boosting: Aggressive. In: Roli, F., Kittler, J. (eds.) MCS 2002. LNCS, vol. 2364. Springer, Heidelberg (2002)
10. Newman, D.J., Hettich, S., Blake, C.L., Merz, C.J.: UCI repository of machine learning databases (1998),
    `http://www.ics.uci.edu/$\sim$mlearn/MLRepository.html`
11. Oza, N.C.: Boosting with averaged weight vectors. In: Windeatt, T., Roli, F. (eds.) MCS 2003. LNCS, vol. 2709, pp. 973–978. Springer, Heidelberg (2003)
12. Torres-Sospedra, J., Hernndez-Espinosa, C., Fernández-Redondo, M.: Combining MF networks: A comparison among statistical methods and stacked generalization. In: Schwenker, F., Marinai, S. (eds.) ANNPR 2006. LNCS (LNAI), vol. 4087, pp. 210–220. Springer, Heidelberg (2006)
13. Torres-Sospedra, J., Hernndez-Espinosa, C., Fernández-Redondo, M.: Designing a multilayer feedforward ensembles with cross validated boosting algorithm. In: IJCNN 2006 proceedings, pp. 2257–2262 (2006)
14. Torres-Sospedra, J., Hernndez-Espinosa, C., Fernndez-Redondo, M.: Designing a multilayer feedforward ensemble with the weighted conservative boosting algorithm. In: IJCNN 2007 Proceedings, pp. 684–689. IEEE, Los Alamitos (2007)
15. Torres-Sospedra, J., Hernndez-Espinosa, C., Fernndez-Redondo, M.: Mixing aveboost and conserboost to improve boosting methods. In: IJCNN 2007 Proceedings, pp. 672–677. IEEE, Los Alamitos (2007)
16. Verikas, A., Lipnickas, A., Malmqvist, K., Bacauskiene, M., Gelzinis, A.: Soft combination of neural classifiers: A comparative study. Pattern Recognition Letters 20(4), 429–444 (1999)
17. Wanas, N.M., Kamel, M.S.: Decision fusion in neural network ensembles. In: Proceedings of the 2001 International Joint Conference on Neural Networks, IJCNN 2001, vol. 4, pp. 2952–2957 (2001)
18. Wolpert, D.H.: Stacked generalization. Neural Networks 5(6), 1289–1301 (1994)
19. Xu, L., Krzyzak, A., Suen, C.Y.: Methods of combining multiple classifiers and their applications to handwriting recognition. IEEE Transactions on Systems, Man, and Cybernetics 22(3), 418–435 (1992)
20. Zimmermann, H.-J., Zysno, P.: Decision and evaluations by hierarchical aggregation of information. Fuzzy Sets and Systems 10(3), 243–260 (1984)

The Mixture of Neural Networks as Ensemble Combiner

Mercedes Fernández-Redondo[1], Joaquín Torres-Sospedra[1],
and Carlos Hernández-Espinosa[1]

Departamento de Ingenieria y Ciencia de los Computadores, Universitat Jaume I,
Avda. Sos Baynat s/n, C.P. 12071, Castellon, Spain
{redondo,jtorres,espinosa}@icc.uji.es

Abstract. In this paper we propose two new ensemble combiners based on the *Mixture of Neural Networks* model. In our experiments, we have applied two different network architectures on the methods based on the *Mixture of Neural Networks*: the *Basic Network* (*BN*) and the *Multilayer Feedforward Network* (*MF*). Moreover, we have used ensembles of *MF* networks previously trained with *Simple Ensemble* to test the performance of the combiners we propose. Finally, we compare the *mixture combiners* proposed with three different mixture models and other traditional combiners. The results show that the mixture combiners proposed are the best way to build Multi-net systems among the methods studied in the paper in general.

1 Introduction

The most important property of an artificial neural network is the ability to correctly respond to inputs which were not used in the learning set. One technique commonly used to increase this ability consists on training some Multilayer Feedforward networks with different weights initialization. Then the mean of the outputs is applied to get the final output. This method, known as *Simple Ensemble* (Fig.1) increases the generalization capability with respect to a single network [17].

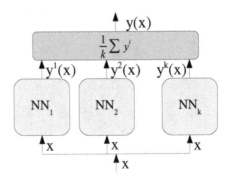

Fig. 1. Simple Ensemble diagram

L. Prevost, S. Marinai, and F. Schwenker (Eds.): ANNPR 2008, LNAI 5064, pp. 168–179, 2008.

The *Mixture of Neural Networks* (Fig.2) is a modular model to build a Multi-Net system which consist on training different neural networks, also called expert networks or experts, with a gating network. The method divides the problem into subproblems, each subproblem tends to be solved by one expert. The gating network is used to weight and combine the outputs of the experts to get the final output of the system.

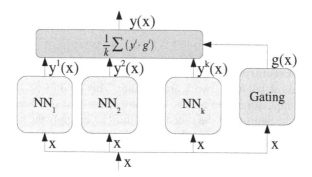

Fig. 2. Mixture of Neural Networks diagram

In a previous paper [16], we analysed the original *Mixture of Neural Networks* model [13] which uses the *Basic Network* as expert and gating networks. Moreover, we successfully proposed a model based on the use of the *Multilayer Feedforward* network.

In this paper we propose two ensemble combiners based on that idea. We think that the the accuracy of an ensemble of *MF* networks can be increased by applying the gating network of the mixture model to combine the ensemble.

In our experiments, we have applied the proposed combiners to ensembles previously trained with *Simple Ensemble (SE)*. We have named them, *Mix-SE-BN* and *Mix-SE-MF*, depending on the method used to train the ensemble (*SE* in this case) and the architecture applied to the gating net (*BN* or *MF*).

Finally, we compare the accuracy of the *mixture models* and the *mixture combiners* on *Simple Ensemble*. Moreover, we also compare the general results of the *mixture combiners* with seven traditional combiners also applied to the ensembles previously trained with *Simple Ensemble*.

This paper is organized as follows. Firstly, some theoretical concepts are briefly reviewed in section 2. Then, the eight databases used and the experimental setup are described in section 3. Finally, the experimental results and their discussion are in section 4.

2   Theory

2.1   Network Architectures

The Basic Network. The *Basic Network* (Figure 3) consists of two layers of neurons that apply the identity function. This network can only solve linear problems [10].

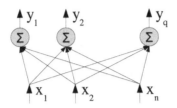

Fig. 3. Basic Network Structure

The Multilayer Feedforward network. The *Multilayer Feedforward* architecture (Figure 4) is the most known network architecture. This kind of networks consists of three layers of computational units. The neurons of the first layer apply the identity function whereas the neurons of the second and third layer apply the sigmoid function. This kind of networks can approximate any function with a specified precision [1,10].

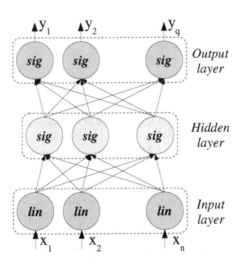

Fig. 4. Multilayer Feedforward Structure

2.2 Simple Ensemble

A simple ensemble can be constructed by training different networks with the same training set, but with different random initialization [3].

Algorithm 1. Simple Ensemble $\{T,V\}$

---

for $i = 1$ to k do
   Traditional Network Training $\{T,V\}$
end for
Save Ensemble Configuration

Algorithm 2. Traditional Network Training$\{T, V\}$

Set initial weights randomly
for $i = 1$ to $iterations$ **do**
    Train the network on the training set T
    Calculate MSE over validation set V
    Save epoch weights and calculated MSE
end for
Select epoch with minimum MSE
Assign best epoch configuration to the network
Save network configuration

In our experiments we have trained the networks for few iterations. In each iteration, the weights have been adapted with *Back-Propagation* over the training set. At the end of each iteration the Mean Square Error (MSE) has been calculated over the Validation set. When the learning process has finished, we assign the weights of the iteration with minimum MSE to the final network. For this reason the original learning set L is divided into two subsets: The Training set T and the Validation set V.

2.3   Mixture of Neural Networks

The *Mixture of Neural Networks* is a Multi-net system which consists on training k expert networks along with a gating network that weights the output of the experts. In the original version of the method [7] the *Basic Network* was applied as expert and gating networks. In the version *Mixture of Multilayer Feedforward Networks* [16], the *Multilayer feedforward* network was applied as expert and gating networks. In both cases, the objective function L, eq. 1, was applied to adapt the weights of the networks.

$$L = \log \left(\sum_{net=1}^{k} g_{net} \cdot \exp \left(-\frac{1}{2} \cdot \left\| d - y^{net} \right\|^2 \right) \right) \qquad (1)$$

Where d refers to the desired output or target.

In reference [16] we can found the basic description of the training algorithm carried out when the *Basic Network* or the *Multilayer Feedforward* network are applied.

Finally, in the experiments carried out in the current paper, we have added a new method based in the networks previously described. We have build a modular system system with the *Mixture of Neural Networks* where the *Multilayer Feedforward* network has been applied to the experts and the *Basic Network* has been applied to the gating network.

The basic algorithm carried out to build a Multi-net system with the *Mixture of Neural Networks* is described in the following algorithm.

Algorithm 3. Mixture of Neural Networks

Random initialization of networks
for $ite = 1$ to $iterations$ **do**
    for each pattern from training set **do**
        for $net = 1$ to k **do**
            Adapt expert weights
        end for
        Adapt gating weights
    end for
    Calculate L_{ite} over Validation set
    Save weights
end for
Select iteration with maximum L (best iteration)
Set best iteration weights to network
Save final configuration

Resuming, we have applied three versions of the *Mixture of Neural Networks* model using the *Basic Network* and the *Multilayer Feedforward*. These *Mixture models* are:

- *Mix-BN-BN*: *BN* as experts and gating net.
- *Mix-BN-MF*: *MF* as experts and *BN* as gating net.
- *Mix-MF-MF*: *MF* as experts and gating net.

2.4   Mixture as Combiner

The *Mixture of Neural Networks* approach can be applied as ensemble combiner if the weights of the ensembles are assigned to the experts and they are keep unchanged during the training process. With this simple procedure the *Mixture of Neural Networks* can be applied to combine the networks of ensemble of neural networks.

In this paper, the *mixture model* is modified in order to get the *mixture combiner*, an efficient ensemble combiner. The *mixture combiner* is described in algorithm 4.

Algorithm 4. Mixture Combiner

Assign ensemble weights to experts
Random initialization of gating network weights
for $ite = 1$ to $iterations$ **do**
    for each pattern from training set **do**
        Adapt gating weights
    end for
    Calculate L_{ite} over Validation set
    Save weights
end for
Select iteration with maximum L (best iteration)
Set best iteration weights to network
Save final configuration

In our experiments, we have only used ensembles previously trained with *Simple Ensemble (SE)*. Moreover, we have applied to the gating network the two network architectures previously described so two different new *mixture combiners* are proposed in this paper. These proposed combiners are:

- *Mix-SE-BN*: *BN* as gating net to combine a *SE*.
- *Mix-SE-MF*: *MF* as gating net to combine a *SE*.

3   Experimental Testing

In our experiments we have trained Multi-net systems of 3, 9, 20 and 40 MF experts with *Simple Ensemble* and the three different *Mixture models* on the eight problems from the *UCI* repository of machine learning [12]. Moreover, we have applied two new combiners based on the *Mixture of Neural Networks* to ensembles previously trained with *Simple Ensemble*.

Finally, we have repeated ten times the whole learning process in order to get a mean performance of the ensemble for each database and an error in the performance calculated by standard error theory.

3.1   Datasets

We have used the following eight different classification problems from the *UCI repository of machine learning databases* [12] to test the performance of methods: *Balance Scale Database* (bala), *Dermatology Database* (derma), *Ecoli Database* (ecoli), *Glass Identification Database* (glas), *The Monk's Problem 1* (mok1), *The Monk's Problem 2* (mok2), *Haberman's Survival Data* (survi) and *Congressional Voting Records Database* (vote).

The training parameters of the networks have been set after performing a deep *trial and error* procedure on a validation set.

4   Results and Discussion

4.1   Results

In this section we present the experimental results we have obtained with the ensembles of MF networks trained with *Simple Ensemble*, the *Mixture models* and the *Mixture combiners* proposed.

Table 1 shows the results we got with *Simple Ensemble*.

Table 1. Simple Ensemble results

Database	3 Nets	9 Nets	20 Nets	40 Nets
bala	96 ± 0.5	95.8 ± 0.5	95.8 ± 0.6	95.9 ± 0.5
derma	97.2 ± 0.7	97.5 ± 0.7	97.3 ± 0.7	97.6 ± 0.7
ecoli	86.6 ± 0.8	86.9 ± 0.8	86.9 ± 0.8	86.9 ± 0.7
glas	94 ± 0.8	94 ± 0.7	94 ± 0.7	94.2 ± 0.6
mok1	98.3 ± 0.9	98.8 ± 0.8	98.3 ± 0.9	98.3 ± 0.9
mok2	88 ± 2	90.8 ± 1.8	91.1 ± 1.1	91.1 ± 1.2
survi	74.3 ± 1.3	74.2 ± 1.3	74.3 ± 1.3	74.3 ± 1.3
vote	95.6 ± 0.5	95.6 ± 0.5	95.6 ± 0.5	95.6 ± 0.5

Tables 2-4 show the results of the *Mixture models* reviewed in this paper (*Mix-BN-BN*, *Mix-BN-MF* and *Mix-MF-MF*).

Table 2. Mix-BN-BN results

Database	3 Nets	9 Nets	20 Nets	40 Nets
bala	90.5 ± 0.9	90.2 ± 1	91 ± 0.7	89.8 ± 0.8
derma	96.8 ± 0.7	97 ± 0.6	97.3 ± 0.6	97.2 ± 0.8
ecoli	52.5 ± 1.1	48 ± 2	69.9 ± 1.5	74.7 ± 1.4
glas	89.4 ± 1	91.2 ± 1.1	90.2 ± 1.3	91 ± 1.1
mok1	88 ± 2	94 ± 3	94 ± 2	94 ± 3
mok2	62.1 ± 1.7	67.5 ± 2	66.8 ± 1.6	68 ± 2
survi	72.3 ± 1.2	72.6 ± 0.9	73.8 ± 0.9	73.6 ± 1.2
vote	95 ± 1.2	96.1 ± 0.6	96.1 ± 0.6	96.5 ± 0.7

Table 3. Mix-MF-BN results

Database	3 Nets	9 Nets	20 Nets	40 Nets
bala	94.1 ± 1	93.9 ± 1.2	94.6 ± 1.1	95.2 ± 0.7
derma	97 ± 0.8	97.2 ± 0.8	97 ± 0.7	96.9 ± 0.8
ecoli	85 ± 0.8	86.5 ± 1	85.9 ± 0.7	84.6 ± 1.3
glas	94.6 ± 1	94.6 ± 1.2	94.2 ± 1.3	95 ± 1.2
mok1	99.3 ± 0.8	99.3 ± 0.8	98.8 ± 0.9	100 ± 0
mok2	77 ± 3	77 ± 2	84 ± 1.8	80.3 ± 1.8
survi	74.6 ± 1.3	74.9 ± 1.2	74.6 ± 1.1	75.1 ± 1.2
vote	96.1 ± 0.6	96.1 ± 0.6	96.1 ± 0.6	95.8 ± 0.6

Table 4. Mix-MF-MF results

Database	3 Nets	9 Nets	20 Nets	40 Nets
bala	95.1 ± 0.6	95 ± 1	94.2 ± 0.9	94.9 ± 0.6
derma	97.2 ± 0.7	96.9 ± 0.9	97 ± 0.8	96.3 ± 1.1
ecoli	85.4 ± 0.6	84.3 ± 0.8	86.3 ± 1	86.5 ± 0.8
glas	95.2 ± 0.7	94.6 ± 1	95.2 ± 1	93.8 ± 1
mok1	98.6 ± 0.9	98.3 ± 0.9	99.5 ± 0.5	98.4 ± 0.8
mok2	90.3 ± 1.2	87.3 ± 1.5	88.5 ± 1.5	90.8 ± 1.6
survi	74.6 ± 1.2	74.8 ± 1.4	74.4 ± 1.2	73.3 ± 1.3
vote	96 ± 0.6	96 ± 0.6	96 ± 0.6	96.1 ± 0.6

Table 5. Mix-SE-BN results

Database	3 Nets	9 Nets	20 Nets	40 Nets
bala	96.2 ± 0.6	96.1 ± 0.7	96.5 ± 0.5	96.6 ± 0.5
derma	97 ± 0.7	97.3 ± 0.7	97.3 ± 0.7	97.3 ± 0.7
ecoli	86.2 ± 0.9	87.4 ± 0.8	87.2 ± 0.6	86.9 ± 0.6
glas	93.8 ± 0.6	94.4 ± 0.6	94.4 ± 0.6	94.8 ± 0.7
mok1	98.5 ± 0.8	99.8 ± 0.3	100 ± 0	100 ± 0
mok2	87 ± 2	91.1 ± 1.5	91.8 ± 1.4	91.1 ± 0.9
survi	74.1 ± 1.6	74.8 ± 1.6	74.3 ± 1.2	74.3 ± 1.2
vote	95.6 ± 0.5	95.6 ± 0.5	95.6 ± 0.5	95.6 ± 0.5

Table 6. Mix-SE-MF results

Database	3 Nets	9 Nets	20 Nets	40 Nets
bala	96.2 ± 0.5	96.2 ± 0.5	95.9 ± 0.8	96.2 ± 0.7
derma	97.2 ± 0.7	97.2 ± 0.7	97.3 ± 0.7	97.3 ± 0.7
ecoli	86.5 ± 0.9	87.5 ± 0.7	86.8 ± 0.8	86.9 ± 0.6
glas	94 ± 0.8	94 ± 0.7	94.4 ± 0.7	94.2 ± 0.6
mok1	98.5 ± 0.8	99.8 ± 0.3	100 ± 0	100 ± 0
mok2	88 ± 2	91.5 ± 1.3	91.8 ± 0.9	91.6 ± 1.4
survi	74.3 ± 1.5	74.4 ± 1.5	74.3 ± 1.3	74.3 ± 1.3
vote	95.6 ± 0.5	95.6 ± 0.5	95.6 ± 0.5	95.6 ± 0.5

The results of the *Mixture combiners* proposed (*Mix-SE-BN* and *Mix-SE-MF*) are in tables 5-6.

4.2   General Measurements

We have also calculated the Increase of Performance (IoP eq.2) and the Percentage of Error Reduction (PER eq.3) of the results with respect to a single network in order to get a general value for the comparison among the studied methods.

The IoP value is an absolute measurement that denotes the increase of performance of the ensemble with respect to a single network.

$$IoP = Error_{SingleNet} - Error_{Ensemble} \tag{2}$$

The PER value is a relative measurement which ranges from 0%, where there is no improvement by the use of an ensemble method with respect to a single network, to 100%.

$$PER = 100 \cdot \frac{Error_{SingleNet} - Error_{Ensemble}}{Error_{SingleNet}} \tag{3}$$

Table 7. Global Results - Mean Increase of Performance

method	3 nets	9 nets	20 nets	40 nets
simple ensemble	9.17	9.61	9.59	9.67
mix-bn-bn	-1.29	-0.1	2.76	3.56
mix-mf-bn	7.61	7.84	8.58	8.27
mix-mf-mf	9.48	8.81	9.32	9.3
mix-se-bn	9.01	9.98	10.06	10
mix-se-mf	9.14	9.94	9.93	9.94
bayesian	9.2	8.91	8.51	8.37
borda	9.12	9.33	9.34	9.53
choquet	8.99	9.61	–	–
dan	8.44	8.55	8.47	8.52
nash	9.18	9.76	9.73	9.81
w.ave	9.08	9.86	9.84	9.25
zimm	9.18	9.71	8.81	7.34

Table 8. Global Results - Mean Percentage of Error Reduction

method	3 nets	9 nets	20 nets	40 nets
simple ensemble	42.51	44.85	44.11	45.45
mix-bn-bn	−11.85	−6.27	13.29	17.48
mix-mf-bn	36.79	38.54	40.24	38.2
mix-mf-mf	43.39	39.75	42.3	40.2
mix-se-bn	41.36	45.97	46.38	46.22
mix-se-mf	42.41	45.39	45.45	45.65
bayesian	41.18	37.14	32.21	29.2
borda	41.21	42.9	42.39	44.26
choquet	41.53	44.18	−	−
dan	37.44	38.81	37.41	39.27
nash	42.7	45.01	45.34	45.45
w.ave	39.63	44.72	44.75	38.69
zimm	42.92	45.11	38.94	25.92

There can also be negative values in both measurements, IoP and PER, which means that the performance of the ensemble is worse than the performance of a single net.

Finally, we have calculated the mean IoP and the mean PER across all databases to get a global measurement to compare the methods presented in the paper. Table 7 shows the results of the mean IoP whereas table 8 shows the results of the mean PER. In these tables we have also included the general results of seven traditional combiners: *Bayesian Combination* (bayesian) [20,11,6], *Borda Count* (borda) [5], *Choquet Integral* (choquet) [2,4,18], *Dinamically Averaged Networks* (dan) [8], *Nash Vote* (nash) [18,19], *Weighted Average* (w.ave) [9] and *Zimmermann's Operator* (zimm) [18,21].

The results of these seven traditional combiners on ensembles trained with *Simple Ensemble* were published in [14,15]. Moreover, we could not apply the *choquet integral* to 20 and 40 network ensembles due to its complexity.

4.3 Discussion

Although some important conclusions can be derived by analysing the results, this discussion will be focused on the comparison of the combiners proposed with the mixture models and with the seven traditional combiners.

In a previous paper we concluded that the idea of applying a sophisticated weighted average based on the gating network of the mixture model should be seriously considered. As we thought, the results show that the mixture combiners we propose in this paper provides the best overall global results. In fact the best overall results is got by the mixture model composed by a simple ensemble as experts and a basic network as gating network *mix-se-bn*.

We can also see that the accuracy of the combiners proposed increases as the number of expert networks increases even if the performance of the ensemble (expert networks) do not increase. The performance of the 20-network ensemble trained with simple ensemble is lower than the 9-network one.

The combiners proposed are more robust than the traditional combiners. For instance, the *Zimmermann's operator* provides the best mean PER for the case of 3-network ensembles but it also provides the worst mean PER for the case of 40-network ensembles. The *Nash vote* also provides good results for the case of 3-network but the accuracy of the combiner in the other cases are better than *Simple Ensemble* but worse than the *Mixture Combiners* proposed.

5   Conclusions

In this paper, we have proposed two combiners based on the *Mixture of Neural Networks*. We have applied them to ensembles of *Multilayer Feedforward* networks previously trained with *Simple ensemble*. In the first combiner, *Mix-SE-BN*, we have applied the *Basic Network* as gating network to weight and combine the outputs provided by the networks of the ensemble previously trained with *Simple Ensemble*. In the second one, *Mix-SE-MF*, we have applied the *Multilayer Feedforward* network as gating network to combine the ensemble previously trained with *Simple ensemble*.

In our experiments we have compared the two new *Mixture combiners* with three different *Mixture models* and seven traditional combiners. In the first mixture model, *Mix-BN-BN*, the *Basic Network* is used as expert and gating networks. In the second, *Mix-MF-BN*, the *Multilayer Feedforward* network is used as expert networks whereas the *Basic Network* is used as gating network. In the last one, *Mix-MF-MF*, the *Multilayer Feedforward* network is used as expert and gating networks.

To compare the combiners proposed with the seven traditional combiners, we have used ensembles of 3, 9, 20 and 40 networks previously trained with *Simple Ensemble*. Then, to compare the Mixture Combiners with the Mixture Models, we have built the *Mixture models* with 3, 9, 20 and 40 experts and a single gating network.

Finally, we have calculated the mean *Increase of Performance* and the mean *Percentage of Error Reduction* with respect to a single *MF* network to compare all the methods.

According the general measurements, the *mixture combiners* on *Simple Ensemble* are the best way to build Multi-Net systems among the models and combiners studied in this paper. In fact, the best results are provided by ensembles of 20 networks trained with *Simple Ensemble* and combined with a *Basic network* as gating network, *Mix-SE-BN*. Moreover, the combiners proposed are more robust than the traditional ones. The *Zimmermann's operator* provides the best and the worst mean PER whereas the *Nash vote* provides, in the majority of cases, better results than *Simple Ensemble* but worse than the *mixture combiners* proposed.

Moreover, the mixture models (*Mix-BN-BN, Mix-MF-BN Mix-MF-MF*) perform worse than *Simple Ensemble* in general. In fact, *Mix-BN-BN* works worse than a single *MF* network in two cases, 3-network and 9-network ensembles. The complexity of the *mixture model* is its main problem, all the networks are trained at the same time and the experts tend to be more cooperative and less competitive. Some authors defend the idea that competitive systems provides good results and are less complex than cooperative systems. The *mixture combiners* are less complex, a previously trained ensemble is assigned to the expert networks and their weights are keep unchanged, only the gating network is trained so adding a new expert only requires the training of the new expert and the gating net.

We can conclude by remarking that the accuracy of an ensemble of *Multilayer feed-forward* networks can be improved by applying the gating network of the *Mixture of Neural Networks* as ensemble combiner.

References

1. Bishop, C.M.: Neural Networks for Pattern Recognition, New York, NY, USA. Oxford University Press, Inc., Oxford (1995)
2. Cho, S.-B., Kim, J.H.: Combining multiple neural networks by fuzzy integral for robust classification. IEEE Transactions on System, Man, and Cybernetics 25(2), 380–384 (1995)
3. Fernndez-Redondo, M., Hernndez-Espinosa, C., Torres-Sospedra, J.: Multilayer feedforward ensembles for classification problems. In: Pal, N.R., Kasabov, N., Mudi, R.K., Pal, S., Parui, S.K. (eds.) ICONIP 2004. LNCS, vol. 3316, pp. 744–749. Springer, Heidelberg (2004)
4. Gader, P.D., Mohamed, M.A., Keller, J.M.: Fusion of handwritten word classifiers. Pattern Recogn. Lett. 17(6), 577–584 (1996)
5. Ho, T.K., Hull, J.J., Srihari, S.N.: Decision combination in multiple classifier systems. IEEE Transactions on Pattern Analysis and Machine Intelligence 16(1), 66–75 (1994)
6. Jacobs, R.A.: Methods for combining experts' probability assessments. Neural Comput. 7(5), 867–888 (1995)
7. Jacobs, R.A., Jordan, M.I., Nowlan, S.J., Hinton, G.E.: Adaptive mixtures of local experts. Neural Computation 3, 79–87 (1991)
8. Jimenez, D.: Dynamically weighted ensemble neural networks for classification. In: Proceedings of the 1998 International Joint Conference on Neural Networks, IJCNN 1998, pp. 753–756 (1998)
9. Krogh, A., Vedelsby, J.: Neural network ensembles, cross validation, and active learning. In: Tesauro, G., Touretzky, D., Leen, T. (eds.) Advances in Neural Information Processing Systems, vol. 7, pp. 231–238. The MIT Press, Cambridge (1995)
10. Kuncheva, L.I.: Combining Pattern Classifiers: Methods and Algorithms. Wiley-Interscience, Chichester (2004)
11. Lam, L., Suen, C.Y.: Optimal combinations of pattern classifiers. Pattern Recogn. Lett. 16(9), 945–954 (1995)
12. Newman, D.J., Hettich, S., Blake, C.L., Merz, C.J.: UCI repository of machine learning databases (1998), http://www.ics.uci.edu/~mlearn/MLRepository.html
13. Sharkey, A.J. (ed.): Combining Artificial Neural Nets: Ensemble and Modular Multi-Net Systems. Springer, Heidelberg (1999)
14. Torres-Sospedra, J., Fernndez-Redondo, M., Hernndez-Espinosa, C.: A research on combination methods for ensembles of multilayer feedforward. In: IJCNN 2005 Proceedings, pp. 1125–1130 (2005)
15. Torres-Sospedra, J., Hernndez-Espinosa, C., Fernndez-Redondo, M.: Combinacin de conjuntos de redes MF. In: SICO 2005 Proceedings, pp. 11–18. Thomson (2005)
16. Torres-Sospedra, J., Hernndez-Espinosa, C., Fernndez-Redondo, M.: Designing a new multilayer feedforward modular network for classification problems. In: WCCI 2006 proceedings, pp. 2263–2268 (2006)
17. Tumer, K., Ghosh, J.: Error correlation and error reduction in ensemble classifiers. Connection Science 8(3-4), 385–403 (1996)
18. Verikas, A., Lipnickas, A., Malmqvist, K., Bacauskiene, M., Gelzinis, A.: Soft combination of neural classifiers: A comparative study. Pattern Recognition Letters 20(4), 429–444 (1999)

19. Wanas, N.M., Kamel, M.S.: Decision fusion in neural network ensembles. In: Proceedings of the 2001 International Joint Conference on Neural Networks, IJCNN 2001, vol. 4, pp. 2952–2957 (2001)
20. Xu, L., Krzyzak, A., Suen, C.Y.: Methods of combining multiple classifiers and their applications to handwriting recognition. IEEE Transactions on Systems, Man, and Cybernetics 22(3), 418–435 (1992)
21. Zimmermann, H.-J., Zysno, P.: Decision and evaluations by hierarchical aggregation of information. Fuzzy Sets and Systems 10(3), 243–260 (1984)

Combining Methods for Dynamic Multiple Classifier Systems

Amber Tomas

The University of Oxford, Department of Statistics
1 South Parks Road, Oxford OX2 3TG, United Kingdom

Abstract. Most of what we know about multiple classifier systems is based on empirical findings, rather than theoretical results. Although there exist some theoretical results for simple and weighted averaging, it is difficult to gain an intuitive feel for classifier combination. In this paper we derive a bound on the region of the feature space in which the decision boundary can lie, for several methods of classifier combination using non-negative weights. This includes simple and weighted averaging of classifier outputs, and allows for a more intuitive understanding of the influence of the classifiers combined. We then apply this result to the design of a multiple logistic model for classifier combination in dynamic scenarios, and discuss its relevance to the concept of diversity amongst a set of classifiers. We consider the use of pairs of classifiers trained on label-swapped data, and deduce that although non-negative weights may be beneficial in stationary classification scenarios, for dynamic problems it is often necessary to use unconstrained weights for the combination.

Keywords: Dynamic Classification, Multiple Classifier Systems, Classifier Diversity.

1   Introduction

In this paper we are concerned with methods of combining classifiers in multiple classifier systems. Because the performance of multiple classifier systems depends both on the component classifiers chosen and the method of combining, we consider both of these issues together. The methods of combining most commonly studied have been simple and weighted averaging of classifier outputs, in the latter case with the weights constrained to be non-negative. Tumer and Ghosh [8] laid the framework for theoretical analysis of simple averaging of component classifiers, and this was later extended to weighted averages by Fumera and Roli [2]. More recently, Fumera and Roli [3] have investigated the properties of component classifiers needed for weighted averaging to be a significant improvement on simple averaging. Although this work answers many questions about combining classifier outputs, it does not provide a framework which lends itself to an intuitive understanding of the problem.

The work presented here we hope goes some way to remedying this situation. We present a simple yet powerful result which can be used to recommend

L. Prevost, S. Marinai, and F. Schwenker (Eds.): ANNPR 2008, LNAI 5064, pp. 180–192, 2008.

a particular method of combination for a given problem and set of component classifiers. We then apply this result to dynamic classification problems. For the purposes of this paper, we define a *dynamic classification problem* as a classification problem where the process generating the observations is changing over time. Multiple classifier systems have been used on dynamic classification by many researchers. A summary of the approaches is given by Kuncheva [5].

The structure of this paper is as follows: in section 2 we present the model for classifier combination that we will be using. We then present our main result in section 3, and discuss its relevance to dynamic classification and classifier diversity. In section 4 we explore the use of component classifier pairs which disagree over the whole feature space, and then in section 5 demonstrate our results on an artificial example.

2   The Model

Because we are interested in dynamic problems, the model we use is time dependent. Elements which are time dependent are denoted by the use of a subscript t. We assume that the population of interest consists of K classes, labelled $1, 2, \ldots, K$. At some time t, an observation $\boldsymbol{x}_t$ and label y_t are generated according to the joint probability distribution $P_t(\boldsymbol{X}_t, Y_t)$. Given an observation $\boldsymbol{x}_t$, we denote the estimate output by the ith component classifier of $\text{Prob}\{Y_t = k | \boldsymbol{x}_t\}$ by $\hat{p}_i(k | \boldsymbol{x}_t)$, for $k = 1, 2, \ldots, K$ and $i = 1, 2, \ldots, M$.

Our final estimate of $P_t(Y_t | \boldsymbol{x}_t)$ is obtained by combining the component classifier outputs according to the multiple logistic model

$$\hat{p}_t(k | \boldsymbol{x}_t) = \frac{\exp(\boldsymbol{\beta}_t^T \boldsymbol{\eta}_k(\boldsymbol{x}_t))}{\sum_{i=1}^M \exp(\boldsymbol{\beta}_t^T \boldsymbol{\eta}_i(\boldsymbol{x}_t))}, \quad k = 1, 2, \ldots, K, \tag{1}$$

where $\boldsymbol{\beta}_t = (\beta_{t1}, \beta_{t2}, \ldots, \beta_{tM})$ is a vector of parameters, the ith component of $\boldsymbol{\eta}_k(\boldsymbol{x}_t), \eta_{ki}(\boldsymbol{x}_t)$, is a function of $\hat{p}_i(k | \boldsymbol{x}_t)$, and $\boldsymbol{\eta}_1(\boldsymbol{x}_t) = 0$ for all $\boldsymbol{x}_t$. In this model we use the same set of component classifiers for all t. Changes in the population over time are modelled by changes in the parameters of the model, $\boldsymbol{\beta}_t$.

Before we can apply (1) to a classification problem, we must specify the component classifiers as well as the form of the functions $\boldsymbol{\eta}_k(\boldsymbol{x}_t), k = 1, 2, \ldots, K$. Specifying the model in terms of the $\boldsymbol{\eta}_k(\boldsymbol{x}_t)$ allows flexibility for the form of the combining rule. In this paper we consider two options:

$$\text{1.} \quad \eta_{ki}(\boldsymbol{x}_t) = \hat{p}_i(k | \boldsymbol{x}_t) - \hat{p}_i(1 | \boldsymbol{x}_t), \quad \text{and} \tag{2}$$

$$\text{2.} \quad \eta_{ki}(\boldsymbol{x}_t) = \log\left(\frac{\hat{p}_i(k | \boldsymbol{x}_t)}{\hat{p}_i(1 | \boldsymbol{x}_t)}\right). \tag{3}$$

Both options allow $\eta_{ki}(\boldsymbol{x}_t)$ to take either positive or negative values. Note that when using option (3), the model (1) can be written as a linear combination of classifier outputs

$$\log\left(\frac{\hat{p}(k | \boldsymbol{x}_t)}{\hat{p}(1 | \boldsymbol{x}_t)}\right) = \sum_{i=1}^M \beta_{ti} \log\left(\frac{\hat{p}_i(k | \boldsymbol{x}_t)}{\hat{p}_i(1 | \boldsymbol{x}_t)}\right). \tag{4}$$

3   Bounding the Decision Boundary

In this section we present our main result. We consider how the decision boundary of a classifier based on (1) is related to the decision boundaries of the component classifiers. The following theorem holds only in the case of 0–1 loss, i.e. when
the penalty incurred for classifying an observation from class j as an observation
from class j' is defined by

$$L(j, j') = \begin{cases} 1 & \text{if } j \neq j' \\ 0 & \text{if } j = j' \end{cases}. \tag{5}$$

In this case, minimising the loss is equivalent to minimising the error rate of the
classifier. At time t, we classify $\boldsymbol{x}_t$ to the class with label $\hat{y}_t$, where

$$\hat{y}_t = \text{argmax}_k \; \hat{p}_t(k|\boldsymbol{x}_t), \tag{6}$$

and $\hat{p}_t(k|\boldsymbol{x}_t)$ is given by (1).

Theorem 1. *When using a 0–1 loss function and non-negative parameter values*
β_t, *the decision boundary of the classifier (6) must lie in regions of the feature*
space where the component classifiers "disagree".

Proof. Assuming 0–1 loss, the decision boundary of the ith component classifier
between the jth and j'th classes is a subset of the set

$$\{\boldsymbol{x} : \hat{p}_i(j|\boldsymbol{x}) = \hat{p}_i(j'|\boldsymbol{x})\}. \tag{7}$$

Define $\mathcal{R}^i_j$ as the region of the feature space in which the ith component classifier
would classify an observation as class j. That is,

$$\mathcal{R}^i_j = \{\boldsymbol{x} : j = \text{argmax}_c \; \hat{p}_i(c|\boldsymbol{x})\}, \; j = 1, 2, \ldots, K. \tag{8}$$

Hence for all $\boldsymbol{x} \in \mathcal{R}^i_j$,

$$\hat{p}_i(j|\boldsymbol{x}) > \hat{p}_i(j'|\boldsymbol{x}), \text{ for } j \neq j'. \tag{9}$$

Define

$$\mathcal{R}^*_j = \cap_i \; \mathcal{R}^i_j. \tag{10}$$

Then for all i, for all $\boldsymbol{x} \in \mathcal{R}^*_j$,

$$\hat{p}_i(j|\boldsymbol{x}) > \hat{p}_i(j'|\boldsymbol{x}). \tag{11}$$

From (11), for $\beta_{ti} \geq 0, i = 1, 2, \ldots, M$, it follows that for all $\boldsymbol{x} \in \mathcal{R}^*_j$,

$$\sum_{i=1}^{M} \beta_{ti}\{\hat{p}_i(j|\boldsymbol{x}) - \hat{p}_i(1|\boldsymbol{x})\} > \sum_{i=1}^{M} \beta_{ti}\{\hat{p}_i(j'|\boldsymbol{x}) - \hat{p}_i(1|\boldsymbol{x})\}. \tag{12}$$

Similarly, from (11), we can show that for $\beta_{ti} \geq 0, i = 1, 2, \ldots, M$, for $\boldsymbol{x} \in \mathcal{R}_j^*$,

$$\sum_{i=1}^{M} \beta_{ti} \log \left(\frac{\hat{p}_i(j|\boldsymbol{x})}{\hat{p}_i(1|\boldsymbol{x})} \right) > \sum_{i=1}^{M} \beta_{ti} \log \left(\frac{\hat{p}_i(j'|\boldsymbol{x})}{\hat{p}_i(1|\boldsymbol{x})} \right). \tag{13}$$

For the classification model (6), the decision boundary between the jth and j'th classes can be written as

$$\{\boldsymbol{x} : \boldsymbol{\beta}_t^T \boldsymbol{\eta}_j(\boldsymbol{x}_t) = \boldsymbol{\beta}_t^T \boldsymbol{\eta}_{j'}(\boldsymbol{x}_t)\}. \tag{14}$$

Therefore, for the definitions of $\boldsymbol{\eta}_j(\boldsymbol{x}_t)$ considered in section 2, we can see from (12) and (13) that $\mathcal{R}_j^*$ does not intersect with the set (14). That is, there is no point on the decision boundary of our final classifier that lies in the region where all component classifiers agree.

Note that from (11) it is easy to show that this result also holds for the combining rules used in [8] and [2], in which case the result can be extended to any loss function. For example, figure 1 shows the decision boundaries of three component classifiers in a two-class problem with two-dimensional feature space. The shaded areas represent the regions of the feature space where all component classifiers agree, and therefore the decision boundary of the classifier must lie outside of these shaded regions.

This result helps us to gain an intuitive understanding of classifier combination in simple cases. If the Bayes boundary does not lie in the region of disagreement of the component classifiers, then the classifier is unlikely to do well. If the

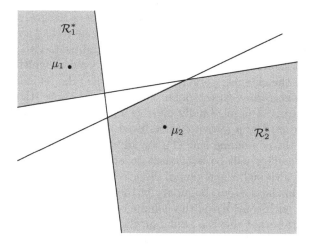

Fig. 1. The decision boundaries of the component classifiers are shown in black, and the regions in which they all agree are shaded grey. $\boldsymbol{\mu}_1$ and $\boldsymbol{\mu}_2$ denote the means of classes one and two respectively. When using non-negative parameter values, the decision boundary of the classifier (6) must lie outside of the shaded regions.

region of disagreement does contain the Bayes boundary (at least in the region of highest probability density), then the smaller this region the closer the decision boundary of the classifier must be to the optimal boundary. However, clearly in practice we do not know the location of the Bayes boundary. If the component classifiers are unbiased, then they should "straddle" the Bayes boundary. If the component classifiers are biased, then the Bayes boundary may lie outside the region of disagreement, and so it is possible that one of the component classifiers will have a lower error rate than a simple average of the classifier outputs. In this case, using a weighted average should result in improved performance over the simple average combining rule. This corresponds to the conclusions of Fumera and Roli [3].

3.1   Relevance to Dynamic Scenarios

If the population of interest is dynamic, then in general so is the Bayes boundary and hence optimal classifier [4]. However, because our model uses the same set of component classifiers for all time points, the region of disagreement is fixed. Therefore, even if the Bayes boundary is initially contained within the region of disagreement, after some time this may cease to be the case. If the Bayes boundary moves outside the region of disagreement, then it is likely the performance of the classifier will deteriorate. Therefore, if using non-negative weights, it is important to ensure the region of disagreement is as large as possible when selecting the component classifiers for a dynamic problem.

3.2   On the Definition of Diversity

Consider defining the diversity of a set of classifiers as the volume of the feature space on which at least two of the component classifiers disagree, i.e. the "region of disagreement" discussed above. Initially this may seem like a reasonable definition. However, it is easy to construct a counter example to its appropriateness. Consider two classifiers c_1 and c_2 on a two class problem which are such that whenever one of the classifiers predicts class one, the other classifier will predict class two. Then according to the definition suggested above, the set of component classifiers $\{c_1, c_2\}$ is maximally diverse. This set is also maximally diverse according to the *difficulty* measure introduced by Kuncheva and Whitaker [6]. However, using the combining rule (1), for all values of the parameters β_{t1} and β_{t2}, the final classifier will be equivalent to either c_1 or c_2 (this is proved in section 4). Thus although the region of disagreement is maximised, there is very little flexibility in the decision boundary of the classifier as β_t varies.

The problem with considering the volume of the region of disagreement as a diversity measure is that this is a bound on the flexibility of the combined classifier. Ideally, a measure of diversity would reflect the actual variation in decision boundaries that it is possible to obtain with a particular set of classifiers and combining rule. However, the region of disagreement is still a useful concept for the design of dynamic classifiers. For a classifier to perform well on a dynamic scenario it is necessary that the region of disagreement is maximised as well as the flexibility of the decision boundary within that region. One way to improve the flexibility of the decision boundary whilst maximising the region of disagreement (and maintaining an optimal level of "difficulty" amongst the component classifiers) is now discussed.

4   Label-Swapped Component Classifiers

Consider again the pair of component classifiers discussed in section 3.2 which when given the same input on a two-class problem will always output different labels. One way in which to produce such a pair of classifiers is to train both classifiers on the same data, except that the labels of the observations are reversed for the second classifier. We refer to a pair of classifiers trained in this way as a *label-swapped pair*.

In this section we consider combining several pairs of label-swapped classifiers on a two-class problem. The region of disagreement is maximised (as each pair disagrees over the entire feature space), and we increase the flexibility of the decision boundary within the feature space by combining several such pairs.

Suppose we combine M pairs of label-swapped classifiers using model (1), so that we have $2M$ component classifiers in total. An observation $\boldsymbol{x}_t$ is classified as being from class 1 if $\hat{p}_t(1|\boldsymbol{x}_t) > \hat{p}_t(2|\boldsymbol{x}_t)$, i.e.

$$\sum_{i=1}^{2M} \beta_{ti}\eta_{2i}(\boldsymbol{x}_t) < 0. \tag{15}$$

Theorem 2. *Suppose $\eta_{2i}(\boldsymbol{x}_t) > 0$ if and only if $\hat{p}_i(2|\boldsymbol{x}_t) > \hat{p}_i(1|\boldsymbol{x}_t)$, and that*

$$\eta_{22}(\boldsymbol{x}_t) = -\eta_{21}(\boldsymbol{x}_t). \tag{16}$$

Then the classifier obtained by using (6) with two label-swapped classifiers c_1 and c_2 and parameters β_{t1} and β_{t2} is equivalent to the classifier c_i, where $i = \text{argmax}_j \beta_{tj}$.

Proof. From (15), with $M = 1$, we see that $\hat{p}_t(1|\boldsymbol{x}_t) > \hat{p}_t(2|\boldsymbol{x}_t)$ whenever

$$(\beta_{t1} - \beta_{t2})\eta_{21}(\boldsymbol{x}_t) < 0. \tag{17}$$

Therefore, $\hat{p}_t(1|\boldsymbol{x}_t) > \hat{p}_t(2|\boldsymbol{x}_t)$ when either

$$\beta_{t1} < \beta_{t2} \text{ and } \eta_{21}(\boldsymbol{x}_t) > 0,$$
$$\text{or} \quad \beta_{t1} > \beta_{t2} \text{ and } \eta_{21}(\boldsymbol{x}_t) < 0,$$

i.e. when

$$\beta_{t1} < \beta_{t2} \text{ and } \hat{p}_2(1|\boldsymbol{x}_t) > \hat{p}_2(2|\boldsymbol{x}_t),$$
$$\text{or} \quad \beta_{t1} > \beta_{t2} \text{ and } \hat{p}_1(1|\boldsymbol{x}_t) > \hat{p}_1(2|\boldsymbol{x}_t).$$

So if $\beta_{t1} > \beta_{t2}$, the combined classifier is equivalent to using only c_1, and if $\beta_{t2} > \beta_{t1}$ the combined classifier is equivalent to using only c_2.

Note that the conditions required by theorem 2 hold for the two definitions of $\eta_2(\boldsymbol{x}_t)$ recommended in section 2, namely

$$\eta_{2i}(\boldsymbol{x}_t) = \log\left(\frac{\hat{p}_i(2|\boldsymbol{x}_t)}{\hat{p}_i(1|\boldsymbol{x}_t)}\right), \text{ and}$$

$$\eta_{2i}(\boldsymbol{x}_t) = \hat{p}_i(2|\boldsymbol{x}_t) - \hat{p}_i(1|\boldsymbol{x}_t).$$

Corollary 1. *For all β_{t1} and β_{t2}, when using label-swapped component classifiers c_1 and c_2, the decision boundary of the combined classifier (6) is the same as the decision boundary of c_1 (and c_2).*

This follows directly from theorem 2.

Now suppose we combine M pairs of label-swapped classifiers and label them such that c_{2i} is the label-swapped partner of c_{2i-1}, for $i = 1, 2, \ldots, M$.

Theorem 3. *Using M pairs of label-swapped classifiers with parameters β_{t1}, $\beta_{t2}, \ldots, \beta_{t,2M}$ is equivalent to the model which uses only classifiers c_1, $c_3, \ldots, c_{2M-1}$ with parameters $\beta_{t1}^*, \beta_{t2}^*, \ldots, \beta_{tM}^*$, where*

$$\beta_{ti}^* \overset{\triangle}{=} \beta_{t,2i-1} - \beta_{t,2i}. \tag{18}$$

Proof. From (15),

$$\hat{p}_t(1|\boldsymbol{x}_t) > \hat{p}_t(2|\boldsymbol{x}_t) \tag{19}$$

when

$$\sum_{i=1}^{2M} \beta_{ti}\eta_{2i}(\boldsymbol{x}_t) < 0, \tag{20}$$

i.e. when

$$(\beta_{t1} - \beta_{t2})\eta_{21}(\boldsymbol{x}_t) + (\beta_{t3} - \beta_{t4})\eta_{23}(\boldsymbol{x}_t)$$
$$+ \ldots + (\beta_{t,2M-1} - \beta_{t,2M})\eta_{2(2M-1)}(\boldsymbol{x}_t) < 0$$

i.e. when

$$\sum_{i=1}^{M} \beta_{ti}^*\eta_{2(2i-1)}(\boldsymbol{x}_t) < 0, \tag{21}$$

where

$$\beta_{ti}^* = \beta_{t,2i-1} - \beta_{t,2i}. \tag{22}$$

Comparing (21) with (15), we can see this is equivalent to the classifier which combines $c_1, c_3, \ldots, c_{2M-1}$ with parameters $\beta_{t1}^*, \beta_{t2}^*, \ldots, \beta_{tM}^*$.

Importantly, although the β_{tj} may be restricted to taking non-negative values, in general the β_{ti}^* can take negative values. Hence we have shown that using label-swapped component classifiers and non-negative parameter estimates is equivalent to a classifier with unconstrained parameter estimates and which does not use label-swapped pairs. However, because in practice we must estimate the parameter values for a given set of component classifiers, using label-swapped classifiers with a non-negativity constraint will not necessarily give the same classification performance as using non-label-swapped classifiers with unconstrained parameter estimates. For example, LeBlanc and Tibshirani [7] and

Breiman [1] reported improved classification performance when constraining the parameters of the weighted average of classifier outputs to be non-negative. The benefit of the label-swapped approach is that it combines the flexibility of unconstrained parameters required for dynamic problems with the potential improved accuracy of parameter estimation obtained when imposing a non-negativity constraint. Clearly then the benefit of using label-swapped classifiers (if any) will be dependent on the algorithm used for parameter estimation.

It is important to note that using label-swapped classifiers leads us to requiring twice as many component classifiers and hence parameters as the corresponding model with unconstrained parameters. In addition, the additional computational effort involved in enforcing the non-negativity constraint means that the label-swapped approach is significantly more computationally intensive than using standard unconstrained estimates.

5   Example

In this section we demonstrate some of our results on an artificial dynamic classification problem. There exist two classes, class 1 and class 2, and observations from each class are distributed normally with a common covariance matrix. The probability that an observation is generated from class 1 is 0.7. At time $t = 0$ the mean of class 1 is $\mu_1 = (1, 1)$, and the mean of class 2 is $\mu_2 = (-1, -1)$. The mean of population 1 changes in equal increments from $(1, 1)$ to $(1, -4)$ over 1000 time steps, so that at time t, $\mu_1 = (1, 1 - 0.005t)$. It is assumed that observations arrive independently without delay, and that after each classification is made the true label of the observation is revealed before the following observation arrives.

Before we can apply the classification model (1) to this problem, we need to decide on how many component classifiers to use, train the component classifiers, decide on the form of $\eta_k(x_t)$ and decide on an algorithm to estimate the parameter values β_t for every t. Clearly each one of these tasks requires careful thought in order to maximise the performance of the classifier. However, because this is not the subject of this paper, we omit the details behind our choices. For the following simulations we used three component classifiers, each of which was trained using linear discriminant analysis on an independent random sample of 10 observations from the population at time $t = 0$. Figure 2 shows the Bayes boundary at times $t = 0$, $t = 500$ and time $t = 1000$, along with the decision boundaries of the three component classifiers. We choose to use $\eta_k(x_t)$ defined by (3) (so for this example the decision boundary of the classifier is also linear), and used a particle filter approximation to the posterior distribution of β_t at each time step. The model for parameter evolution used was

$$\beta_{t+1} = \beta_t + \omega_t, \tag{23}$$

where ω_t has a normal distribution with mean $\mathbf{0}$ and covariance matrix equal to 0.005 times the identity matrix. 300 particles were used for the approximation. An observation x_t was classified as belonging to class k if

$$k = \operatorname{argmax}_j \hat{E}_{\beta_t} [\hat{p}_t(j|x_t)]. \tag{24}$$

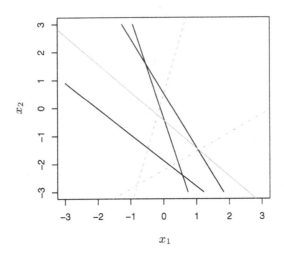

Fig. 2. The decision boundaries of the component classifiers (black) and the Bayes boundary (grey) at times $t = 0$ (solid), $t = 500$ (dashed) and $t = 1000$ (dot-dashed)

Denote by Err_{-i} the error rate of the ith component classifier on the training data of the other component classifiers, for $i = 1, 2, 3$. The value of β_{0i} was chosen to be proportional to $1 - \mathrm{Err}_{-i}$ for $i = 1, 2, 3$. Each simulation involved repeating the data generation, classification and updating procedure 100 times, and the errors of each run were averaged to produce an estimate of the error rate of the classifier at every time t.

We repeated the simulation three times. In the first case, we constrained the parameters β_t to be non-negative. A smooth of the estimated average error rate is shown in figure 3(a), along with the Bayes error (grey line) and the error of the the component classifier corresponding to the smallest value of Err_{-i} (included to demonstrate the deterioration in performance of the "best" component classifier at time $t = 0$, dashed line). The error rate of the classifier is reasonably close to the Bayes error for the first 200 updates, but then the performance deteriorates. After $t = 200$, the Bayes boundary has moved enough that it can no longer be well approximated by a linear decision boundary lying in the region of disagreement.

In the second case, we use the same three classifiers as above, but include their label-swapped pairs. We hence have six component classifiers in total. A smooth of the estimated error rate is shown in figure 3(b), and it is clear that this classifier does not succumb to the same level of decreased performance as seen in figure 3(a).

Thirdly, we used the same set of three component classifiers but without constraining the parameter values to be non-negative. The resulting error rate, shown in figure 3(c), is very similar to that using label-swapped classifiers.

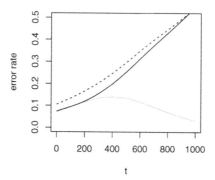

(a) Non-negative parameter values

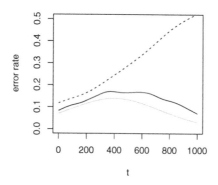

(b) Label-swapped classifiers

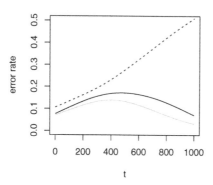

(c) Unconstrained parameter values

Fig. 3. Smoothed average error rates for the three scenarios described (solid black line) along with the Bayes error (grey) and error rate of the component classifier with the lowest estimated error rate at time $t = 0$ (dashed black line)

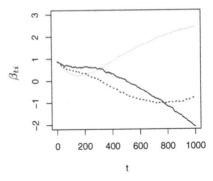

(a) Unconstrained parameter values

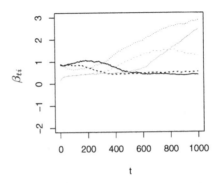

(b) Label-swapped classifiers: $\boldsymbol{\beta}_t$

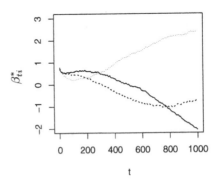

(c) Label-swapped classifiers: $\boldsymbol{\beta}_t^*$

Fig. 4. Average expected parameter values β_i, for $i = 1$ (solid line), $i = 2$ (dashed) and $i = 3$ (dotted). In figure 4(b) the grey and black lines of each type correspond to a label-swapped pair.

In figure 4, we show the average expected parameter values returned by the updating algorithm in cases two and three. Clearly the values of β^*_{ti} shown in figure 4(c) are very similar to the unconstrained parameter values in figure 4(a), which explains the similarity of classification performance between the label-swapped and unconstrained cases. Furthermore, we can see from figure 4 that negative parameter values become necessary after about 200 updates, again explaining the behaviour seen in figure 3(a).

This example shows that it is important to consider the region of disagreement in dynamic classification problems. Furthermore, we found no clear difference in performance between the classifier using label-swapped component classifiers with non-negative parameter values, and the classifier using unconstrained parameter estimates.

6   Conclusions

When using a combining model of the form (1) or a linear combiner with non-negative parameter values, it can be useful to consider the region of disagreement of the component classifiers. This becomes of even greater relevance when the population is believed to be dynamic, as the region of disagreement is a bound on the region in which the decision boundary of the classifier can lie. If the Bayes boundary lies outside the region of disagreement, then it is unlikely that the classifier will perform well. In stationary problems it may be beneficial to constrain the region of disagreement. However, in dynamic scenarios when the Bayes boundary is subject to possibly large movement, it seems most sensible to maximise this region. This can be done for a two-class problem by using label-swapped classifiers with non-negative parameter estimates, or more simply and efficiently by allowing negative parameter values. Which of these approaches results in better classification performance is likely to depend on the parameter estimation algorithm, and should be further investigated.

References

1. Breiman, L.: Stacked Regressions. Machine Learning 24, 49–64 (1996)
2. Fumera, G., Roli, F.: Performance Analysis and Comparison of Linear Combiners for Classifier Fusion. In: Caelli, T.M., Amin, A., Duin, R.P.W., Kamel, M.S., de Ridder, D. (eds.) SPR 2002 and SSPR 2002. LNCS, vol. 2396, pp. 424–432. Springer, Heidelberg (2002)
3. Fumera, G., Roli, F.: A Theoretical and Experimental Analysis of Linear Combiners for Multiple Classifier Systems. IEEE Trans. Pattern Anal. Mach. Intell. 27(6), 942–956 (2005)
4. Kelly, M., Hand, D., Adams, N.: The Impact of Changing Populations on Classifier Performance. In: KDD 1999: Proc. 5th ACM SIGKDD International Conference on Knowledge Discovery and Data Mining, San Diego, California, United States, pp. 367–371. ACM, New York (1999)
5. Kuncheva, L.I.: Classifier Ensembles for Changing Environments. In: Roli, F., Kittler, J., Windeatt, T. (eds.) MCS 2004. LNCS, vol. 3077, pp. 1–15. Springer, Heidelberg (2004)

6. Kuncheva, L.I., Whitaker, C.J.: Measures of Diversity in Classifier Ensembles and Their Relationship with the Ensemble Accuracy. Machine Learning 51, 181–207 (2003)
7. Le Blanc, M., Tibshirani, R.: Combining Estimates in Regression and Classification. Technical Report 9318, Dept. of Statistics, Univ. of Toronto (1993)
8. Tumer, K., Ghosh, J.: Analysis of Decision Boundaries in Linearly Combined Neural Classifiers. Pattern Recognition 29, 341–348 (1996)

Researching on Multi-net Systems Based on Stacked Generalization

Carlos Hernández-Espinosa, Joaquín Torres-Sospedra,
and Mercedes Fernández-Redondo

Departamento de Ingenieria y Ciencia de los Computadores, Universitat Jaume I,
Avda. Sos Baynat s/n, C.P. 12071, Castellon, Spain
{espinosa,jtorres,redondo}@icc.uji.es

Abstract. Among the approaches to build a Multi-Net system, *Stacked Generalization* is a well-known model. The classification system is divided into two steps. Firstly, the level-0 generalizers are built using the original input data and the class label. Secondly, the level-1 generalizers networks are built using the outputs of the level-0 generalizers and the class label. Then, the model is ready for pattern recognition. We have found two important adaptations of *Stacked Generalization* that can be applyied to artificial neural networks. Moreover, two combination methods, *Stacked* and *Stacked+*, based on the *Stacked Generalization* idea were successfully introduced by our research group. In this paper, we want to empirically compare the version of the original *Stacked Generalization* along with other traditional methodologies to build Multi-Net systems. Moreover, we have also compared the combiners we proposed. The best results are provided by the combiners *Stacked* and *Stacked+* when they are applied to ensembles previously trained with *Simple Ensemble*.

1 Introduction

Perhaps, the most important property of a neural network is the generalization capability. The ability to correctly respond to inputs which were not used in the training set.

It is clear from the bibliography that the use of an ensemble of neural networks increases the generalization capability, [13] and [7], for the case of *Multilayer Feedforward* and other classifiers. The two key factors to design an ensemble are how to train the individual networks and how to combine them.

Although most of the methods to create a Multi-Net System are based on the *Ensemble approach* (*Boosting, Bagging, Cross-Validation*) [2,4] or in the *Modular approach* (*Mixture of Neural Networks*) [8,12], we have also analyzed other complex methodologies and models to perform an exhaustive comparison. *Stacked Generalization* is one of the most known alternatives to the traditional methods previously mentioned.

Stacked Generalization was introduced by Wolpert in 1994 [14]. Firstly, a set of cross-validated generalizers called *level-0 generalizers* are trained with the original input data and class label. Then, a set of generalizers called *level-1 generalizers* are trained using the information provided by *level-0* generalizers along with the class label. Unfortunately, *Stacked Generalization* can not be directly applied to artificial

L. Prevost, S. Marinai, and F. Schwenker (Eds.): ANNPR 2008, LNAI 5064, pp. 193–204, 2008.
© Springer-Verlag Berlin Heidelberg 2008

neural networks because it can not be applied to methods that require being trained before classification.

Some authors like Ghorbani & Owrangh [3] and Ting & Witten [9,10] have proposed some versions based on the Wolpert's model which can be directly applied to neural networks. Moreover, we proposed in [11] two combination methods based on *Stacked Generalization* model, *Stacked* and *Stacked+*. Unfortunately there is not any comparison among them.

The original *Stacked Generalization* can be incorrectly identified as a method to combine an ensemble. In the original model specifications, the process to create the combiners highly depends on the process to create the experts so the *stacked models* can not be considered as a traditional ensemble. Moreover, the *stacked combiners* we proposed should not be considered as pure 2-leveled models.

The *Stacked Generalization* model applied to neural networks is shown in figure 1. We can see that the *level-0 generalizers* are called *Expert Networks* (*EN*) whereas the *level-1 generalizers* are called *Combination Networks* (*CN*).

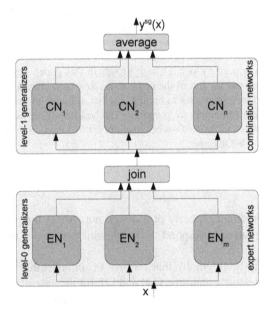

Fig. 1. Stacked Generalization Model

In this paper, we want to compare Ghorbani & Owrangh and Ting & Witten models with other ensemble models and with combiners *Stacked Stacked+*. To perform this comparison, ten databases from the UCI repository have been chosen.

This paper is organized as follows. Firstly, some theoretical concepts are briefly reviewed in section 2. Then, the databases used and the experimental setup are described in section 3. Finally, the experimental results and their discussion are in section 4.

2   Theory

2.1   The Multilayer Feedforward Network

The *Multilayer Feedforward* architecture is the most known network architecture. This kind of networks consists of three layers of computational units. The neurons of the first layer apply the identity function whereas the neurons of the second and third layer apply the sigmoid function. It has been proved that *MF* networks with one hidden layer and threshold nodes can approximate any function with a specified precision [1] and [5]. Figure 2 shows the diagram of the *Multilayer Feedforward* network.

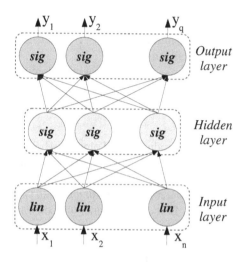

Fig. 2. Multilayer Feedforward Structure

In our experiments we have trained the networks with the following algorithm.

Algorithm 1. Neural Network Training $\{T, V\}$

Set initial weights values
for $i = 1$ to *iterations* **do**
  Train the network with the patterns from the training set T
  Calculate MSE over validation set V
  Save epoch weights and calculated MSE
end for
Select epoch with minimum MSE
Assign best epoch configuration to the network
Save network configuration

2.2   Traditional Multi-net Models

Simple Ensemble. A simple ensemble can be constructed by training different networks with the same training set, but with different random initialization [2].

Bagging. This method consists on generating different datasets drawn at random with replacement from the original training set [2].

Adaptive Boosting. In *Adaptive Boosting*, also known as *Adaboost*, the successive networks are trained with a training set selected randomly from the original set, but the probability of selecting a pattern changes depending on the correct classification of the pattern and on the performance of the last trained network [2].

Cross-Validation. We have used two different versions of k-fold cross-validation: *CVC* [2] and *CVCv2* [4]. In CVC, the training set is divided into k subsets being $k-1$ subsets used to train the network. In this case, the subset which is left out is not used in the training and validation process. In *CVCv2*, cross-validation is applied to the learning set to get k different subsets. In this case, $k-1$ subsets are used to train the network and the left one is used for validation.

In both cases, we can construct k classifiers with different sets by changing the subset that is left out.

2.3   Stacked Generalization

Stacked Generalization was introduced by Wolpert [14]. Some authors have adapted the Wolpert's method to use with neural networks like Ghorbani & Owrangh [3] and Ting & Witten [9,10].

The training in *Stacked Generalization* is divided into two steps. In the first one, the expert networks are trained. In the second one, the combination networks are trained with the outputs provided by the experts. However, there are some constraints related to the datasets used to train the expert and combination networks. The method is fully described in [14]. In fact, it has been suggested that *Stacked Generalization* is a sophisticated version of *Cross-Validation*.

Unfortunately, the method proposed by Wolpert can not be directly applied to generalizers that requires being trained before classificating patterns. Although this drawback, there are some authors that have adapted *Stacked Generalization* to specific classifiers. There are two versions that have to be taken into account: The version proposed by Ghorbani & Owrangh [3] and the version proposed by Ting & Witten [9,10]. Those authors described their procedure to create the different training sets for the experts and combination networks.

Stacked Generalization - Version 1. Ting & Witten proposed a version of *Stacked Generalization* that can be applied to the *Multilayer Feedforward* architecture [9,10]. The training set was randomly splitted into k equal subsets: $T = \{T_1, T_2, ..., T_k\}$. Then, $T^{-j} = \{T - T_j\}$ was used to train the experts networks and the experts output on T_j were used to train the combination networks.

Stacked Generalization - Version 2. Ghorbani & Owrangh proposed a version of *Stacked Generalization* that was applied directly to Artificial Neural Networks [3]. They applied cross-validation to create the different training sets of the experts by randomly splitting the training set into k equal subsets: $T = \{T_1, T_2, ..., T_k\}$. With this procedure,

k different classifiers can be built with different training sets by changing the subset that is left out as in *CVC*. Then, the outputs of the experts on T were used to train the combination networks.

Combining networks with Stacked and Stacked+. Moreover, *Stacked Generalization* can be used as a combiner of ensembles of neural networks as we proposed in [11] in which we introduced *Stacked* and *Stacked+*.

In those combiners, the networks of an ensemble previously trained with *Simple Ensemble* were used as expert networks. Then, the outputs provided by the ensemble on the original training set were used to train the combination networks. Additionally, in the combiner *stacked+*, the original input data was also used to train the combination network.

3   Experimental Setup

To test the performance of the methods described in this paper, we have used ensembles of 3, 9, 20 and 40 expert networks. In the case of two versions of *Stacked Generalization* we have used classification systems with 3, 9, 20 and 40 expert networks that have been combined by a single combination network. Finally, in the case of *Stacked* and *Stacked+*, we have used ensembles of 3, 9, 20 and 40 networks previously trained with *Simple Ensemble*, these networks have been combined by a single combination network.

In addition, the whole learning process have been repeated ten times, using different partitions of data in training, validation and test sets. With this procedure we can obtain a mean performance of the ensemble and the error calculated by standard error theory.

3.1   Databases

We have used ten different classification problems from the *UCI repository of machine learning databases* [6] to test the performance of the methods and combiners reviewed in this paper. These databases are:

- *Arrhythmia Database* (aritm)
- *Dermatology Database* (derma)
- *Ecoli Database* (ecoli)
- *Solar Flares Database* (flare)
- *Image segmentation Database* (img)
- *Ionosphere Database* (ionos)
- *Pima Indians Diabetes Database* (pima)
- *Haberman's Survival Data* (survi)
- *Vowel Database* (vowel)
- *Wisconsin Breast Cancer Database* (wdbc)

All the training parameters (Hidden units, Adaptation step, Momentum rate and Number of iterations of *Back-propagation*) have been set after an exhaustive trial and error procedure. Due to the lack of space, the parameters are omitted.

4   Results and discussion

4.1   Results

The main results of our work are presented in this subsection. Tables 1-5 show the performance of the ensembles of 3, 9, 20 and 40 networks trained with the traditional ensemble methods described in subsection 2.2.

Table 1. Simple Ensemble results

	3 Nets	9 Nets	20 Nets	40 Nets
aritm	73.4 ± 1	73.8 ± 1.1	73.8 ± 1.1	73.8 ± 1.1
derma	97.2 ± 0.7	97.5 ± 0.7	97.3 ± 0.7	97.6 ± 0.7
ecoli	86.6 ± 0.8	86.9 ± 0.8	86.9 ± 0.8	86.9 ± 0.7
flare	81.8 ± 0.5	81.6 ± 0.4	81.5 ± 0.5	81.6 ± 0.5
img	96.5 ± 0.2	96.7 ± 0.3	96.7 ± 0.2	96.8 ± 0.2
ionos	91.1 ± 1.1	90.3 ± 1.1	90.4 ± 1	90.3 ± 1
pima	75.9 ± 1.2	75.9 ± 1.2	75.9 ± 1.2	75.9 ± 1.2
survi	74.3 ± 1.3	74.2 ± 1.3	74.3 ± 1.3	74.3 ± 1.3
vowel	88 ± 0.9	91 ± 0.5	91.4 ± 0.8	92.2 ± 0.7
wdbc	96.9 ± 0.5	96.9 ± 0.5	96.9 ± 0.5	96.9 ± 0.5

Table 2. Bagging results

	3 Nets	9 Nets	20 Nets	40 Nets
aritm	74.7 ± 1.6	75.9 ± 1.7	75.9 ± 1.7	74.7 ± 1.5
derma	97.5 ± 0.6	97.7 ± 0.6	97.6 ± 0.6	97.6 ± 0.6
ecoli	86.3 ± 1.1	87.2 ± 1	87.1 ± 1	86.9 ± 1.1
flare	81.9 ± 0.6	82.4 ± 0.6	82.2 ± 0.5	82 ± 0.6
img	96.6 ± 0.3	96.7 ± 0.3	97 ± 0.3	97.1 ± 0.3
ionos	90.7 ± 0.9	90.1 ± 1.1	89.6 ± 1.1	90 ± 1.1
pima	76.9 ± 0.8	76.6 ± 0.9	77 ± 1	77 ± 1.1
survi	74.2 ± 1.1	74.4 ± 1.5	74.6 ± 1.7	74.2 ± 1.3
vowel	87.4 ± 0.7	90.8 ± 0.7	91.3 ± 0.6	91.2 ± 0.8
wdbc	96.9 ± 0.4	97.3 ± 0.4	97.5 ± 0.4	97.4 ± 0.3

Tables 6-7 show the results we have obtained with the combiners *Stacked* and *Stacked+* on ensembles previoulsly trained with *Simple Ensemble*.

Tables 8-9 show the results related to the first version of *Stacked Generalization*. These tables show the performance of the experts combined as an ensemble and the performance of the whole model.

In a similar way, tables 10-11 show the results related to the second version of *Stacked Generalization*.

Table 3. Adaptive Boosting results

	3 Nets	9 Nets	20 Nets	40 Nets
aritm	71.8 ± 1.8	73.2 ± 1.6	71.4 ± 1.5	73.8 ± 1.1
derma	98 ± 0.5	97.3 ± 0.5	97.5 ± 0.6	97.8 ± 0.5
ecoli	85.9 ± 1.2	84.7 ± 1.4	86 ± 1.3	85.7 ± 1.4
flare	81.7 ± 0.6	81.1 ± 0.7	81.1 ± 0.8	81.1 ± 0.7
img	96.8 ± 0.2	97.3 ± 0.3	97.29 ± 0.19	97.3 ± 0.2
ionos	88.3 ± 1.3	89.4 ± 0.8	91.4 ± 0.8	91.6 ± 0.7
pima	75.7 ± 1	75.5 ± 0.9	74.8 ± 1	73.3 ± 1
survi	75.4 ± 1.6	74.3 ± 1.4	74.3 ± 1.5	73 ± 2
vowel	88.4 ± 0.9	94.8 ± 0.7	96.1 ± 0.7	97 ± 0.6
wdbc	95.7 ± 0.6	95.7 ± 0.7	96.3 ± 0.5	96.7 ± 0.9

Table 4. CVC results

	3 Nets	9 Nets	20 Nets	40 Nets
aritm	74 ± 1	74.8 ± 1.3	74.8 ± 1.3	73.4 ± 1.9
derma	97.3 ± 0.7	97.6 ± 0.6	97.3 ± 0.6	97.3 ± 0.6
ecoli	86.8 ± 0.8	87.1 ± 1	86.5 ± 1	86.8 ± 0.9
flare	82.7 ± 0.5	81.9 ± 0.6	81.7 ± 0.7	81.7 ± 0.7
img	96.4 ± 0.2	96.6 ± 0.2	96.8 ± 0.2	96.6 ± 0.2
ionos	87.7 ± 1.3	89.6 ± 1.2	89.6 ± 1.3	88.3 ± 1
pima	76 ± 1.1	76.9 ± 1.1	76.2 ± 1.3	76.6 ± 1
survi	74.1 ± 1.4	75.2 ± 1.5	73.8 ± 0.9	74.6 ± 1
vowel	89 ± 1	90.9 ± 0.7	91.9 ± 0.5	92.2 ± 0.8
wdbc	97.4 ± 0.3	96.5 ± 0.5	97.4 ± 0.4	96.8 ± 0.5

Table 5. CVCv2 results

	3 Nets	9 Nets	20 Nets	40 Nets
aritm	76.1 ± 1.6	75.4 ± 1.5	74.3 ± 1.2	77 ± 0.8
derma	98 ± 0.3	97.3 ± 0.5	97.5 ± 0.6	97.2 ± 0.6
ecoli	86.8 ± 0.9	86.6 ± 1	86.6 ± 1.1	86.3 ± 0.9
flare	82.5 ± 0.6	82.1 ± 0.5	81.8 ± 0.4	82.2 ± 0.5
img	96.9 ± 0.3	97 ± 0.3	97.03 ± 0.17	96.7 ± 0.3
ionos	89.7 ± 1.4	90.4 ± 1.3	91 ± 0.9	92 ± 1
pima	76.8 ± 1	76.8 ± 1.1	76.7 ± 0.8	76.1 ± 0.9
survi	74.1 ± 1.2	73 ± 1	73.6 ± 1	73.4 ± 1.2
vowel	89.8 ± 0.9	92.7 ± 0.7	93.3 ± 0.6	92.9 ± 0.7
wdbc	96.7 ± 0.3	96.8 ± 0.3	95.9 ± 0.6	96 ± 0.5

Table 6. Simple Ensemble - Stacked Combiner results

	3 Nets	9 Nets	20 Nets	40 Nets
aritm	75.4 ± 1.4	75.1 ± 1.2	73.8 ± 1.3	73.9 ± 1.4
derma	97.2 ± 0.7	97.5 ± 0.7	97.5 ± 0.7	97.6 ± 0.7
ecoli	86.6 ± 0.9	86.8 ± 1.1	86.8 ± 0.9	86.8 ± 1.1
flare	81.4 ± 0.6	81.1 ± 0.5	81.4 ± 0.6	81.3 ± 0.8
img	96.5 ± 0.2	96.6 ± 0.3	97 ± 0.2	97.2 ± 0.2
ionos	92 ± 0.8	92.9 ± 1	92.7 ± 1.1	92.4 ± 1
pima	76.1 ± 1	76.1 ± 1.1	76.4 ± 0.9	75.9 ± 0.9
survi	74.4 ± 1.4	73.8 ± 1.5	73.8 ± 1.3	74.1 ± 1.2
vowel	89.4 ± 0.8	92.3 ± 0.5	93.3 ± 0.6	94.2 ± 0.8
wdbc	97.1 ± 0.5	97.2 ± 0.4	97.2 ± 0.5	97.2 ± 0.5

Table 7. Simple Ensemble - Stacked+ Combiner results

	3 Nets	9 Nets	20 Nets	40 Nets
aritm	74.4 ± 1.4	73.6 ± 1.7	74.7 ± 1.1	74.5 ± 1.3
derma	97.2 ± 0.7	97.3 ± 0.7	97.5 ± 0.7	97.6 ± 0.7
ecoli	86.8 ± 1	86.3 ± 1.2	86.8 ± 1.1	86.8 ± 1
flare	81.9 ± 0.4	81.7 ± 0.7	81.5 ± 0.7	81.1 ± 0.7
img	96.7 ± 0.3	96.8 ± 0.3	97 ± 0.3	96.8 ± 0.2
ionos	92 ± 0.9	92.7 ± 1	92.9 ± 1.2	92.4 ± 1.2
pima	76.1 ± 1	75.7 ± 1	75.9 ± 1.2	75.9 ± 1
survi	73.9 ± 1.4	73.6 ± 1.4	73.8 ± 1.2	73.9 ± 1.4
vowel	89.8 ± 0.8	92.3 ± 0.6	93.3 ± 0.7	94.1 ± 0.7
wdbc	97.2 ± 0.5	97.4 ± 0.5	97.3 ± 0.5	97.3 ± 0.5

Table 8. Stacked Generalization Version 2 Results - Expert networks as Ensemble

	3 Nets	9 Nets	20 Nets	40 Nets
aritm	72.5 ± 1.5	72.4 ± 1.5	72.4 ± 1.5	72.4 ± 1.5
derma	96.6 ± 0.7	96.8 ± 0.6	96.6 ± 0.9	96.3 ± 0.9
ecoli	86 ± 1	86 ± 0.9	85.7 ± 1	85.4 ± 1.1
flare	82 ± 0.6	81.8 ± 0.6	81.8 ± 0.6	81.9 ± 0.6
img	96.5 ± 0.2	96.6 ± 0.2	96.6 ± 0.3	96.6 ± 0.2
ionos	88 ± 1.3	89 ± 1.2	88.9 ± 1.2	89.1 ± 1.1
pima	77.4 ± 0.8	77.2 ± 0.7	77.4 ± 0.7	77.2 ± 0.8
survi	74.9 ± 1.4	74.9 ± 1.4	75.1 ± 1.4	75.3 ± 1.4
vowel	86.7 ± 0.7	88.6 ± 0.6	89.7 ± 0.7	89.8 ± 0.4
wdbc	96.9 ± 0.6	97 ± 0.6	97 ± 0.6	97 ± 0.6

Table 9. Stacked Generalization Version 2 Results - Whole model

	3 Nets	9 Nets	20 Nets	40 Nets
aritm	74.8 ± 1.5	74 ± 2	75 ± 2	74.9 ± 1.9
derma	96.6 ± 0.9	96.5 ± 1	96.5 ± 1	96.6 ± 1.1
ecoli	85 ± 1.2	84.3 ± 1	85.4 ± 1.3	84.6 ± 1.1
flare	81.8 ± 0.7	81.8 ± 0.7	82 ± 0.8	81.8 ± 0.7
img	96.9 ± 0.2	97.1 ± 0.3	97.1 ± 0.3	97.2 ± 0.2
ionos	89.4 ± 1.2	90.6 ± 0.9	90.6 ± 0.7	91.1 ± 1
pima	76.3 ± 1.1	76.2 ± 1.1	76.5 ± 0.9	75.9 ± 1.4
survi	73.1 ± 1.2	73.4 ± 0.6	73.1 ± 1	73.4 ± 1.1
vowel	87 ± 0.5	90.7 ± 0.6	92.6 ± 0.8	92.7 ± 0.7
wdbc	96.7 ± 0.6	96.6 ± 0.5	96.6 ± 0.5	96.5 ± 0.6

Table 10. Stacked Generalization Version 2 Results - Expert networks as Ensemble

	3 Nets	9 Nets	20 Nets	40 Nets
aritm	74.5 ± 1.3	74.3 ± 1	74.3 ± 1.4	74 ± 1
derma	97.5 ± 0.7	97.8 ± 0.6	97.3 ± 0.6	97.3 ± 0.4
ecoli	87.5 ± 0.9	87.2 ± 1	87.2 ± 0.9	86.6 ± 1.2
flare	82.1 ± 0.5	81.9 ± 0.7	82.1 ± 0.6	81.7 ± 0.7
img	96 ± 0.3	96.7 ± 0.2	96.9 ± 0.3	96.7 ± 0.3
ionos	88.9 ± 0.9	89.1 ± 0.9	89.7 ± 1.6	88.9 ± 1.5
pima	76.1 ± 1.2	76.8 ± 0.9	77.4 ± 1	75.8 ± 0.6
survi	74.1 ± 1.4	74.1 ± 1.5	75.1 ± 1.1	74.4 ± 1.3
vowel	86.9 ± 0.5	90.6 ± 0.5	91.1 ± 0.6	91.5 ± 0.7
wdbc	96.8 ± 0.5	97 ± 0.5	96.9 ± 0.5	96.7 ± 0.6

Table 11. Stacked Generalization Version 2 Results - Whole Stacked Model

	3 Nets	9 Nets	20 Nets	40 Nets
aritm	74.9 ± 1.3	74.5 ± 1	72.9 ± 1.3	75.6 ± 1
derma	97.5 ± 0.7	97.6 ± 0.7	97.3 ± 0.6	97.3 ± 0.4
ecoli	86.5 ± 0.9	86.3 ± 0.9	84.4 ± 1.1	86 ± 1.1
flare	81.6 ± 0.9	82.1 ± 0.8	81.4 ± 0.9	81.7 ± 0.7
img	97 ± 0.2	96.8 ± 0.3	96.8 ± 0.2	97 ± 0.2
ionos	90 ± 0.9	91 ± 1	90.1 ± 1.2	89.3 ± 1.5
pima	76.4 ± 1	76.7 ± 1	76.5 ± 1.4	74.5 ± 1
survi	73.9 ± 1.1	72.6 ± 1.1	72.8 ± 1.5	73.8 ± 1
vowel	88.1 ± 0.6	92.2 ± 0.5	92.8 ± 0.7	93.6 ± 0.6
wdbc	96.5 ± 0.5	96.9 ± 0.4	96.5 ± 0.6	96.6 ± 0.6

4.2   General Measurements

We have calculated the mean Increase of Performance (IoP eq.1) and the mean Percentage of Error Reduction (PER eq.2) across all databases to get a global measurement to

Table 12. General Results

Mean Increase of Performance

method	3 nets	9 nets	20 nets	40 nets
Simple Ensemble	0.69	1.01	1.04	1.16
Bagging	0.84	1.44	1.51	1.34
CVC	0.66	1.24	1.13	0.95
CVCv2	1.26	1.33	1.3	1.51
Adaboost	0.3	0.85	1.14	1.26
stacked on simple ensemble	1.13	1.44	1.51	1.59
stacked+ on simple ensemble	1.12	1.26	1.58	1.57
SG Ver.1 - Experts	0.28	0.56	0.64	0.62
SG Ver.2 - Experts	0.57	1.08	1.31	0.9
SG Ver.1 - Whole model	0.29	0.67	1.05	1.01
SG Ver.2 - Whole model	0.76	1.19	0.67	1.07

Mean Percentage of Error Reduction

method	3 nets	9 nets	20 nets	40 nets
Simple Ensemble	5.58	8.38	8.08	9.72
Bagging	6.85	12.12	13.36	12.63
CVC	6.17	7.76	10.12	6.47
CVCv2	10.25	10.02	7.57	7.48
Adaboost	1.32	4.26	9.38	12.2
stacked on simple ensemble	8.48	12.04	13.43	14.6
stacked+ on simple ensemble	9.4	11.8	14	13.89
SG Version 1 - Experts	0.7	3.46	3.52	2.78
SG Version 2 - Experts	3.44	9.22	9.48	6.31
SG Version 1 - Whole model	1.4	3.91	6.07	5.94
SG Version 2 - Whole model	5.57	10.14	4.86	7.82

compare the methods (Table 12). A negative value on these measurements means that the the ensemble performs worse than a single network.

$$IoP = Error_{SingleNet} - Error_{Ensemble} \qquad (1)$$

$$PER = 100 \cdot \frac{Error_{SingleNet} - Error_{Ensemble}}{Error_{SingleNet}} \qquad (2)$$

4.3 Discussion

Before the results discussion, the main results have been resumed in table 13 in which the best performance for each database is shown along with the method and number of networks we got it with.

In the resumed table, we can see that any version of the *Stacked Generalization* model do not appear on it. Although there are two cases in which the experts of *Stacked Generalization Ver.2* as ensemble is the best method, these experts are trained as in

Table 13. Best Resutls

	Performance	Method	Networks
aritm	77 ± 0.8	CVCv2	40
derma	98 ± 0.3	CVCv2	3
ecoli	87.5 ± 0.9	SG Version 2 - Experts	3
flare	82.7 ± 0.5	CVC	3
img	97.3 ± 0.3	Adaboost	9
ionos	92.9 ± 1	Stacked	9
pima	77.4 ± 1	SG Version 2 - Experts	20
survi	75.4 ± 1.6	Adaboost	3
vowel	97 ± 0.6	Adaboost	40
wdbc	97.5 ± 0.4	Bagging	20

CVC. We can extract from the table that the best individual results are provided by *Adaboost* and *Cross-Validation*.

Finally, we can see in the global measurements tables that the best results are got by *Cross-Validation* methods, Bagging and the combiners *Stacked* and *Stacked+*.

5   Conclusions

In this paper we have elaborated a comparison among traditional methodologies to build ensembles, two combiners based on *Stacked Generalization* and two stacked models proposed by Ting & Witten (SG Ver.1) by and Ghorbani & Owrangh (SG Ver.2).

Firstly, we can notice that *Bagging* and *Cross-Validation* are the best methods among the results related to the traditional ensemble methods according to the general results.

Secondly, we can see that the stacked combiners, *Stacked* and *Stacked+*, considerably improve the results got by the ensembles trained with *Simple Ensemble*. Moreover, these combiners on *Simple Ensemble* provide better results than some traditional ensemble methods. The best overall general measurements is provided by applying *Stacked* to a 40-network ensemble trained with *Simple Ensemble*.

Thirdly, we can see that the second version of the original *Stacked Generalization* model is better than first one. We can also see that the experts trained with the second version are also better. In fact, the first version is not on the top of the best performing methods, being similar to *Adaboost*.

Finally, comparing the results related to the first and second version of the original *Stacked Generalization* model, we can see that the performance of the whole stacked models is similar to the performance of their experts as ensemble. There is not a high increase of performance by the use of the combination network. Incredibly, the increase of performance got by the combiners *Stacked* and *Stacked+* with respect to their experts is considerable higher.

In conclusion, *Cross-Validation*, *Adaboost* and the combiners *Stacked* and *Stadked+* on *Simple Ensemble* are the best ways to build a *Multi-Net system*. *Adaboost* and the *Cross-Validation* methods got the best individual results whereas the general results show that the best Multi-Net system is provided by the 40-network version of *Simple*

Ensemble combined with *Stacked*. The two versions of the original *Stacked Generalization* model do not get neither the best individual performance for a database nor the best global results. Moreover, there are some cases in which the two versions of the original *Stacked Generalization* perform worse than a *Simple Ensemble*.

References

1. Bishop, C.M.: Neural Networks for Pattern Recognition. Oxford University Press, Inc., New York (1995)
2. Fernndez-Redondo, M., Hernndez-Espinosa, C., Torres-Sospedra, J.: Multilayer feedforward ensembles for classification problems. In: Pal, N.R., Kasabov, N., Mudi, R.K., Pal, S., Parui, S.K. (eds.) ICONIP 2004. LNCS, vol. 3316, pp. 744–749. Springer, Heidelberg (2004)
3. Ghorbani, A.A., Owrangh, K.: Stacked generalization in neural networks: Generalization on statistically neutral problems. In: Proceedings of the International Joint conference on Neural Networks, IJCNN 2001, Washington D.C., USA, pp. 1715–1720. IEEE, Los Alamitos (2001)
4. Hernndez-Espinosa, C., Torres-Sospedra, J., Fernndez-Redondo, M.: New experiments on ensembles of multilayer feedforward for classification problems. In: IJCNN 2005 proceedings, pp. 1120–1124 (2005)
5. Kuncheva, L.I.: Combining Pattern Classifiers: Methods and Algorithms. Wiley-Interscience, Chichester (2004)
6. Newman, D.J., Hettich, S., Blake, C.L., Merz, C.J.: UCI repository of machine learning databases (1998),
   `http://www.ics.uci.edu/~mlearn/MLRepository.html`
7. Raviv, Y., Intratorr, N.: Bootstrapping with noise: An effective regularization technique. Connection Science, Special issue on Combining Estimators 8, 356–372 (1996)
8. Sharkey, A.J. (ed.): Combining Artificial Neural Nets: Ensemble and Modular Multi-Net Systems. Springer, Heidelberg (1999)
9. Ting, K.M., Witten, I.H.: Stacked generalizations: When does it work? In: International Joint Conference on Artificial Intelligence proceedings, vol. 2, pp. 866–873 (1997)
10. Ting, K.M., Witten, I.H.: Issues in stacked generalization. Journal of Artificial Intelligence Research 10, 271–289 (1999)
11. Torres-Sospedra, J., Hernndez-Espinosa, C., Fernndez-Redondo, M.: Combining MF networks: A comparison among statistical methods and stacked generalization. In: Schwenker, F., Marinai, S. (eds.) ANNPR 2006. LNCS (LNAI), vol. 4087, pp. 210–220. Springer, Heidelberg (2006)
12. Torres-Sospedra, J., Hernndez-Espinosa, C., Fernndez-Redondo, M.: Designing a new multilayer feedforward modular network for classification problems. In: WCCI 2006 proceedings, pp. 2263–2268 (2006)
13. Tumer, K., Ghosh, J.: Error correlation and error reduction in ensemble classifiers. Connection Science 8(3-4), 385–403 (1996)
14. Wolpert, D.H.: Stacked generalization. Neural Networks 5(6), 1289–1301 (1994)

Real-Time Emotion Recognition from Speech Using Echo State Networks

Stefan Scherer, Mohamed Oubbati, Friedhelm Schwenker, and Günther Palm

Institute of Neural Information Processing, Ulm University, Germany
{stefan.scherer,mohamed.oubbati,friedhelm.schwenker,
guenther.palm}@uni-ulm.de

Abstract. The goal of this work is to investigate real-time emotion recognition in noisy environments. Our approach is to solve this problem using novel recurrent neural networks called echo state networks (ESN). ESNs utilizing the sequential characteristics of biologically motivated modulation spectrum features are easy to train and robust towards noisy real world conditions. The standard Berlin Database of Emotional Speech is used to evaluate the performance of the proposed approach. The experiments reveal promising results overcoming known difficulties and drawbacks of common approaches.

1 Introduction

The present innovations in affective computing aim to provide simpler and more natural interfaces for human-computer interaction applications. Detecting and recognizing the emotional status of a user is important in designing and developing efficient and productive human-computer interaction interfaces [2]. The efficiency gain is well founded on the fact that in healthy human to human interaction emotion is essential in every bit of communication. For example, while explaining something to another person, one could communicate understanding with a simple smile, with no need to say "I understand, what you are telling me" [15]. Hence, emotion analysis and processing is a multi-disciplinary topic, which has been emerging as a rich research area in recent times [2,4,6,13,18]. The visual cues, such as facial expressions and hand gestures are the natural indicators of emotions. However, these require additional hardware and computational resources for processing. Alternatively, speech can be used for emotion recognition which is not only simple to process, but can also be incorporated into the existing speech processing applications [2,6,17]. Most commonly used features are pitch, energy and speech spectral based features [13]. In this work, a novel approach based on long term modulation spectrum of speech is used to detect the emotions close to real-time using a recurrent neural network called echo state network (ESN).

One of the main issues in designing an automatic emotion recognition system is the selection of the features that can represent the corresponding emotions. In [12], pitch and linear predictive coding (LPC) features were used as input

L. Prevost, S. Marinai, and F. Schwenker (Eds.): ANNPR 2008, LNAI 5064, pp. 205–216, 2008.

to an artificial neural network (ANN). After detecting the start and end points of the utterances, a 300 dimensional vector was used, which resulted in classification rates of around 50% detecting eight different emotions. In earlier work multi classifier systems (MCS) were trained with three feature types, comprising modulation spectrum, as in this work, relative spectral transform - perceptual linear prediction (RASTA-PLP), and perceived loudness features, in a MCS to recognize seven different emotions with an accuracy of more than 70% [18]. The Mel Frequency Cepstral Coefficients (MFCC) based features were used in [11], which were obtained with a window of 25 ms sampled every 10 ms. The Hidden Markov Model (HMM) was then used for training each of the four targeted emotions. After training a combination of all the phoneme class based HMMs on the TIMIT database, for each of the emotions, the classification performance reached around 76%. In [3], k-nearest neighbor (KNN) algorithm was applied to classify four emotions, which resulted in 65 % accuracy. The pitch based statistics, such as contour of pitch, maximum, minimum, and slope were considered as features. Broadly speaking, differences in the various approaches arise from the overall goal (recognizing single vs. multiple emotions), the specific features used, and the classification framework. The anger vs. neutral emotion classification was studied, particularly in the context of interactive voice response systems with specific application to call centers in [19]. 37 prosody features related to pitch, energy, and duration were used as features, and for classification neural networks, support vector machines (SVM), and KNN were applied. With the most significant feature set of 19 features, the best recognition accuracy about 90% was achieved using SVMs. In another study, agitation vs. calm emotion classification was performed with KNN, ANN, and set of experts [14]. "Agitation" included happiness, anger, and fear, and "calm" comprised neutral, and sadness emotional states. Pitch, energy, speaking rate, formants, and their bandwidths were used as features, which resulted in an accuracy of 77%.

However, commonly used features and classifiers are sensitive towards noise. In this work, a system overcoming these issues is targeted. Furthermore, classifiers still require time to classify the utterances as they rely on statistics of the features and are computationally intensive. Here we use a special characteristic of long term modulation spectrum, which reflects syllabic and phonetic temporal structures of speech [5,7]. Recently, a novel recurrent neural network (RNN) called echo state network (ESN) is developed. An ESN has an easy training algorithm, where only the output weights are to be adjusted. The basic idea of ESN is to use a Dynamic Reservoir (DR), which contains a large number of sparsely interconnected neurons with non-trainable weights. The previously mentioned features are used as inputs to an ESN classifier. Since, the only weights that need to be adjusted in an ESN are the output weights, the training is not computationally expensive using the direct pseudo inverse calculation instead of gradient descent training. The performance is close to real-time, as the decisions are made on short-segments of the signal (100 ms), rather than over the entire utterance. This is of great advantage since emotions are constantly changing and aggregating statistics of pitch or other similar features may not suffice [17]. An example of a scenario where emotions change rapidly can be found in [15]. In this scenario a tennis player feels a piercing pain in his lower back and he first turns around clenching his fist and feeling angry, but as he sees that a woman in a wheelchair hit him his feelings changed to sadness and sympathy. This small example illustrates the possible rapid changes of how we value situations. Therefore, it is necessary to build a system that does not need to aggregate information over several seconds, but is able to classify emotions close to real time. In this work we present a feature extraction system that extracts after a lead time of 400 ms feature vectors with a frequency of 25 Hz, which is sufficient for emotion recognition in many applications.

The paper is organized and presented in four sections: Section 2 gives an overview of the database used for experiments, Sect. 3 describes the feature extraction, Sect. 4 introduces the echo state networks used for classifictation, Sect. 5 presents the experiments and results, and finally Sect. 6 concludes.

2   Database Description

The Berlin Database of Emotional Speech is used as a test bed for our approach. This corpus is a collection of around 800 utterances spoken in seven different emotions: anger, boredom, disgust, fear, happiness, sadness, and neutral [1]. The database is publicly available at http://pascal.kgw.tu-berlin.de/emodb/. Ten professional actors (five male and five female) read the predefined utterances in an anechoic chamber, under supervised conditions. The text was taken from everyday life situations, and did not include any emotional bias. The utterances are available at a sampling rate of 16 kHz with a 16 bit resolution and mono channel. A human perception test to recognize various emotions with 20 participants resulted in a mean accuracy of around 84% [1].

3   Feature Extraction

Short term analysis of the speech signal, such as extracting spectral features from frames not more than several milliseconds, dominates speech processing for many years. However, these features are strongly influenced by environmental noise and are therefore unstable. In [8], it is suggested to use the so called modulation spectrum of speech to obtain information about the temporal dynamics of the speech signal to extract reliable cues for the linguistic context. Since emotion in speech is often communicated by varying temporal dynamics in the signal the same features are used to classify emotional speech in the following experiments [17].

The proposed features are based on long term modulation spectrum. In this work, the features based on slow temporal evolution of the speech are used to represent the emotional status of the speaker. These slow temporal modulations of speech emulate the perception ability of the human auditory system. Earlier studies reported that the modulation frequency components from the range

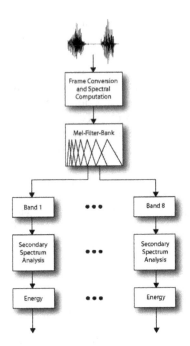

Fig. 1. Schematic description for feature extraction

between 2 and 16 Hz, with dominant component at around 4 Hz , contain important linguistic information [5,7,10]. Dominant components represent strong rate
of change of the vocal tract shape. This particular property, along with the other
features has been used to discriminate speech and music [16]. In this work, the
proposed features are based on this specific characteristic of speech, to recognize
the emotional state of the speaker.

The block diagram for the feature extraction for a system to recognize emotions is shown in Fig. 1. The fast Fourier transform (FFT) for the input signal
$x(t)$ is computed over N points with a shift of n samples, which results in a $\frac{N}{2}$
dimensional FFT vector. Then, the Mel-scale transformation, motivated by the
human auditory system, is applied to these vectors. The Mel-filter bank with
eight triangular filters $H_i[k]$, is defined by:

$$H_i[k] = \begin{cases} \frac{2(k-b_i)}{(d_i-b_i)(c_i-b_i)} & b_i \leq k \leq c_i \\ \\ \frac{2(d_i-k)}{(d_i-b_i)(d_i-c_i)} & c_i \leq k \leq d_i \end{cases} , \tag{1}$$

where $i = 1, ..., 8$ indicates the index of the i-th filter. b_i and d_i indicate the frequency range of filter H_i and the center frequency c_i is defined as $c_i = (b_i+d_i)/2$.
These ranges are equally distributed in the Mel-scale, and the corresponding frequencies b_i and d_i are listed in Table 1. For $k < b_i$ and $k > d_i$ $H_i[k] = 0$.

Table 1. Start and end frequencies of the triangular Mel-filters

Band	Start Freq. (Hz)	End Freq. (Hz)
1	32	578
2	257	964
3	578	1501
4	966	2217
5	1501	3180
6	2217	4433
7	3180	6972
8	4433	8256

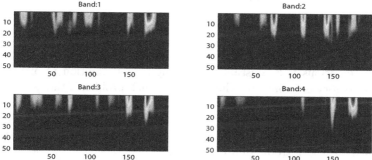

Fig. 2. Modulation spectrum for the first four bands of a single angry utterance. The x-axis represents the time scale, in frames and the y-axis, the frequency in Hz.

For each of the bands, the modulations of the signal are computed by taking FFT over the P points, shifted by p samples, resulting in a sequence of $\frac{P}{2}$ dimensional modulation vectors. Most of the prominent energies can be observed within the frequencies between 2 - 16 Hz. Figure 2 illustrates the modulation spectrum based energies for a single angry utterance, for the values $N = 512$, $n = 160$, $P = 100$ and $p = 1$ for the first four bands. For the classification task following values were used: $N = 1600$, $n = 640$, $P = 10$, $p = 1$. Since the signal is sampled with 16 kHz, N corresponds to 100 ms and n to 40 ms resulting in a feature extraction frequency of 25 Hz. According to the window size P a lead time of 400 ms is necessary. Therefore, one feature vector in the modulation spectrum takes 400 ms into account with an overlap of 360 ms, due to p.

4 Echo State Networks

Feed forward neural networks have been successfully used to solve problems that require the computation of a static function, i.e. a function whose output depends only upon the current input. In the real world however, many problems cannot be solved by learning a static function because the function being computed may produce different outputs for the same input if it is in different states. Since

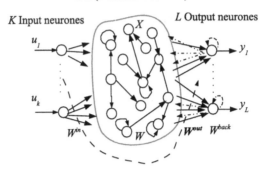

Fig. 3. Basic architecture of ESN. Dotted arrows indicate connections that are possible but not required.

expressing emotions is a constantly changing signal, emotion recognition falls into this category of problems. Thus, to solve such problems, the network must have some notion of how the past inputs affect the processing of the present input. In other words, the network must have a memory of the past input and a way to use that memory to process the current input. This limitation can be rectified by the introduction of feedback connections in the network. The class of Neural Networks which contain feedback connections are called RNNs. In principle RNNs can implement almost arbitrary sequential behavior, which makes them promising for adaptive dynamical systems. However, they are often regarded as difficult to train. Using ESNs only two steps are necessary for training: First, one forms a DR, with input neurons and input connections, which has the echo state property. The echo state property says: "if the network has been run for a very long time, the current network state is uniquely determined by the history of the input and the (teacher-forced) output." [9]. According to experience, it is better to ensure that the internal weight matrix has maximum eingenvalue $|\lambda_{max}| < 1$. Second, one attaches output neurons to the network and trains suitable output weights.

As presented in (Fig. 3), we consider a network with K inputs, N internal neurons and L output neurons. Activations of input neurons at time step n are $U(n) = (u_1(n), \ldots, u_k(n))$, of internal units are $X(n) = (x_1(n), \ldots, x_N(n))$, and of output neurons are $Y(n) = (y_1(n), \ldots, y_L(n))$. Weights for the input connection in a (NxK) matrix are $W^{in} = (w_{ij}^{in})$, for the internal connection in a (NxN) matrix are $W = (w_{ij})$, and for the connection to the output neurons in an L x $(K + N + L)$ matrix are $W^{out} = (w_{ij}^{out})$, and in a (NxL) matrix $W^{back} = (w_{ij}^{back})$ for the connection from the output to the internal units.

The activation of internal and output units is updated according to:

$$X(n + 1) = f(W^{in}U(n + 1) + WX(n) + W^{back}Y(n)) \qquad (2)$$

where $f = (f_1, \ldots, f_N)$ are the internal neurons output sigmoid functions. The outputs are computed according to:

$$Y(n+1) = f^{out}(W^{out}(U(n+1), X(n+1), Y(n)))$$ (3)

where $f^{out} = (f_1^{out}, \ldots, f_L^{out})$ are the output neurons output sigmoid functions. The term $(U(n+1), X(n+1), Y(n))$ is the concatenation of the input, internal, and previous output activation vectors. The idea of this network is that only the weights for connections from the internal neurons to the output (W^{out}) are to be adjusted.

Here we present briefly an off-line algorithm for the learning procedure:

1. Given I/O training sequence $(U(n), D(n))$
2. Generate randomly the matrices (W^{in}, W, W^{back}), scaling the weight matrix W such that it's maximum eingenvalue $|\lambda_{max}| \leq 1$.
3. Drive the network using the training I/O training data, by computing

$$X(n+1) = f(W^{in}U(n+1) + WX(n) + W^{back}D(n))$$ (4)

4. Collect at each time the state $X(n)$ as a new row into a state collecting matrix M, and collect similarly at each time the sigmoid-inverted teacher output $tanh^{-1}D(n)$ into a teacher collection matrix T.
5. Compute the pseudo inverse of M and put

$$W^{out} = (M^+T)^t$$ (5)

t: indicates transpose operation.

For exploitation, the trained network can be driven by new input sequences and using the equations (2) and (3).

5   Experiments and Results

All the experiments were carried out on the German Berlin Database of Emotional Speech, which is described in Sect. 2. The utterances comprising anger and neutral are specifically used with regard to the most important task in call center applications, where it is necessary to recognize angry customers for the system to react properly.

In the first experiment especially the real-time recognition capability of the ESN is tested. After thorough tuning the following parameters revealed optimal results. A randomly initialized network of 500 neurons was used. The connectivity within the network was 0.5 which indicates that 50% of the connections were set within the network. Additionally, the spectral width λ_{max} was set to 0.2. Figure 4 (a) shows the cumulative sums over frame wise decisions of three different utterances taken from the first fold of the 10 fold cross validation experiment. The vertical bars represent borders of single utterances taken from the database. The ESN needs some time to adapt to the new portrayed emotion

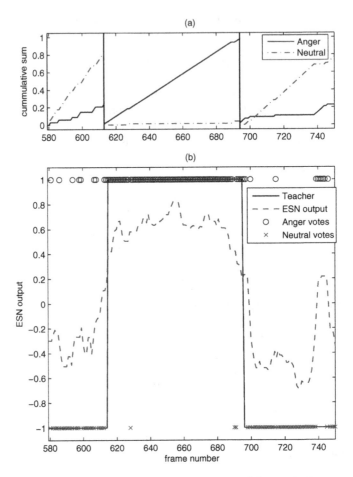

Fig. 4. Example results of ESN generalization behavior: (a) cumulative sum over frame wise decisions; (b) median filtered ESN output with superimposed plotted frame wise decisions and teacher signal

in the following utterance. However, after only a few frames the ESN achieves adoption and recognizes the correct emotion fast. The output of the ESN counts as a vote for anger if the sign of the output is + and a vote for neutral if it is −. It is seen that the cumulative sum of the correct emotion is above the other only a few frames after an emotional shift and hardly any errors are made in most of the cases. At every point in time a possible shift from neutral to angry and vice versa is possible. Only once in the 10 folds of the cross validation the correct emotion is overruled by around 52% of the votes, leading to an accuracy of more than 99%. In every other case the correct emotion wins at least at the end of an utterance, most of the cases behave similarly to the three in Fig. 4 (a). In Fig. 4 (b) the dashed line corresponds to the median filtered ESN output signal. The median filter used had the magnitude 10. However, the circles and

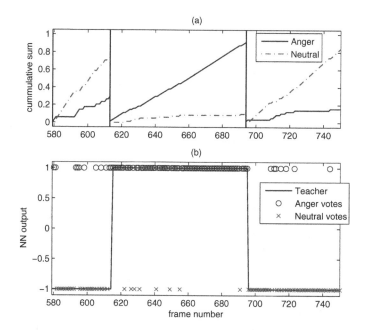

Fig. 5. Example results of NN generalization behavior: (a) cumulative sum over frame wise decisions; (b) frame wise decisions with superimposed plotted teacher signal

crosses correspond to the frame wise decisions not the median filtered output in order to show the real-time performance of the network. Additionally, the plot of the ESN output is superimposed by the teacher signal. Each circle or cross that does not lie on the solid line resembles an error of the network on a single frame. These errors may occur, but oscillating outputs could just be filtered out and only if a series of the same decisions follow each other an emotion will be recognized in future applications.

In Fig. 5 (a) the cumulative sums over the frame wise decisions of a simple NN classifier are shown. In most of the cases the NN errs more often than the ESN. This may be the result of the ESN's capability to take earlier frames into account. Due to the recurrent architecture of the ESN it is possible to "keep several frames in mind". In a similar way HMMs are capable to process sequences of frames. The cumulative sums of the wrong emotion overruled the correct emotion in the 10 fold cross validation 6 times using the NN classifier. Furthermore, the calculation of the nearest neighbor is far more computationally expensive as the output for a single frame of the ESN. In Fig. 5 (b) again circles or crosses not lying on the line resembling the teacher signal count as errors. To be able to calculate the outputs in real-time it would be necessary to reduce the search space for the NN classifier. For example, by using a clustering method such as Learning Vector Quantization (LVQ) or K-Means. However, using these methods may result in more errors.

Table 2. Frame wise error rate of the two classifiers according to differing conditions of noise

Type of Noise	Frame wise error rates	
	ESN	NN
no noise	0.15	0.21
coffee machine	0.14	0.23
office	0.16	0.22
vacuum cleaner	0.16	0.23
inside car	0.18	0.26
city/street	0.18	0.27

In a second experiment we added different amounts of noise to the audio signals in order to check whether the ESN or the NN are capable of dealing with noise and how the classification performance develops. Theoretically the modulation spectrum features should be quite stable towards noise, since only voice relevant frequencies pass through the Mel filtering. Additionally, it is possible to compensate noise using ESNs [9]. Table 2 shows the recognition results adding noise to the audio signal. The error rates correspond to the average rate of misclassified frames. In the first row no additional noise was added. In the following rows the amount of noise slowly increases from a quiet coffee machine up to noise recorded in Helsinki on a sidewalk. The noise of the coffee machine is mostly due to dripping water. The office environment corresponds to people typing, chatting in the background, and copying some papers. The vacuum cleaner is a very constant but loud noise. Inside the car mostly the engine of the own car is heard. The last type of noise comprises passing trucks, cars, motorbikes, and people passing by. All the noise was added to the original signal before feature extraction. The audio material was taken from the homepage of the "Freesound Project"[1]. It is seen that the results using modulation spectrum based features are quite robust using both classifiers. However, the recognition stays more stable using the ESN, which confirms the abilities of the ESN to compensate noise to a certain amount.

6   Conclusions

This paper presented an approach towards recognizing emotions from speech close to real-time. Features motivated by the human auditory system were used as input for an ESN. The real-time recognition performance of the network in one of the most important tasks in automatic call centers is impressive. Furthermore, the utilized features as well as the ESN are very stable towards noise of different types, like cars, office environment, or a running vacuum cleaner. The data was tested using a standard emotional speech database. However, the recognition task was rather limited and could be extended further to recognize

---

[1] Data is freely available: http://freesound.iua.upf.edu/

various other emotions such as happiness, fear, disgust, and sadness, and towards the classification of real-world data. In earlier work, these emotions were recognized using additional feature types and multi classifier systems [18]. In this work emotions can not be recognized in real-time, which is a major drawback as it was illustrated in an example in this paper. However, this could be improved using ESNs. These issues will be studied in the future.

Acknowledgements

This work is supported by the competence center Perception and Interactive Technologies (PIT) in the scope of the Landesforschungsschwerpunkt project: "Der Computer als Dialogpartner: Perception and Interaction in Multi-User Environments" funded by the Ministry of Science, Research and the Arts of Baden-Württemberg.

References

1. Burkhardt, F., Paeschke, A., Rolfes, M., Sendlmeier, W., Weiss, B.: A database of german emotional speech. In: Proceedings of Interspeech 2005 (2005)
2. Cowie, R., Douglas-Cowie, E., Tsapatsoulis, N., Votsis, G., Kollias, S., Fellenz, W., Taylor, J.G.: Emotion recognition in human-computer interaction. IEEE Signal Processing Magazine 18(1), 32–80 (2001)
3. Dellaert, F., Polzin, T., Waibel, A.: Recognizing emotion in speech. In: Proceedings of ICSLP, pp. 1970–1973 (1996)
4. Devillers, L., Vidrascu, L., Lamel, L.: Challanges in real-life emotion annotation and machine learning based detection. Neural Networks 18, 407–422 (2005)
5. Drullman, R., Festen, J., Plomp, R.: Effect of reducing slow temporal modulations on speech reception. Journal of the Acousic Society 95, 2670–2680 (1994)
6. Fragopanagos, N., Taylor, J.G.: Emotion recognition in human-computer interaction. Neural Networks 18, 389–405 (2005)
7. Hermansky, H.: Auditory modeling in automatic recognition of speech. In: Proceedings of Keele Workshop (1996)
8. Hermansky, H.: The modulation spectrum in automatic recognition of speech. In: Proceedings of IEEE Workshop on Automatic Speech Recognition and Understanding (1997)
9. Jaeger, H.: Tutorial on training recurrent neural networks, covering bppt, rtrl, ekf and the echo state network approach. Technical Report 159, Fraunhofer-Gesellschaft, St. Augustin Germany (2002)
10. Kanederaa, N., Araib, T., Hermansky, H., Pavele, M.: On the relative importance of various components of the modulation spectrum for automatic speech recognition. Speech Communications 28, 43–55 (1999)
11. Lee, C.M., Yildirim, S., Bulut, M., Kazemzadeh, A., Busso, C., Deng, Z., Lee, S., Narayanan, S.S.: Emotion recognition based on phoneme classes. In: Proceedings of ICSLP 2004 (2004)
12. Nicholson, J., Takahashi, K., Nakatsu, R.: Emotion recognition in speech using neural networks. Neural Computing and Applications 9, 290–296 (2000)

13. Oudeyer, P.-Y.: The production and recognition of emotions in speech: features and algorithms. International Journal of Human Computer Interaction 59(1-2), 157–183 (2003)
14. Petrushin, V.: Emotion in speech: recognition and application to call centers. In: Proceedings of Artificial Neural Networks in Engineering (1999)
15. Picard, R.W.: Affective Computing. MIT Press, Cambridge (2000)
16. Scheirer, E., Slaney, M.: Construction and evaluation of a robust multifeature speech/music discriminator. In: Proceedings of ICASSP, vol. 1, pp. 1331–1334 (1997)
17. Scherer, K.R., Johnstone, T., Klasmeyer, G.: Affective Science. In: Handbook of Affective Sciences - Vocal expression of emotion, pp. 433–456. Oxford University Press, Oxford (2003)
18. Scherer, S., Schwenker, F., Palm, G.: Classifier fusion for emotion recognition from speech. In: Proceedings of Intelligent Environments 2007 (2007)
19. Yacoub, S., Simske, S., Lin, X., Burns, J.: Recognition of emotions in interactive voice response systems. In: Proceedings of Eurospeech 2003 (2003)

Sentence Understanding and Learning of New Words with Large-Scale Neural Networks

Heiner Markert, Zöhre Kara Kayikci, and Günther Palm

University of Ulm
Institute of Neural Information Processing
89069 Ulm, Germany
{heiner.markert,zoehre.kara,guenther.palm}@uni-ulm.de
http://www.informatik.uni-ulm.de/neuro/

Abstract. We have implemented a speech command system which can understand simple command sentences like "Bot lift ball" or "Bot go table" using hidden Markov models (HMMs) and associative memories with sparse distributed representations. The system is composed of three modules: (1) A set of HMMs is used on phoneme level to get a phonetic transcription of the spoken sentence, (2) a network of associative memories is used to determine the word belonging to the phonetic transcription and (3) a neural network is used on the sentence level to determine the meaning of the sentence. The system is also able to learn new object words during performance.

Keywords: Associative Memories, Hidden Markov Models, Hebbian Learning, Speech Recognition, Language Understanding.

1 Introduction

A variety of speech recognition systems are currently in use in applications such as command & control, data entry, and document preparation. In this study, we have applied neural associative memories to a speech processing system in a hybrid approach with hidden Markov models (HMMs) [1][2]. The system is able to recognize spoken command sentences such as "Bot show plum" or "Bot pick blue plum". Those sentences are processed in three stages: At the first stage, the auditory signal is transformed into a sequence of corresponding triphones via a HMM based triphone recognizer. At the second stage, the generated stream of triphones is forwarded to a word recognition module, which consists of a number of binary heteroassociative memories and is able to recognize single words. The module determines the best matching words from the triphone data. As last processing step, the words are forwarded to a sentence recognition module which parses the sentence with respect to a given grammar. After successful processing, each word in the sentence is labeled with its grammatical role, giving a relatively straightforward representation of the meaning of the sentence.

The system is able to learn new object words. Learning is triggered by the special phrase "This is" followed by a novel word, e.g. "This is lemon". After learning, the

L. Prevost, S. Marinai, and F. Schwenker (Eds.): ANNPR 2008, LNAI 5064, pp. 217–227, 2008.
© Springer-Verlag Berlin Heidelberg 2008

system is able to understand this new object word like any other object and can successfully process sentences like "Bot show lemon" or "Bot pick yellow lemon".

2 Neural Associative Memories

A neural associative memory (NAM) is a single layer neural network which maps a set of input patterns to a set of corresponding output patterns. The patterns are binary and sparsely coded. A NAM can be represented as a synaptic connectivity matrix. Pairs of patterns are stored in the binary matrix using clipped binary Hebbian learning [3][4]. In case of autoassociation, input and output patterns are identical. Autoassociation allows for pattern completion, fault tolerance and long-term activation of patterns via feedback links. Heteroassociation is used to map one pattern onto another (e.g. to map a pattern "X" to a representation "Y"). They are also often used to translate between different neural coding schemes of different populations.

We have decided to use Willshaws model of associative memory [5][6] as the basic model. Different retrieval strategies are employed in different parts of the system. One of these strategies is one step retrieval with threshold [4][7], where the threshold is set to a global value:

$$Y_j^k = 1 \Leftrightarrow \left(X^k W\right)_j \geq \Theta, \tag{1}$$

where Θ is the global threshold, X is the input, Y is the output and W is the matrix of synaptic strengths. A special case of this strategy is the Willshaw strategy, where the threshold is set to the number of ones in the binary input vector X.

In more complex memories we also use the so-called spike counter model [8]. With this model, the network is simulated in global steps, where in each global step a complete retrieval is calculated by each of the associative memories. In contrast to Willshaws model [5][6] which interprets incoming activation as membrane potential values, incoming activation levels per neuron are interpreted as value of the membrane potentials derivative in the spike counter model. Thus, if a neuron receives strong input, its membrane potential is rising fast, while with low input, the membrane potential is increasing only slowly. In this simple model, the membrane potential is modified linearly. The neurons spike whenever the membrane potential reaches a given threshold and the membrane potential is reset to zero after the spike is emitted. In each retrieval, every neuron is allowed to spike at most twice and the retrieval is stopped when no more neuron is able to spike (which happens either when each neuron has spiked twice or there is no more neuron left that receives any positive postsynaptic input). The spike patterns are then forwarded through the heteroassociative connections and the next global time step starts.

3 The System

The system is composed of a chain of three modules. The first module is a triphone recognizer based on hidden Markov models, which is responsible for generating a sequence of triphones corresponding to the spoken command sentence. The second module is an isolated word recognizer from which single words are retrieved. The last

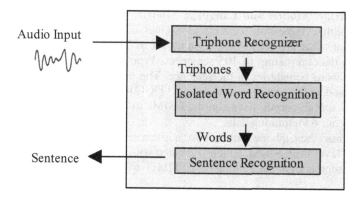

Fig. 1. Overview of the language processing system. The three main components of the system are connected in a feed-forward manner. Besides the triphone recognizer, all boxes are implemented as neural networks. Communication between the boxes is based on neural representations that allow ambiguities to be forwarded between the different units of the system. In case the word recognizer cannot decide between "ball" and "wall", a superposition of both words is sent to the sentence recognizer which can then use additional contextual input to resolve the ambiguity (see examples in text below).

module in the system is the sentence recognition network which analyzes the retrieved words with respect to simple grammar rules. Fig. 1 shows a block diagram of the system.

The triphone recognizer module, based on hidden Markov models, receives audio input via a microphone and converts it to a stream of triphones. This stream is then transfered to the isolated word recognizer module which extracts words from the stream of triphones. Afterwards, the output of the word recognizer is sent to the sentence recognition module which parses the stream of words against a given grammar and assigns grammatical roles to the words. The last two modules are networks of neural associative memories.

In the following, the three parts are described in more detail.

3.1 Triphone Recognizer

Speech Material. To train triphone HMMs, the training data set of TIMIT continuous speech corpus without the "SA"-type sentences has been used [9]. TIMIT is manually labeled and includes time-aligned, manually verified phonetic and word segmentations. The original set of 61 phonemes was reduced to a set of 39 phonemes in this study. The TIMIT training data consists of 462 speakers and 1716 sentences.

We added our own recordings to the training material in order to adapt the system to our scenario. A set of 105 different sentences with a vocabulary of 43 words has been recorded from four different speakers of our institute. From this data, 70 sentences were used for training and the remaining 35 sentences were used for testing. For training and testing the same speakers were used. The total number of words in the test set of our own data is 504, the training set consists of 1068 words in total.

Hidden Markov Models and Language Models. The HMMs used in the triphone recognizer utilize three-state continuous 8-Gaussian triphone models, i.e, a set of 7421 triphones that are seen in the training set of TIMIT speech corpus and in our own speech data consisting of 105 sentences. Word internal models are used to determine the word boundaries in the sentence. The models are trained using the decision tree based state tying procedure using HTK [2]. The triphone models are trained with TIMIT speech corpus. To adapt the HMMs to our scenario, the models are also trained with our own training data.

To recognize the triphone sequence for a given sentence, a bigram language model on triphone level is also used, which is created with respect to the triphone transcriptions of the sentences in the training set of TIMIT speech corpus and our speech data.

3.2 Isolated Word Recognition

The isolated word recognition module consists of five heteroassociative memories. Fig. 2 shows the general structure of the module. Each box denotes a heteroassociative memory, the arrows correspond to auto- or heteroassociative connections. Due to the large number of triphones, we decided to use diphones as basic processing units in the neural network for isolated word recognition. This requires a translation of the triphones from the HMMs into diphones before they are processed by the word recognition network. The total number of diphones used in the memories is 1600.

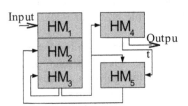

Fig. 2. Isolated word recognition module. It consists of five heteroassociative memories (depicted as gray boxes) that are connected by auto- or heteroassociations respectively (arrows). The input diphones extracted from the HMMs is sent to HM1, while the output words are taken from HM4 after a complete word has been processed (see text for details).

The word boundaries are determined with respect to diphones in the input sequence: The HMMs detect diphones due to a language model which knows about common di- and triphones. In our experiments word boundaries were recognized quite stable using this approach.

The tasks of the heteroassociative memories HM_1 through HM_5 are described in more detail in the following:

HM_1 is a matrix of dimension $L \times n$, where L is the length of the input code vector and n is the number of diphones, ($n=1600$ in our system). The memory receives input diphones from the HMMs and presents them to the word recognition network as 1 out of n code.

HM_2 has the same structure as HM_1, but receives diphones expected as input in the next step (see HM_5 below).

HM_3 is a matrix of dimension $n \times n$. It stores diphone transitions of the words in the vocabulary.

The memories HM_1, HM_2 and HM_3 can be regarded as one combined memory HM_*. In each retrieval step, HM_1 represents the diphones from the HMMs, while HM_2 and HM_3 represent the diphones predicted by the network from the previous input. The outputs of the three memories are summed up such that the influence of HM_2 and HM_3 is reduced compared to that of HM_1, giving the acoustic input a higher priority. After summation, a global threshold is applied.

The network generates a list of already heard diphones and presents it to memory HM_4. Meanwhile, the currently heard diphone(s) are presented to memory HM_5.

HM_4 is a matrix of dimension $n \times M$, where M is the number of output units, (M=200). The memory is used to store all words known to the system and to activate those that match the current list of diphones, generating a list of word hypotheses. During retrieval, the activated words are forwarded to memory HM$_5$. The output pattern is a randomly generated 5 out of 200 code vector.

HM_5 is used to predict the diphones expected in the next step. It takes the word hypotheses from HM_4 and the currently heard diphone into account and tries to predict which diphones are to be expected in the next step.

HM_5 is organized in columns. Each column belongs to one specific diphone. For each word containing that diphone, the corresponding column stores a heteroassociation from the word representation (input) to the diphone following the column-specific diphone in that word (output). By exciting only those columns that belong to diphones matching the current output of memory HM_*, HM_5 generates a prediction which diphones are about to enter the network in the next time step.

In praxis, the input to HM_5 is not a single word but a superposition of possible words generated in HM_4, thus, the output is usually not unique but also a superposition of possible diphones. Furthermore it is possible that the same diphone occurs twice in a word and thus the output can be ambiguous even if the memory is addressed with a single word.

Retrieval. When input is given to the network as a sequence of triphones from the HMMs, e.g. "sil b+ow b-ow+t ow-t sh+ow sh-ow b+l b-l+uw l-uw p+l p-l+ah l-ah+m ah-m sil" for the spoken sentence "bot show blue plum", the first step is to divide it into subsequences for isolated words with respect to diphones, e.g. "sil b+ow b-ow+t ow-t / sh+ow sh-ow ...", where "/" denotes the word boundaries. Afterwards, the subsequence of triphones is decomposed into diphones, because the following parts of the network are processing diphones to limit the use of memory resources.

For the subsequence "b+l l+uw uw", firstly the diphone "b+l" enters the network. For the first diphone of each word, the memories HM_2 and HM_3 are not activated and the output of HM_* is the input diphone itself. The resulting diphone is then forwarded to HM_4 to generate a superposition of the words which contain the diphone "b+l". For the first diphone "b+l", the words "blue", "black" and "table" are activated simultaneously. The output words from HM_4 are then sent to HM_5 to predict the next possible diphones.

Fig. 3 illustrates the module when processing the last two diphones. In the next step, HM_1 receives the second diphone "l+uw", HM_5 predicts the next diphones and

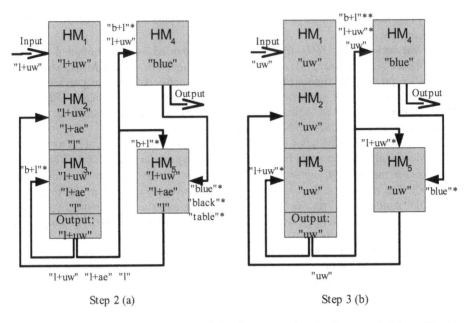

Fig. 3. The isolated word recognition module when processing (a) the second diphone "l+uw" and (b) the last phoneme "uw". The text on the arrows denotes the input that the areas receive via the corresponding connection. * means that the input is from the previous step (delay), ** means delay by two time steps.

forwards them to HM_2. HM_3 takes the output diphones of HM_* in the previous step as input. The resulting diphone "l+uw" in HM_* is added to the diphone in the first retrieval step and forwarded to HM_4. In HM_4, the word "blue" is activated. In the following step, the last phoneme "uw" is processed by the network in the same way.

After retrieving a complete word, the module is reset to its initial state in order to start with the next word in the input stream. In particular, to "reset" the module means to delete the list of already processed diphones that is fed into HM_4. After successfully retrieving a word it is forwarded to the sentence recognition system.

The experimental result on our test set (see Section 3.1) shows that our system slightly outperforms the adapted HMMs: the system recognized 98% of the words (2% word level error), while a set of adapted HMMs recognized 96% of the words correctly. Due to the fault-tolerance property of associative memories, the system is able to deal with spurious inputs, such as incorrectly recognized subword units. Thus, the correct word can be retrieved in spite of possible a corresponding noisy or incomplete subword-unit transcriptions to a certain degree.

3.3 Sentence Recognition

The sentence recognition network parses the stream of words detected by the isolated word recognizer with respect to a certain grammar. Fig. 4 gives a short overview of this module.

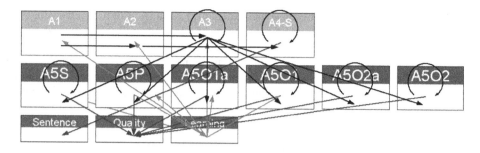

Fig. 4. The sentence recognition network. Each box corresponds to one autoassociative memory, arrows denote heteroassociations. Boxes with circular arrows use the autoassociative short-term memory (e.g. they stabilize patterns in time).

The module is composed of several autoassociative memories (depicted as boxes in Fig. 4) that are connected by heteroassociations (arrows in Fig. 4). Heteroassociations are used to exchange activation between the associative memories and to translate between the different neural representations used in different autoassociative memories. After a sentence is successfully processed, the sentence module has determined the type of sentence (e.g. Subject-predicate-object, SPO) and assigned to each word its grammatical role.

The memories in the sentence module use the so called "spike counter model" of associative memory [10]. Each memory stores a set of different patterns (assemblies). To allow for easy display of the network state, each pattern has an associated name stored in a simple look-up table. Whenever a network activity is displayed or a name of a pattern is mentioned, we really mean that the underlying group of neurons is activated.

The different memories in the module serve the following purposes:

A1 is the input memory. Activation from the isolated word recognizer arrives here. This memory is only required to allow for developing and testing the module individually. A1 holds all words known to the system in the 5 out of 200-code from the isolated word recognition module.

A2 and A3 distinguish between semantic elements (words, A3) and syntactical elements like word boundary signals and sentence end markers (arbitrary patterns, A2) respectively.

A4-S is a sequence memory. Sequences are realized by an additional hetero-association from the memory onto itself that stores the state transitions of the sequence elements. The heteroassociation is delayed and weaker than the auto-association, so normally an active pattern is stable. With short inhibition, the autoassociation can be inhibited and the heteroassociation, which is still effective because of its higher delay, will activate the next pattern in the sequence.

A4-S holds the grammatical information, i.e. the sentence types that are known to the system. It stores sequences like S->P->O->OK_SPO or S->P->A->O->A->O, where S means subject, P means predicate, A adjective and O object.

A5S, A5P, A5O1a, A5O1, A5O2a, A5O2: These memories hold the subject, predicate, attribute to first object, first object, attribute to second object and second object of the sentence, respectively. They are subsequently filled while the sentence is parsed against the grammar. Because these memories use a short term mechanism, the words assigned to them are active for a longer period of time, allowing to access the information later, e.g. for action planning.

The other fields (Sentence, Quality, Learning) are additional status information fields that can distinguish between only two states. The Sentence-memory activates a "complete"-state when a complete sentence was correctly parsed, the Quality-memory activates a "good"-pattern when there were no ambiguities on single word level (see below) and the Learning-box activates a special learn signal when a "this is"-sentence is recognized.

4 Disambiguation

The system is able to deal with ambiguities on the single word level. When a word was not uniquely understood by the HMM (e.g. an ambiguity between "ball" and "wall"), the isolated word recognition module is not able to assign a unique interpretation to the stream of triphones generated by the HMMs. In this case, a superposition of several words that all match the stream closely is activated and forwarded to the sentence recognition network. The sentence recognizer is able to deal with that ambiguity and keeps it active until further context information can be used to resolve the ambiguity. The ambiguous sentence "bot lift bwall", where "bwall" is an ambiguity between ball and wall, can be resolved to "bot lift ball" in the sentence processing network because a wall is not a liftable object. To achieve this disambiguation a bidirectional link between A5P and A5O1 is used, which supports matching pairs of predicates and objects.

5 Online Learning

New object words can be learned by the system during performance without further training the triphone models or changing the structure of the system. The online learning performance highly depends on the performance of HMMs that need to be trained with enough speech data and have a comprehensive language model in order to enable the HMMs to generate a plausible phonetic representation for novel words. Learning of a new word is initiated by a sentence of the type "This is cup", where "cup" is the word that has to be learned. "This is" arouses the system to learn a new object word.

While learning of a novel word, in the isolated word recognition module, HM_1 and HM_2 are not updated, whereas HM_3 is updated according to the sequence of diphones of the novel word. To store the new object word in HM_4 and HM_5, a new word vector is randomly generated and stored in the associative memories.

The representation of the novel word is then forwarded to the sentence recognition network. The associative memories used here employ the so called "spike counter model" [10] which allows for automatic detection of novel address patterns. The modules are able to measure the retrieval quality, where a low quality indicates that the address pattern did not match any of the stored patterns closely. If this is the case and an additional learn signal is applied to the memory, it generates a new pattern and stores it autoassociatively. If however the address pattern matches a stored pattern closely, no new pattern is generated even if a learn signal is applied. This ensures that already known stimuli will be recognized. After learning a new pattern, the heteroassociations between the concerned memories are updated according to a simple binary Hebbian learning rule.

In the special case of learning new object words, the sentence recognition network uses the phrase "This is" to determine that a new word is probably going to be learned. If "This is" was heard, the Learning-memory (see Fig. 4) activates a special pattern responsible for learning new object words. If this pattern is active, the neurons emit a special learn signal to all memories in the sentence module that are concerned with object words.

After learning a new object word, the system is able to deal with it in exactly the same way as it deals with the vocabulary it knows from initialization. If e.g. "cup" was learned during performance by the command "this is cup", the system can afterwards understand sentences like "bot show cup", "bot lift cup" etc.

At the current state of our work, in order to demonstrate the online learning capability of the system, we initially stored 40 out of 43 words in the system and the remaining 3 words were used for learning during performance. This is due to the fact that we did not yet record enough speech material for our own speakers to train the HMMs such that they can generate plausible phonetic representation for arbitrary new words.

6 Discussion

We have presented a system that is able to understand spoken language and to transform low level subsymbolic auditory data into corresponding symbolic representations that allow for easy deduction of the meaning of the sentence.

The system is composed of three rather independent main components, the HMM triphone recognizer, the isolated word recognizer and the grammar parser (also called sentence recognizer). Currently, this three components are interconnected in a feed forward manner due to the early stage of development of especially the isolated word recognizer. In particular a feedback connection from the grammar parser to the isolated word recognizer would be beneficial as a connection would allow the word recognizer to focus on words that fit for example grammatical or semantical constraints arising while the current sentence is parsed. Although we did not yet implement this kind of connection, we plan to add feedback connections in the near future.

The system is expected to scale well with the size of the vocabulary. It is well known that associative memories store information efficiently in terms of storage capacity [6][7].

The word recognition network additionally allows for different coding schemes and subword units to be used. In particular, the latter can be either phonemes, context dependent phonemes such as diphones or triphones, semi-syllables or syllables. The dimensionality of the memories will then scale with respect to the number of subword units. For the vocabularies of small size, it is convenient to work with phonemes or diphones in heteroassociative memories instead of triphones (e.g. for a given set of 40 phonemes, the corresponding set of diphones is composed of up to 1600 diphones whereas the total number of corresponding triphones is around 10000). On the other hand, for large vocabularies it is more convenient to use triphones or semi-syllabels.

The sentence understanding network can operate using random k-out-of-n codes for word representation and the dimensionality of the memories can be chosen such that high storage capacity can be reached for a given vocabulary. This means in particular, that memory requirements grow less then linearly with the number of words to be stored. For increased, more complex grammars, population A4-G has to be scaled. Almost the same arguments apply here. Due to the efficient tree-like structure of the grammatical representation, the network scales rather well with respect to a more complex grammar also. The computational complexity in terms of computing time scales basically linearly with the number of neurons in the system, as long as the overall spike activity does not increase significantly (i.e., if the number of active one entries in the code vectors is held constant). As additional sparseness also increases the storage capacity, the above constraint is however fulfilled in practical systems. Note further that the number of neurons does increase less than linearly with the size of the vocabulary, this is also true for the computational time required.

To show the correct semantical understanding of parsed sentences by the system, the model is embedded into a robot [11]. Therefore, the system is extended by a neural action planning part, some simple motor programs (e.g. to pan and tilt the camera) and a RBF-based object recognition system. The scenario is a robot standing in front of a white table with fruits lying on it, and the robot has to perform actions corresponding to spoken commands. We can demonstrate that the robot is able to perform simple actions like e.g. showing a plum by centering it in its camera image after a corresponding command like "bot show plum" was given.

In the robotics scenario described above, the system is not only able to learn new object words but also to bind the words to visual representations (this happens in the action planning network) and also to learn new visual object representations and bind them to object words (new or already known objects). This allows e.g. to generate synonyms by introducing new object words and to bind them to an already known visual object.

If we compare the system represented in this study with a HMM based word recognition system in terms of learning of new words, HMMs have to make some modifications to the pronouncing dictionary and task grammar at each time to add a new word. In contrast, the presented system is capable of learning new word representations by simply adding new patterns to associative memories; it does not need structural and time-consuming modifications as long as the associative memories are not overloaded.

Our approach to language understanding should be understood in the context of a larger model that integrates language understanding, visual object recognition, and action planning in a functional, large-scale neural network in a robotics scenario Parts of the presented work have been developed earlier within the MirrorBot project of the European Union[1], and results on sentence understanding [10] and the robotics application [11] have been published earlier. Most other approaches deal only with one of the aspects mentioned above at a time. Closely related to our work are the approaches of Arbib [12], Roy [13], Kirchmar and Edelmann [14] and of Billard and Hayes [15].

---

[1] See http://www.his.sunderland.ac.uk/mirrorbot/

However, to our knowledge this is the first robot control architecture including simple language understanding, visual object recognition and action planning, that is realized completely by neural networks [11] and that is able to resolve ambiguities and to learn new words during performance [10]. It also represents the first real-time functional simulation of populations of spiking neurons in more than ten cortex areas in cooperation.

References

1. Rabiner, L., Juang, B.H.: Fundamentals of Speech Recognition. Prentice-Hall, Inc, Englewood Cliffs (1993)
2. Young, S., et al.: The HTK Book (for HTK Version 3.2.1). Cambridge University Engineering Department, Cambridge (2002)
3. Hebb, D.O.: The Organization of Behaviour. John Wiley, New York (1949)
4. Schwenker, F., Sommer, F., Palm, G.: Iterative Retrieval of Sparsely Coded Associative Memory Patterns. Neural Networks 9, 445–455 (1996)
5. Willshaw, D., Buneman, O., Longuet-Higgins, H.: Non-holographic Associative Memory. Nature 222, 960–962 (1969)
6. Palm, G.: On Associative Memory. Biological Cybernetics 36, 19–31 (1980)
7. Palm, G.: Memory Capacities of Local Rules for Synaptic Modification. A Comparative Review, Concepts in Neuroscience 2, 97–128 (1991)
8. Knoblauch, A., Palm, G.: Pattern Separation and Synchronization in Spiking Associative Memories and Visual Areas. Neural Networks 14, 763–780 (2001)
9. TIMIT Acoustic-Phonetic Continuous Speech Corpus. National Institute of Standards and Technology, Speech Disc 1-1.1, NTIS Order No. PB91-505065 (1990)
10. Markert, H., Knoblauch, A., Palm, G.: Modelling of syntactical processing in the cortex. Biosystems 89, 300–315 (2007)
11. Fay, R., Kaufmann, U., Knoblauch, A., Markert, H., Palm, G.: Combining Visual Attention, Object Recognition and Associative Information Processing in a Neurobotic System. In: Wermter, S., Palm, G., Elshaw, M. (eds.) Biomimetic Neural Learning for Intelligent Robots. LNCS (LNAI), vol. 3575, pp. 118–143. Springer, Heidelberg (2005)
12. Arbib, M.A., Billard, A., Iacoboni, M., Oztop, E.: Synthetic brain imaging: grasping, mirror neurons and imitation. Neural Networks 13(8/9), 931–997 (2000)
13. Roy, D.: Learning visually grounded words and syntax for a scene description task. Comput. Speech Lang. 16(3), 353–385 (2002)
14. Kirchmar, J.L., Edelman, G.: Machine psychology: autonomous behavior, perceptual categorization and conditioning in a brain-based device. Cereb. Cortex 12(8), 818–830 (2002)
15. Billard, A., Hayes, G.: DRAMA, a connectionist architecture for control and learning in autonomous robots. Adapt. Behav. J. 7(1), 35–64 (1999)

Multi-class Vehicle Type Recognition System

Xavier Clady[1], Pablo Negri[1], Maurice Milgram[1], and Raphael Poulenard[2]

[1] Université Pierre et Marie Curie-Paris 6, CNRS FRE 2907
Institut des Systèmes Intelligents et Robotique
[2] LPR Editor - Montpellier
xavier.clady@upmc.fr

Abstract. This paper presents a framework for multiclass vehicle type (Make and Model) identification based on oriented contour points. A method to construct a model from several frontal vehicle images is presented. Employing this model, three voting algorithms and a distance error allows to measure the similarity between an input instance and the data bases classes. These scores could be combined to design a discriminant function. We present too a second classification stage that employ scores like vectors. A nearest-neighbor algorithm is used to determine the vehicle type. This method have been tested on a realistic data set (830 images containing 50 different vehicle classes) obtaining similar results for equivalent recognition frameworks with different features selections [12]. The system also shows to be robust to partial occlusions.

1 Introduction

Many vision based Intelligent Transport Systems are dedicated to detect, track or recognize vehicles in image sequences. Three main applications can be distinguished. Firstly, embedded cameras allow to detect obstacles and to compute distances from the equiped vehicle [15]. Secondly, road monitoring measures traffic flow [2], notifies the health services in case of an accident or informes the police in case of a driving fault. Finally, Vehicle based access control systems for buildings or outdoor sites have to authentify incoming (or outcoming) cars [12]. The first application has to classify region-of-interest (ROI) in two classes: vehicles or background. Vehicles are localized in an image with 2D or 3D bounding box [10,15]. The second one can use geometric models in addition to classify vehicles in some categories such sedans, minivans or SUV. These 2D or 3D geometric models are defined by deformable or parametric vehicle templates [5,6,7].

Rather than these two systems, the third one uses often only the recognition of a small part of vehicle : the license plate. It is enough to identify a vehicle, but in practice the vision based number plate recognition system can provide a wrong information, due to a poor image quality or a fake plate. Combining such systems with others process dedicated to identify vehicle type (brand and model) the authentication can be increased in robustness (see fig. 1). This paper adresses the identification problem of a vehicle type from a vehicle greyscale frontal image: the input of the system is an unknown vehicle class, that the system has to determine from a data base.

L. Prevost, S. Marinai, and F. Schwenker (Eds.): ANNPR 2008, LNAI 5064, pp. 228–239, 2008.

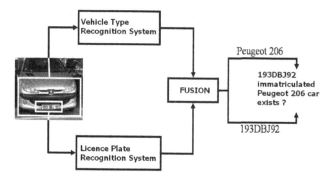

Fig. 1. The fusion system

Few papers deal with a similar problem. In a recognition framework for rigid object recognition, Petrovic and Cootes [12] tested various features for vehicle type classification. Their decision module is based on two distance measures (with or without Principal Component Analysis pre-stage): the dot product $d = 1 - f_1 f_2$ and the Euclidean measure $d = |f_1 - f_2|$, where f_i is the feature vectors. The dot product gives slighthly outperforming results. Best results are obtained with gradients based representations. These results can be explained because the vehicle rigid structure is standardized by the manufacturer for each model. The relevant information contained in contour edge and orientation is independent of the vehicle color. Daniel T.Munroe et al [14] studied machine learning classification techniques applied on features vectors (extracted with a Canny edge detector). L. Dlagnekov [3] used Scale Invariant Feature Transforms (SIFT) to compute and match keypoints. Zafar et al. [18] used a similar algorithm. David A. Torres[16] extended the work of Dlagnekov by replacing the SIFT features with features which characterize contour lines. In [9], Kazemi et al investigated use of Fast Fourier Transforms, Discrete Wavelet Transforms and Discrete Curvelet Transforms based image features. All these works used gradient or contour based features.

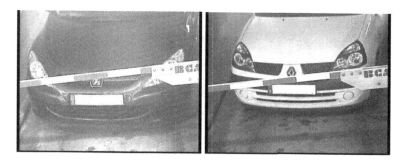

Fig. 2. Real vehicle images with the tollgate presence

In this paper, a multiclass recognition system is developed using the oriented-contour pixels to represent each vehicle class. The system analyses a vehicle frontal view identifying the instance as the most similar model class in the data base. The classification is based on a voting process and a Euclidean edge distance. The algorithm have to deal with partial occlusions. Tollgates hide a part of the vehicle (see fig. 2) and making inadequate the appearance-based methods. In spite of tollgate presence, our system doesn't have to change the training base or apply time-consuming reconstruction process.

In section 2, we explain how we define a model for every class in the data base using the oriented-contour points. Section 3 employs this model to obtain *scores* measuring the similarity between the input instance and the data bases classes. These scores could be combined to design a discriminant function. We present too a second classification stage that employ scores like vectors. A nearest-neighbor algorithm is used to classify the vehicle type. Results of our system are presented in the section 4. We finish with conclusions and perspectives.

2   Model Creation

During the initial phase of our algorithm, we produce a model for all the K vehicle types classes composing the system knowledge. The list of classes the system is capable to recognize is called Knowledge Base (**KnB**). In our system, the Knowledge Base will be the **50** vehicle type classes.

2.1   Images Databases

All ours experiments have been carried out on the Training Base (**TrB**) and on the Test Base (**TsB**). The **TrB** samples (291 images) are used to produce the oriented-contour point models of the vehicle classes. While the **TsB** samples (830 images) are utilized to evaluate the performance of the classification system. In figure 3, the upper row shows samples from **TrB** and in the bottom row, the figure shows the corresponding vehicle type class of the **TsB**. These databases are composed of frontal vehicle views, captured in different car parks, under different light conditions and different points of view.

Fig. 3. In the upper row, the figure shows samples from **TrB**. In the bottom row, the figure shows the corresponding vehicle type class of the **TsB**.

2.2   Prototype Image

We create a canonical rear-viewed vehicle image I from the four corner points of the license plate {**A,B,C,D**} (see fig. 4). The image templates are called prototypes and in the present work are 600 * 252 pixels (rows * columns). A ROI defined by the points {**A,B,C,D**} is independent of the vehicle location in the image and the scale (fig.4.a). In order to correct the orientation of the original image (see example in fig.3), an affine transformation moves original points {**A,B,C,D**} to the desired {**A',B',C',D'**} reference position, considering the vehicle grille and the license plate in the same plane. A license plate recognition system provides the corners of the vehicle license plate.

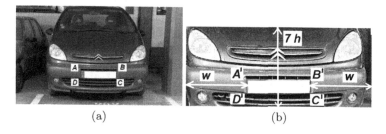

(a)   (b)

Fig. 4. (a) original image, (b) prototype I

The Sobel operator is used to compute the gradient's magnitude and orientation of the greyscale prototype I ($|\nabla g_I|, \phi_I$). An oriented-contours points matrix $\mathbf{E}_I$ is obtained using an histogram based threshold process. Each edge point $\mathbf{p}_i$ of $\mathbf{E}_I$ is considered as a vector in $\Re^3$: $\mathbf{p}_i=[x,y,o]'$, where (x,y) is the point position, and o is the gradient orientation of $\mathbf{p}_i$ [11]. We sample the gradient orientations to N bins. To manage the cases of vehicles of the same type but with different colors, the modulus π is used instead of the modulus 2π [1]. In the present application, $N = 4$.

2.3   Model Features

Oriented-Contour points features array. Each class in the **KnB** is represented by n prototypes in the **TrB**. This quantity n varies from class to class, having some defined with one prototype only.

Superposing the n prototypes of the class k, we find an array of the redundant oriented-contour points. This feature array of Oriented-Contour based points models this class in the **KnB**. The algorithm operates the n prototypes of the class k in the **TrB** by couples (having $C_{n,2}$ couples at all). Let be $(\mathbf{E}_i, \mathbf{E}_j)$ a couple of Oriented-Contour Points matrix of the prototypes 1 and 2 from the k class. We define an 600x252xN accumulator matrix A_{ij} and the vote process is as follow: a) taking a point $\mathbf{p}_i$ of $\mathbf{E}_i$, we seek in $\mathbf{E}_j$ the nearest point $\mathbf{p}_j$ with the same gradient orientation; b) the algorithm increments the accumulator A_{ij}

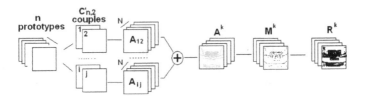

Fig. 5. Model creation

in the middle point of $\overline{\mathbf{p}_i\mathbf{p}_j}$ at the same gradient orientation; c) the procedure is repeated for all the points $\mathbf{p}_i$ of $\mathbf{E}_i$. Considering the addition of all A_{ij} we obtain the accumulator array A^k: $A^k = \sum_{i,j} A_{ij}$. The most voted points $\mathbf{a}_m = [x,y,o]$ of A^k are selected iteratively. We impose a distance of 5 pixels between the $\mathbf{a}_m$ in order to obtain a homogeneous distribution of the model points. We store $\mathbf{a}_m$ in a feature array $\mathbf{M}^k$. The array $\mathbf{M}^k$ contains the Oriented-Contour Points that are rather stable through the n samples of the class k.

When $n = 1$, the accumulator matrix A^k cannot be computed: the feature array $\mathbf{M}^k$ is then determined from the maximum values of the gradient magnitude $|\nabla g_I|$.

Weighted Matrix. The Chamfer distance is applied to determine the distance from every picture element to the given $\mathbf{M}^k$ set (fig. 6). This figure shows the four R_i^k Chamfer region matrix (one for each gradient orientation) obtained after threshold the Chamfer chart matrix D_i^k with the distances smaller than r.

Two weighted regions arrays W_+^k and W_-^k will be created for each class k. W_+^k is based on the R^k region matrix where each pixel point has a weight related to the discrimination power of the corresponding oriented-contour points. Pixel

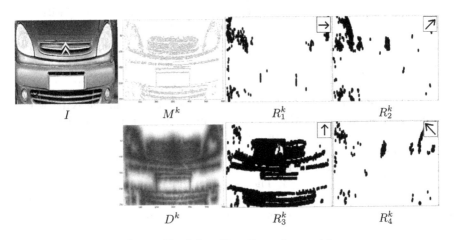

Fig. 6. Obtaining Chamfer region matrix

points rarely present in the others classes obtain highest weights. We give lowest weights to the points present in the majority of the Knowledge Base classes.

$$W_+^k = \frac{1}{K-1} \sum_{i,i\neq k} (R^k - R^i \cap R^k)$$

W_-^k gives a negative weight to the points of the other models which are not present in the matrix R^k of the model k. Pixel points that are present in most of the other classes obtain highest weight values. In the other hand, pixel points present in few classes get lowest weight values.

$$W_-^k = -\frac{1}{K-1} \sum_{i,i\neq k} (R^i - R^i \cap R^k)$$

The K classes in the **KnB** are modelled by $\{\mathcal{M}_1, ..., \mathcal{M}_K\}$, where each $\mathcal{M}_k = \{\mathbf{M}^k, W_+^k, W_-^k\}$.

3   Classification

This section develops the methods to classify the samples providing from the $\mathcal{TfB}$ using the models $\mathcal{M}_k$. A new instance t is evaluated on the classification function $G(t) = ArgMax\{g_1(t), ..., g_K(t)\}$ using the *winner-take-all* rule. The example t is labelled by $k \in \mathcal{K}$ from the highest score of the g_k. Two types of matching scores compose the g_k (see fig. 7). The first obtains a score based on three kind of votes (positive, negative and class votes) for each class k. The second score evaluates the distance between the oriented-contour points of the model $\mathbf{M}^k$ to the oriented-contour points of t.

Obtaining the image prototype of the sample t from the Test Base, we calculate the oriented-contour points matrix $\mathbf{E}_t$ (section 2.2). Considering the large number of points in $\mathbf{E}_t$, we have to choose a limited set of T points. The value

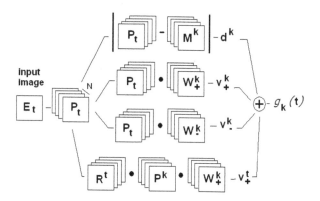

Fig. 7. Obtaining the discriminant function

of T is a compromise between the computing time and a good rate of correct classifications (in our algorithm, $T = 3500$). To select these points, we construct a sorted list of the prototype positions (x, y, o). We sort in decreasing order, the values of the weighted arrays W_+^i $i = 1, ..., K$, placing the discriminant pixels (highest values) in the firsts positions of the list. Looking iteratively if the pixels in the list are present in $\mathbf{E}_t$, we pick up the T points, and place them in $\mathbf{P}_t$.

3.1 Designing the Discriminant Function

Positive votes. The methodology consists in accumulating votes for the class k, whenever a point of $\mathbf{P}_t$ falls in a neighbourhood of a $\mathbf{M}^k$ point. We define the neighbourhood of the point $\mathbf{M}^k$ as a circle of radius r around the point of interest. This neighbourhood representation is modelled in the Chamfer regions R_i^k. Moreover, each point of $\mathbf{P}_t$ votes for the class k with a different weight depending on its value in the matrix W_+^k.

The nonzero points of the dot product of $\mathbf{P}_t$ and W_+^k correspond to the points of $\mathbf{P}_t$, that belong to a neighbourhood of the $\mathbf{M}^k$'s points. Thereafter, we calculate the amount of positive votes in equation 1 where $[\bullet]$ is the dot product.

$$v_+^k = \sum_x \sum_y \sum_o \mathbf{P}_t \bullet W_+^k \qquad (1)$$

Negative votes. The negative votes take into account the points of $\mathbf{P}_t$ that did not fall into the neighbourhood of the $\mathbf{M}_k$ points. We punish the class k by accumulating these points weighted by the matrix W_-^k. The amount of negative votes is defined as:

$$v_-^k = \sum_x \sum_y \sum_o \mathbf{P}_t \bullet W_-^k$$

Votes to test. We calculate the votes from the models to the sample test. In short, the method is the same as the one detailed in the preceding section. We first build the chart of Chamfer Distances for $\mathbf{E}_t$. We keep the regions around the oriented-contour points of $\mathbf{E}_t$ which are at a distance lower than r pixels in the matrix R^t. Then, randomly selecting T points from the array $\mathbf{M}_k$, we obtain a representation of this set in an array $\mathbf{P}^k$. Each point of the matrix $\mathbf{P}^k$ is weighted by the matrix W_+^k. Total votes from the class k to the sample test t are calculated as:

$$v_+^t = \sum_x \sum_y \sum_o R^t \bullet \mathbf{P}^k \bullet W_+^k$$

Distance Error. The last score is the error measure of matching the $\mathbf{P}_t$ points with their nearest point in $\mathbf{M}_k$. Calculating the average of all the minimal distances, we obtain the error distance d^k [4]:

$$H(\mathbf{P}_t, \mathbf{M}_k) = max(h(\mathbf{P}_t, \mathbf{M}_k), h(\mathbf{M}_k, \mathbf{P}_t))$$

with :

$$h(\mathbf{P}_t, \mathbf{M}_k) = mean_{a \in \mathbf{P}_t}(min_{b \in \mathbf{M}_k} \|a - b\|)$$

Furthermore, values in the error vector have to be processed by a decreasing function considering that in the vote vectors we search for the maximum and for the error vector we search for the minimum.

3.2   Classification Strategies

We have developed two strategies for classification. The first combines the scores in a discriminant function. The second creates voting vector spaces from the scores : the decision is based on a nearest-neighbor process.

First Strategy : Discriminant Function. The four matching scores $\{v_+^k, v_-^k, v_+^t, d^k\}$ are combined in a discriminant function $g_k(t)$ matching the sample test t to the class k. A pseudo-distance of Mahalanobis normalizes the scores: $\bar{v} = (v - \mu)/\sigma$, where (μ, σ) are the mean and the standard deviation of v. The discriminant function is defined as a fusion of scores:

$$g_k(t) = \alpha_1 \, \bar{v}_+^k + \alpha_2 \, \bar{v}_-^k + \alpha_3 \, \bar{v}_+^k + \alpha_4 \, \bar{d}^k \tag{2}$$

The α_i are coefficients which weight each classifier. In our system, we give the same value for all α_i.

Finally, given the test sample t, its class label k is determined from:

$$k = G(t) = ArgMax\{g_1(t), ...g_K(t)\}$$

Second Strategy : Voting Spaces. We construct vector spaces with the results from the voting process. We define:

- $v(t) = \left[v_{+k}^{mh}, v_{-k}^{mh}, v_{+t}^{mh}, d_k^{mh}\right]_{k=1..K}$ as a vector in a $200(=4 * K)$ dimension space, called Ω_{wf} (wf = without fusion).
- $v^{PCAX}(t) = \left[v_{+k}^{mh}, v_{-k}^{mh}, v_{+t}^{mh}, d_k^{mh}\right]_{k=1..K}^{PCAX}$ as a vector in a X dimension space, called Ω_{wf}^{PCAX}, (with a Principal Component Analysis pre-stage).
- $g(t) = [g_k(t)]_{k=1..K}$ as a vector in a 50 $(=K)$ dimension space, called Ω_f (f = with fusion).

In these spaces, given the test sample t, its class label is determined as the nearest-neighbor class. It needs reference samples. We use a cross-validation process : the test database is decomposed in two equal parts. The first is used as references. The second is used for the test.

4   Results

With the first strategy, the system correctly identifies 80,2% of 830 test samples. The mean of the recognition rates per class is 69,4%.

The second strategy obtains better results (in mean, with 100 randomly different repartitions):

- in the first space, Ω_{wf}, we obtain 93,1% for the correctly identification rate (83,5% for the mean of the recognition rates per class).

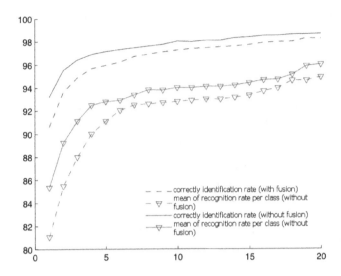

Fig. 8. CMC curves in the Ω_{wf} space (solid line) and in the Ω_f space (dashed line)

- in a second space, Ω_{wf}^{PCA50}, we obtain 86,2% for the correctly identification rate (78,8% for the mean of the recognition rates per class).
- in the last space Ω_f, we obtain 90,6% for the correctly identification rate (86,4% for the mean of the recognition rates per class).

The figure 8 shows the *Cumulative Match Characteristic* curves (CMC[1]). We clearly see that the second strategy in the first space (without fusion and without PCA) gives better results, but with a higher computational cost (due to a high dimensional space). The figure 9 shows us that, without fusion, we have to keep a space dimension higher than 100. Furthermore, a better algorithm performance could be obtained by choosing optimized values for the α_i in the equation 2[2]. Moreover, the recognition rate depends on the used reference samples proportion (see figure 10).

Another test simulates the presence of a tollgate at four different locations; in a car park access control system it is difficult to define the relative vertical position between the barrier and the vehicle even if the license plate is always visible. The results for each tollgate position are showed in figure 11. The better recognition are obtained if the virtual tollgate hides the upper part of the images: a lot of noise points are extracted from this part (see figure 13). These points

---

[1] A Cumulative Match Characteristic (CMC) curve plots the probability of identification against the returned 1:N candidate list size. It shows the probability that a given user appears in different sized candidate lists. The faster the CMC curve approaches 1, indicating that the user always appears in the candidate list of specified size, the better the matching algorithm.

[2] A training algorithm method could be used, but we have to capture more frontal view vehicle samples.

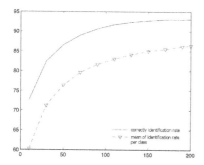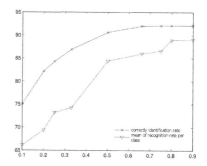

Fig. 9. Recognition rates related to the X dimension in the Ω_{wf}^{PCAX}

Fig. 10. Recognition rates related to the samples proportion used as reference, in Ω_f

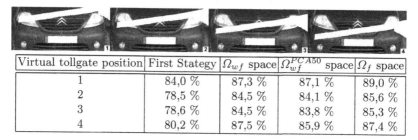

Virtual tollgate position	First Stategy	Ω_{wf} space	Ω_{wf}^{PCA50} space	Ω_f space
1	84,0 %	87,3 %	87,1 %	89,0 %
2	78,5 %	84,5 %	84,1 %	85,6 %
3	78,6 %	84,5 %	83,8 %	85,3 %
4	80,2 %	87,5 %	85,9 %	87,4 %

Fig. 11. The four positions of a virtual tollgate and the recognition rates

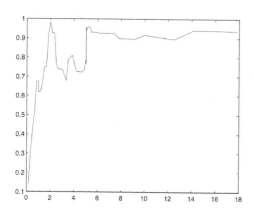

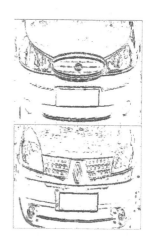

Fig. 12. Mean of the recognition rates par class, related to the images number used in the model creation.

Fig. 13. Results of the contour extraction process

perturb the recognition system. They are filtered if the number of images used in the model creation is sufficient (> 5) as we can see in the figure 12.

5   Conclusions

This article has presented a voting system for the multiclass vehicle type recognition based on Oriented-Contour Points Set. Each vehicle class is composed from one or many grayscale frontal images of one vehicle type (make and model). A discriminant function combines the scores provided from three voting based classifiers and an error distance. A second strategy consists in considering the scores as elements of a vector. A nearest-neighbor process is used to determine the vehicle type. We have tested this method on a realistic data sets of 830 frontal images of cars. The results showed that the method is robust to a partial occlusion of the patterns. The second strategy obtains better results, particularly if the fusion scores are used. Our recognition rate is over 90%. Without occlusion, our system obtains similar results as others works [12]. Future works will be oriented to reduce the influence of the images number used in the model creation process : it could be interesting to recognize the type of a vehicle with only one reference image per class.

References

1. Cootes, T., Taylor, C.: On representing edge structure for model matching. In: Conference on Vision and Pattern Recognition, Hawai, USA, December 2001, vol. 1, pp. 1114–1119 (2001)
2. Douret, J., Benosman, R.: A multi-cameras 3d volumetric method for outdoor scenes: a road traffic monitoring application. In: International Conference on Pattern Recognition, pp. III: 334–337 (2004)
3. Dlagnekov, L.: Video-based car surveillance: License plate make and model recognition. Masters Thesis, University of California at San Diego (2005)
4. Dubuisson, M., Jain, A.: A modified hausdorff distance for object matching. In: International Conference on Pattern Recognition, vol. A, pp. 566–569 (1994)
5. Dubuisson-Jolly, M., Lakshmanan, S., Jain, A.: Vehicle segmentation and classification using deformable templates. IEEE Transactions on Pattern Analysis and Machine Intelligence 18(3), 293–308 (1996)
6. Ferryman, J.M., Worrall, A.D., Sullivan, G.D., Baker, K.D.: A generic deformable model for vehicle recognition. In: British Machine Vision Conference, pp. 127–136 (1995)
7. Han, D., Leotta, M.J., Cooper, D.B., Mundy, J.L.: Vehicle class recognition from video-based on 3d curve probes. In: VS-PETS, pp. 285–292 (2005)
8. Hond, D., Spacek, L.: Distinctive descriptions for face processing. In: British Machine Vision Conference, University of Essex, UK (1997)
9. Kazemi, F.M., Samadi, S., Pooreza, H.R., Akbarzadeh-T, M.R.: Vehicle Recognition Based on Fourier, Wavelets and Curvelet Transforms - a Comparative Study. In: IEEE International conference on Information Technology, ITNG 2007 (2007)
10. Lai, A.H.S., Fung, G.S.K., Yung, N.H.C.: Vehicle type classification from visual-based dimension estimation. In: IEEE International System Conference, pp. 201–206 (2001)

11. Olson, C.F., Huttenlocher, D.P.: Automatic target recognition by matching oriented edge pixels. IEEE Transactions on Image Processing 6(1), 103–113 (1997)
12. Petrovic, V.S., Cootes, T.F.: Analysis of features for rigid structure vehicle type recognition. In: British Machine Vision Conference, vol. 2, pp. 587–596 (2004)
13. Petrovic, V.S., Cootes, T.F.: Vehicle Type Recognition with Match Refinement. In: International Conference on Pattern Recogntion, vol. 3, pp. 95–98 (2004)
14. Munroe, D.T., Madden, M.G.: Multi-Class and Single-Class Classification Approaches to Vehicle Model Recognition from Images. In: AICS (2005)
15. Sun, Z., Bebis, G., Miller, R.: On-road vehicle detection: A review. IEEE Transactions on Pattern Analysis and Machine Intelligence 28(5), 694–711 (2006)
16. Torres, D.A.: More Local Structure for Make-Model Recognition (2007)
17. Zafar, I., Acar, B.S., Edirisinghe, E.A.: Vehicle Make and Model Identification using Scale Invariant Transforms. In: IASTED (2007)
18. Zafar, I., Edirisinghe, E.A., Acar, B.S., Bez, H.E.: Two Dimensional Statistical Linear Discriminant Analysis for Real-Time Robust Vehicle Type Recognition. In: SPIE, vol. 6496 (2007)
19. Zhao, W., Chellappa, R., Phillips, P.J., Rosenfeld, A.: Face recognition: A literature survey. ACM Computing Surveys 35(4), 399–458 (2003)

A Bio-inspired Neural Model for Colour Image Segmentation

Francisco Javier Díaz-Pernas, Míriam Antón-Rodríguez,
José Fernando Díez-Higuera, and Mario Martínez-Zarzuela

Department of Signal Theory, Communications and Telematics Engineering
Telecommunications Engineering School, University of Valladolid, Valladolid, Spain
{pacper,mirant,josdie,marmar}@tel.uva.es

Abstract. This paper describes a multi-scale neural model to enhance regions and extract contours of colour-texture image taking into consideration the theory for visual information processing in the early stages of human visual system. It is composed of two main components: the Colour Opponent System (COS) and the Chromatic Segmentation System (CSS). The structure of the CSS architecture is based on BCS/FCS systems, so the proposed architecture maintains the essential qualities of the base model such as illusory contours extraction, perceptual grouping and discounting the illuminant. Experiments performed show the good visual results obtained and the robustness of the model when processing images presenting different levels of noise.

Keywords: image analysis; image segmentation; neural network; multiple scale model; Boundary Contour System; Feature Contour System; enhancing colour image regions; colour-opponent processes.

1 Introduction

Image segmentation is a difficult yet very important task in many image analysis or computer vision applications. Segmentation subdivides an image into its constituent parts or object. It depends on the global characteristics of an image, which is similar to the judgment of human perception. The proposed model covers the processing of the segmentation stage in a biological way, processing colour and textural information for intensifying regions and extracting perceptual boundaries as a previous processing, prior to region labelling, in order to form up the segmented image. This image enhancing makes labelling easier during the segmentation process and also makes it more efficient by reducing the uncertainty of the images' region allocation.

The skill of identifying, grouping and distinguishing among textures and colours is inherent to the human visual system. For the last few years many techniques and models have been proposed in the area of textures and colour analysis [5], resulting in a detailed characterisation of both parameters. Many of these initiatives, however, have used geometric models, omitting the human vision physiologic base. The main difference, and advantage, of the human visual system is the context dependence [8]; this implies that its processing cannot be simulated by means of geometric techniques

L. Prevost, S. Marinai, and F. Schwenker (Eds.): ANNPR 2008, LNAI 5064, pp. 240–251, 2008.

that obviate this information. A clear example of such a feature is the illusory contour formation, in which context data is used to complete the received information, which is partial or incomplete in many cases.

The architecture described in this work is based on the BCS/FCS neural model, composed of the Boundary Contour System and the Feature Contour System introduced by [7] and [3]. This model suggests a neural dynamics for perceptual segmentation of monochromatic visual stimuli and offers a unified analysis process for different data referring to monocular perception, grouping, textural segmentation and illusory figures perception. The BCS system obtains a map of image contours based on contrast detection processes, whereas the FCS performs diffusion processes with luminance filling-in within those regions limited by high contour activities. Consequently, regions that show certain homogeneity and are globally independent are intensified. Later versions of BCS/FCS models, [9] [15] have considered multi-scale processing, in which the system is sensitive to boundaries and surfaces of different sizes. Recent publications within the Theory of Neural Dynamics of the Visual Perception focus on the modelling of retinal processes in order to solve the problem of consistent interpretation of surface lightness under varying illumination.

In this manner, the main improvement introduced to the original model hereby in this paper, resides in offering a complete colour image processing neural architecture for extracting contours and enhancing the homogeneous areas in an image. In order to do this, the neural architecture develops processing stages, coming from the original RGB image up to the segmentation level, following analogous behaviours to those of the early mammalian visual system. This adaptation has been performed by trying to preserve the original BCS/FCS model structure and its qualities, establishing a parallelism among different visual information channels and modelling physiological behaviours of the visual system processes. Therefore, the region enhancement is based on the feature extraction and perceptual grouping of region points with similar and distinctive values of luminance, colour, texture and shading information.

The paper is organized as follows. In Section 2, we review the segmentation algorithm. Section 3 studies its performance over input images presenting different noise levels; and finally a conclusion is drawn in Section 4.

2 Proposed Neural Model

The architecture of the proposed model (Fig. 1) comprises of two main components, designated respectively Colour Opponent System (COS) and Chromatic Segmentation System (CSS).

The COS module transforms the chromatic components of the input signals (RGB) into a bio-inspired codification system, made up of two opponent chromatic channels, L-M and S-(L+M), and an achromatic channel. Resulting signals from COS are used as inputs for the CSS module where the contour map extraction and two intensified region images, corresponding to the enhancement of L-M and S-(L+M) opponent chromatic channels, are generated in multiple scale processing.

The final output is constituted by three components: an image contour map and two intensified region images corresponding to the enhancement of aforementioned L-M and S-(L+M) opponent chromatic channels.

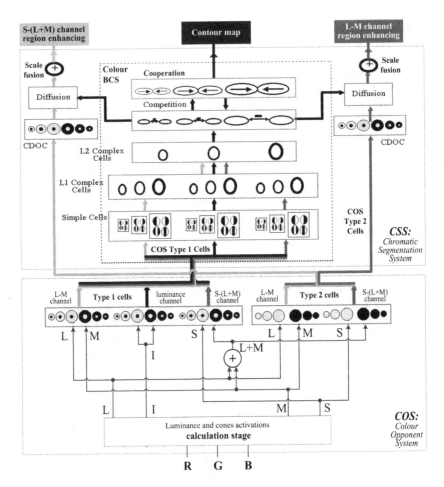

Fig. 1. Proposed model architecture. At the bottom, the detailed COS module structure: on the left, it shows type 1 cells whereas on the right, elements correspond to type 2 opponent cells. At the top, the detailed structure of the CSS based on the BCS/FCS model.

2.1 Colour Opponent System (COS)

The COS module performs colour opponent processes based on opponent mechanisms that are present on the retina and on the LGN of the mammalian visual system [10] [17]. Firstly, luminance (I signal) and activations of the long (L), middle (M), short (S) wavelength cones and (L+M) channel activation (Y signal) are generated from R, G and B input signals. The I signal is computed as a weighted sum [5]; the L, M and S signals are obtained as the transformation matrix [11].

In the COS stage, two kinds of cells are suggested; type 1 and type 2 cells (see Fig. 1). These follow opponent profiles intended for detecting contours (type 1, simple opponency) and colour diffusion (type 2 cells initiate double opponent processes).

Type 1 Opponent Cells

Type 1 opponent cells perform a colour codification system represented by the existence of opponent L-M, S-(L+M), and luminance channels (see Fig. 1). These cells are modelled through two centre-surround multiple scale competitive networks, and form the ON and OFF channels composed of ON-centre OFF-surround and OFF-centre ON-surround competitive fields, respectively. These competitive processes establish a gain control network over the inputs from chromatic and luminance channels, maintaining the sensibility of cells to contrasts, compensating variable illumination, and normalising image intensity [6] [8]. The equations governing the activation of type 1 cells (1) have been taken from the Contrast Enhancement Stage in [9] [15], but adapted to compute colour images.

$$y_{ij}^{g+} = \left[\frac{AD^+ + BS_{ij}^c - CS_{ij}^{sg}}{A + S_{ij}^c + S_{ij}^{sg}}\right]^+ ; \qquad y_{ij}^{g-} = \left[\frac{AD^- + CS_{ij}^{sg} - BS_{ij}^c}{A + S_{ij}^{sg} + S_{ij}^c}\right]^+ \qquad (1)$$

$$S_{ij}^c = \sum_{pq} e_{i+p,j+q}^c G_{pq}^c ; \qquad S_{ij}^{sg} = \sum_{pq} e_{i+p,j+q}^{sg} G_{pq}^{sg} \qquad (2)$$

with A, B, C and D as model parameters, $[w]^+ = max(w,0)$, e^c as central signal, e^s as peripheral signal (see Table 1), the superscript $g=0,1,2$ for small, medium and large scales. The weight functions have been defined as normalised Gaussian functions for central (G^c) and peripheral (G^{sg}) connectivity.

Table 1. Inputs of different channels on type 1 opponent cells

	L-M Opponency	S-(L+M) Opponency	Luminance
e^c	L_{ij}	S_{ij}	I_{ij}
e^{sg}	M_{ij}	Y_{ij}	I_{ij}

Type 2 Opponent Cells

The type 2 opponent cells initiate the double opponent process that take place in superior level, chromatic diffusion stages (see Fig. 1). The double opponent mechanisms are fundamental in human visual colour processing [10].

The receptive fields of type 2 cells are composed of a unique Gaussian profile. Two opponent colour processes occur, corresponding L-M and S-(L+M) channels (see Fig. 1). Each opponent process is modelled by a multiplicative competitive central field, presenting simultaneously an excitation and an inhibition caused by different types of cone signals (L, M, S and Y as sum of L and M). These processes are applied over three different spatial scales in the multiple scale model shown. Equations (3) model the behaviour of these cells, ON and OFF channels, respectively.

$$x_{ij}^{g+} = \left[\frac{AD^+ + BS_{ij}^{+g}}{A + S_{ij}^{Eg}}\right]^+ ; \qquad x_{ij}^{g-} = \left[\frac{AD^- + BS_{ij}^{-g}}{A + S_{ij}^{Eg}}\right]^+ \qquad (3)$$

$$S_{ij}^{+g} = \sum_{pq} G_{pq}^g \left(e_{i+p,j+q}^{(1)} - e_{i+p,j+q}^{(2)} \right); \; S_{ij}^{-g} = \sum_{pq} G_{pq}^g \left(e_{i+p,j+q}^{(2)} - e_{i+p,j+q}^{(1)} \right) \quad (4)$$

$$S_{ij}^{Eg} = \sum_{pq} G_{pq}^g \left(e_{i+p,j+q}^{(1)} + e_{i+p,j+q}^{(2)} \right) \quad (5)$$

with A, B, C and D as model parameters, $[w]^+ = max(w,0)$, $e^{(1)}$ and $e^{(2)}$ being the input signals of the opponent process (see Table 2). The weight functions have been defined as normalised Gaussians with different central connectivity (G^g) for the different spatial scales g=0, 1, 2 (see Table 4).

Table 2. Inputs for different type 2 cells channels

	L-M Opponency	S-(L+M) Opponency
$e^{(1)}$	L_{ij}	S_{ij}
$e^{(2)}$	M_{ij}	Y_{ij}

2.2 Chromatic Segmentation System (CSS)

As previously mentioned, the Chromatic Segmentation System bases its structure on the modified BCS/FCS model [9] [15], adapting its functionality for colour image processing. The CSS module (see Fig. 1) consists of the Colour BCS stage and two chromatic diffusion stages, processing one chromatic channel each.

Colour BCS stage
The Colour BCS stage constitutes our colour extension of the original BCS model. It processes visual information from three parallel channels, two chromatic and a luminance channels to obtain a unified contour map. Analogous to the original model, the Colour BCS module has two differentiated phases: the first one (simple and complex cells) extracts real contours from the output signals of the COS and the second is represented by a competition and cooperation loop, in which real contours are completed and refined, thus generating contour interpolation and illusory contours (see Fig. 1). Colour BCS preserves all of the original model perceptual characteristics such as perceptual grouping, emergent features and illusory perception.

The achieved output coming from the competition stage is a contour map of the original image. This output is transmitted to the diffusion stages where it will act as a control signal serving as a barrier in chromatic diffusions.

Simple cells are in charge of extracting real contours from each of the chromatic and luminance channels. In this stage, the filters from the original model have been replaced by two pairs of Gabor filters with opposite polarity, due to their high sensibility to orientation, spatial frequency and position [4] [13]. Their presence has been proved on the simple cells situated at V1 area of visual cortex [16]. Fig. 2 shows a visual representation of Gabor filter pair profiles.

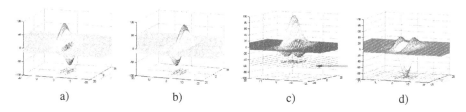

a)   b)   c)   d)

Fig. 2. Receptive fields of the filters used to model simple cells. a) Anti-symmetric light-dark receptive field. b) Anti-symmetric dark-light receptive field. c) Symmetric receptive field with central excitation. d) Symmetric receptive field with central inhibition.

The complex cell stage, using two cellular layers, fuses information from simple cells giving rise to a map which contains real contours for each of the three scales used (see Fig. 1). Detected real contours are passed into a cooperative-competitive loop. This nonlinear feedback network detects, regulates, and completes boundaries into globally consistent contrast positions and orientations, while it suppresses activations from redundant and less important contours, thus eliminating image noise. The loop completes the real contours in a consistent way generating, as a result, the illusory contours [9] [15].

Cooperation is carried out by dipole cells, which have been placed just before cortical cells in V2 area. These cells have been used to model processes such as illusory contour generation, neon colour spreading or texture segregation [7]. Dipole cells act like long-range statistical AND gates, providing active responses if they perceive enough activity over both dipole receptive fields lobes (left and right). Thus, this module performs a long-range orientation-dependent cooperation in such a way that dipole cells are excited by collinear (or close to collinearity) competition outputs and inhibited by perpendicularly oriented cells. This property is known as spatial impermeability and prevents boundary completions towards regions containing substantial amounts of perpendicular or oblique contours. The equations used in competitive and cooperative stages are taken from the original model [9].

Chromatic diffusion stages

As mentioned above, the chromatic diffusion stage has undergone changes that entailed the introduction of Chromatic Double Opponency Cells (CDOC), resulting in a new stage in the segmentation process.

In human visual system, double opponency occurs in visual striate cortex cells, contained in blobs [10]. The model for these cells has the same receptive field as COS type 1 opponent cells (centre-surround competition), but their behaviour is quite a lot more complex since they are highly sensitive to chromatic contrasts. Double opponent cell receptive fields are excited on their central region by COS type 2 opponent cells, and are inhibited by the same cell type. We apply a greater sensibility to contrast as well as a more correct attenuation toward illumination effects, therefore bringing a positive solution to the noise-saturation dilemma.

The mathematical pattern that governs the behaviour of CDOC cells is the one defined by (1) and successive equations, by varying only their inputs, that is, the outputs of the COS type 2 opponent cells for each chromatic channel (see Table 3).

Table 3. Inputs of included Chromatic Double Opponency Cells

	L-M Opponency		S-(L+M) Opponency	
e^c	$(L^+\text{-}M^-)_{ij}$	$(L^-\text{-}M^+)_{ij}$	$(S^+\text{-}Y^-)_{ij}$	$(S^-\text{-}Y^+)_{ij}$
e^{sg}	$(L^+\text{-}M^-)_{ij}$	$(L^-\text{-}M^+)_{ij}$	$(S^+\text{-}Y^-)_{ij}$	$(S^-\text{-}Y^+)_{ij}$

Chromatic diffusion stages perform four nonlinear and independent diffusions for L-M (ON and OFF) and S-Y (ON and OFF) chromatic channels. These diffusions are controlled by means of a final contour map obtained from the competition-cooperation loop while the outputs of CDOC are the signals being diffused. At this stage, each spatial position diffuses its chromatic features in all directions except those in which a boundary is detected. When boundary signals take part, they inhibit diffusion obtaining differentiated activities at each of their sides (thus separating regions with different features) [15]. By means of this process, image regions that are surrounded by closed boundaries tend to obtain uniform chromatic features, even in noise presence, and therefore producing the enhancement of the regions detected in the image. The equations that model the diffusive filling-in can be found in [9].

Scale fusion constitutes the last stage of this pre-processing architecture. A simple linear combination of the three scales (6), obtains suitable visual results at this point.

$$V_{ij} = A_0(F_{ij}^{01} - F_{ij}^{02}) + A_1(F_{ij}^{11} - F_{ij}^{12}) + A_2(F_{ij}^{21} - F_{ij}^{22}) \tag{6}$$

where A_0, A_1 and A_2 are linear combination parameters, F_{ij}^{gt} represents diffusion outputs, with g indicating the spatial scale ($g=0,1,2$) and t denoting the diffused double opponent cell, 1 for ON and 2 for OFF.

3   Tests and Results

This section introduces our tests' simulations over the proposed architecture. Selected colour images are shown in Fig. 3. The format of the images is RGB, with 24 bits per pixel. Each of them shows diverse features that, altogether, validate the model in a general way and also against changing requirements (scale, chromatic, texture...).

Fig. 3. Colour images included in the tests: Mandrill, Aerial, and Pyramid

The offset model parameters were adjusted to obtain the best visual results (see Table 4).

Table 4. Parameter values of the proposed architecture for test simulations

Type 1 COS cells L-M, S-(L+M) and Luminance Opponencies				Type 2 COS cells L-M and S-(L+M) Opponencies			
A	1000.0	σ_c	0.3	A	10000.0	σ_s	0.3
C	1.0	σ_{ss}	0.5	B	10.0	σ_m	0.8
B	1.0	σ_{sm}	1.0	D^+	10.0	σ_l	1.6
D^+, D^-	1.0	σ_{sl}	1.8	D^-	10.0		
Simple cells L-M, S-(L+M) and Luminance Opponencies				Complex cells L-M, S-(L+M) and Luminance Opponencies			
F_s	12.0	σ_{ss}	8.0	γ	1.0	ξ	0.01
F_m	8.0	σ_{sm}	12.0	κ	1.0	υ	0.01
F_l	5.0	σ_{sl}	15.0				
Competition cells				Cooperation cells			
A	3.0	σ_s	4.0	A	3.0	$C_{ws}, C_{wm},$	4.0
B	1.0	σ_m	8.0	B	1.0	C_{wl}	
C	0.2	σ_l	16.0	C_{ls}	0.2	T	2.0
G_f	1.0	σ_k	45.0	C_{lm}	1.0	μ	11.0
G_b	15.0			C_{ll}	15.0	β	0.8
CDOC cells L-M and S-(L+M) Opponencies				Diffusion stage		Scale fusion	
A	10000.0	σ_{ss}	1.2	D	150.0	A_0	8.0
D^+	1.0	σ_{sm}	5.2	δ	20000.0	A_1	2.0
D^-	1.0	σ_{sl}	10.8	ε	30.0	A_2	1.0
σ_c	0.3	σ_{ss}	1.2				

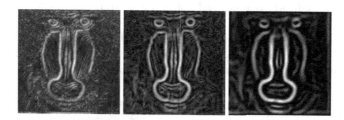

Fig. 4. Final boundary maps at three spatial scales, obtained from the competition stage

In Fig. 4, the final boundary maps obtained from the Colour BCS module are depicted. Small scale performs higher precision and contrast sensitivity even though it presents higher noise level (e.g. mandrill hair), while large scale obtains lesser precision and contrast sensitivity but also higher noise filtering. The BCS colour module has extracted the mandrill's nose contour in a continuous way. This is due to the interaction among the competition and cooperation processes which generate mechanisms of contour interpolation and illusory contour extraction.

In order to evaluate the contour extraction capabilities of the described model, we compared it to the well known Canny extractor [1], using the cvCanny() function from Intel Computer Vision Library, OpenCv [12]. Parameters were assigned using best visual result criteria. For mandrill's image we use: threshold1=2000,

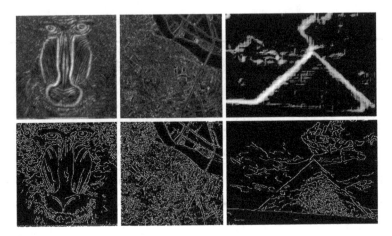

Fig. 5. Comparison with Canny's contour extraction. Top row: Our model, Bottom row: Canny.

threshold2=1000, and aperture size=5; for Aerial image: t1=200, t2=200, and a=3; and for Pyramid image: t1=10000, t2=5000, and a=7.

In Fig. 5, top row shows the contour's structure of our model while bottom row shows the output of the Canny extractor. It can be note that Canny's could not extract the mandrill's nose as a continuous contour, unlike our model. The Aerial image processing shows how the model responds to very detailed images with fine contours, demonstrating a great level of precision at small scales. Finally, the Pyramid image allows us to observe the behaviour of the model when processing large scale, obtaining well defined contours and achieving a strong elimination of the noise, which exists in Canny extraction.

Fig. 6 depicts the final outputs of the model after fusing information from the diffusion stage with the three spatial scales (see Table 4). As it can be noted, chromatic and textural features appear now levelled within each enhanced region. Our

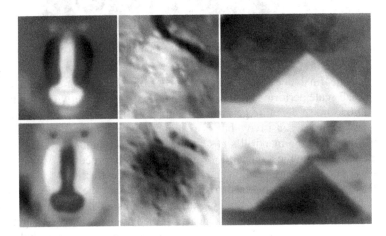

Fig. 6. Final outputs of the model. Top row: L-M channel. Bottom row: S-Y channel.

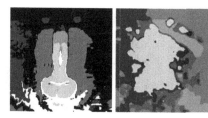

Fig. 7. Labelling of the featured model's output using a Fuzzy ART Network. Supervision parameter (ρ) is 0.9. Left: Mandrill, 7 categories/regions. Right: Aerial, 9 categories/regions.

Fig. 8. Comparison with other segmentation schemas. Left: Original RGB image labelling. Right: Results from the pyramidal segmentation.

model has enhanced the homogeneous regions of Mandrill and Aerial images. The enhancement could be used to estimate towns or places' population from aerial images. The Pyramid image, however, contains a heterogeneous sky with clouds which are distinguished as different enhanced regions by the model. This image also obtains a good visual result from a textured region such as the pyramid's surface.

In order to obtain segmentation, after enhancing regions of the image, the immediate following process will be the region labelling, that is, making all the points belonging the same region have the same region label. In fact, enhancing regions makes the labelling process easier and so more efficient. So as to prove this statement, a Fuzzy ART categorisation model [2] was chosen to run the labelling process. For each point in the image, a four component pattern, composed of the output values from chromatic diffusion stages and their complement coding, is taken, thus the segmentation image is generated by labelling all of the points in accordance with their pattern category, giving a different grey level to each category for result visualization. Fig. 7 shows the categorisation for the Mandrill and Aerial images. In both cases the Fuzzy ART supervision value ρ was 0.9; 7 and 9 categories were created respectively.

To show how the results coming from our architecture favour the segmentation process in comparison with other segmentation models, we used the Aerial image from Fig. 3. Fig. 8-left shows the RGB original image labelling using the same Fuzzy ART model, with the same value for the supervision parameter ($\rho=0.9$). This categorisation created 10 categories. Aerial image was also segmented through a pyramidal process (Fig. 8-right) using the cvPyrSegmentation() function from the OpenCv library [12]. The parameters' values were assigned using the best visual results criterion (t1=70, t2=30). By comparing Fig. 7-right with Fig. 8, it can be easily observed that the introduced model takes into account the colour and textural

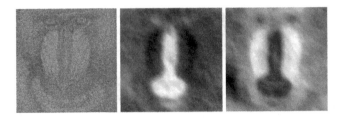

Fig. 9. Results for image with 30% of additive Gaussian noise on each channel. It can be seen that the enhancement results are visually satisfactory even with a high level of noise.

information and this is the reason why it detects a wide central region, in a compatible way with human visual perception, unlike the two other methods. Hence, it can be inferred that the presented architecture obtains far more satisfactory results than the pyramidal method.

In order to validate the architecture against adverse circumstances, some Gaussian noise has been added to each of the R, G, and B input channels. This kind of noise is usually introduced by acquisition devices on industrial applications. Consequently, noise is individually measured as a percentage of noise power over the input channel total power (noise + signal). When additive Gaussian noise is applied to input images, the system responses are satisfying up to the point when a 30% noise level is reached on each channel. As shown in Fig. 9, obtained segmentation give good visual results although detected regions are not as homogeneous as with clean images, they appear notoriously defined. It is important to stress the better performance at larger scales when noise increases, as their weight becomes higher on fusion stages.

4 Conclusion

This work presents a multiple scale neural model for extracting contours and enhancing regions in colour images. The model processes chromatic information to obtain a coherent region enhancing. BCS/FCS systems have been used as a baseline due to their broad validation and acceptance. Furthermore, these systems have been extended with the addition of the stages performing chromatic processing. The use of a multiple scale model has demonstrated a great level of flexibility and adaptation to multiple images, which lays the foundations to obtain a more general purpose architecture for colour image segmentation.

Although it increases the system's complexity and the computational load (more parallel processing and more stages are added), colour processing through type 1 and 2 COS cells and CDOC cells provides significant improvements on the segmentation and discriminates better among regions with similar luminance values but different chromatic features. Furthermore, the architecture shows a great level of parallelism and therefore computation load can be shared among different processors.

We included test simulations in order to validate the model, obtaining really satisfactory visual results. When comparing it to Canny's contour extractor we have observed that the perceptual contour extraction used by our model displays features not present in Canny's like, for example, illusory contours detection and perceptual grouping which helps us to obtain results much more in line with those of human visual perception. The results shown have covered different areas in which computer vision systems have become a helpful solution and a means of automating tools such as face recognition systems or aerial images analysis.

Finally, this paper demonstrates that the described architecture displays a visually satisfactory response against the standard Gaussian noise present in most image acquisition devices.

References

1. Canny, J.: A computational approach to edge detection. IEEE Trans. Pat. Anal. Mach. Intell. 8(6), 679–698 (1986)
2. Carpenter, G.A., Grossberg, S., Rosen, D.B.: Fuzzy ART: Fast Stable Learning and Categorization of Analog Patterns by an Adaptive Resonance System. Neural Networks 4, 759–771 (1991)
3. Cohen, M.A., Grossberg, S.: Neural dynamics of brightness perception: features, boundaries, diffusion, and resonance. Perception and Psychophysics 36, 428–456 (1984)
4. Daugman, J.G.: Two-dimensional spectral analysis of cortical receptive field profiles. Vision Research 20, 847–856 (1980)
5. Gonzalez, R.C., Woods, R.E.: Digital Image Processing, 2/E. Prentice Hall, Englewood Cliffs (2002)
6. Grossberg, S.: Contour enhancement, short term memory, and constancies in reverberating neural networks. Studies in Applied Mathematics 52, 217–257 (1973)
7. Grossberg, S.: Outline of a theory of brightness, colour, and form perception. In: Degreef, E., van Buggenhault, J. (eds.) Trends in mathematical psychology. North Holland, Amsterdam (1984)
8. Grossberg, S., Mingolla, E.: Neural dynamics of perceptual grouping: textures, boundaries, and emergent segmentations. In: Grossberg, S. (ed.) The adaptive brain II, ch. 3. North Holland, Amsterdam (1988)
9. Grossberg, S., Mingolla, E., Williamson, J.: Synthethic aperture radar processing by a multiple scale neural system for boundary and surface representation. Neural Networks 8, 1005–1028 (1995)
10. Hubel, D.H.: Eye, Brain and Vision. Scientific American Library 22, 70 (1995)
11. Hubel, D.H., Livingstone, M.S.: Color and contrast sensitivity in lateral geniculate body and Primary Visual Cortex of the Macaque Monkey. The Journal of Neuroscience 10(7), 2223–2237 (1990)
12. Intel Corporation, Open Source Computer Vision Library (2006), http://www.intel.com/technology/computing/opencv/
13. Landy, M.S., Bergen, J.R.: Texture segregation and orientation gradient. Vision Research 31(4), 679–693 (1991)
14. Mirmehdi, M., Petrou, M.: Segmentation of color textures. IEEE Trans. Pattern Analysis and Machine Intelligence 22(2), 142–159 (2000)
15. Mingolla, E., Ross, W., Grossberg, S.: A neural network for enhancing boundaries and surfaces in synthetic aperture radar images. Neural Networks 12, 499–511 (1999)
16. Pollen, D.A., Ronner, S.F.: Visual cortical neurons as localized spatial frequency filters. IEEE Transactions on Systems, Man, and Cybernetics SMC-13(15), 907–916 (1983)
17. Wilson, H.R., Levi, D., Maffei, L., Rovamo, J., De Valois, R.: The Perception of Form: Retina to Striate Cortex. In: En Spillmann, L., Werner, J.S. (eds.) Visual Perception: The Neurophysiological Foundations, ch.10. Academic Press, San Diego (1990)

Mining Software Aging Patterns by Artificial Neural Networks

Hisham El-Shishiny, Sally Deraz, and Omar Bahy

IBM Cairo Technology Development Center
P.O.B. 166 Ahram, Giza, Egypt
shishiny@eg.ibm.com,sally@eg.ibm.com,obadr024@uottawa.ca

Abstract. This paper investigates the use of Artificial Neural Networks (ANN) to mine and predict patterns in software aging phenomenon. We analyze resource usage data collected on a typical long-running software system: a web server. A Multi-Layer Perceptron feed forward Artificial Neural Network was trained on an Apache web server dataset to predict future server swap space and physical free memory resource exhaustion through ANN univariate time series forecasting and ANN nonlinear multivariate time series empirical modeling. The results were benchmarked against those obtained from non-parametric statistical techniques, parametric time series models and other empirical modeling techniques reported in the literature.

Keywords: Data Mining, Artificial Neural Network, Pattern Recognition, Software Aging.

1 Introduction

It has been observed that software applications executing continuously over a long period of time, such as Web Servers, show a degraded performance and increasing rate of failures [5]. This phenomenon has been called software aging [4]. This may be due to memory leaks, unreleased file-locks and round-off errors. Currently, researchers are looking for methods to counteract this phenomenon by what is so called software rejuvenation methods such as applying a form of preventive maintenance. This could be done by, for example, occasionally stopping the software application, cleaning its internal state and then restarting [9] to prevent unexpected future system outages. This allows for scheduled downtime at the discretion of the user, which suggests an optimal timing of software rejuvenation.

In this work, we investigate the use of Artificial Neural Networks (ANN) univariate time series forecasting and ANN nonlinear multivariate time series empirical modeling to mine and predict software aging patterns in a typical long-range software system: a web server, in order to assess ANN suitability for the analysis of the software aging phenomenon. ANN are used to forecast swap space and free physical memory of an Apache web server and results are cross benchmarked against those reported in the literature based on parametric and non-parametric statistical techniques and other empirical modeling techniques.

L. Prevost, S. Marinai, and F. Schwenker (Eds.): ANNPR 2008, LNAI 5064, pp. 252–262, 2008.
© Springer-Verlag Berlin Heidelberg 2008

This research aims at providing some empirical evidence on the effectiveness of artificial neural networks on modeling, mining and predicting software aging patterns, and the ultimate goal is an optimization model that uses the prediction of resources exhaustion as well as further information for deriving the best rejuvenation schedule.

The rest of this paper is organized as follows: in Section 2, we review related work and in section3, the data collected is described. The adopted Neural Network approach is illustrated in section 4. Finally, conclusion and future work are presented in section 5.

2   Related Work

The software aging problem is currently approached either by building analytical models for system degradation such as probability models, linear and nonlinear statistical models, expert systems and fractal base models [1, 3, 7], or by empirically studying the software systems based on measurements. Few attenpts were reported on the use of Wavelet Networks in software aging [10, 12].

The rate to which software ages is usually not constant, but depends on the time-varying system workload. Therefore, time series models are usually fitted to the data collected to help predicting the future resource usage. Attributes subject to software aging are monitored and related data is collected aiming at predicting the expected exhaustion of resources like real memory and swap space. Then, non-parametric statistical techniques and parametric time series models are employed to analyze the collected data and estimate time to exhaustion via extrapolation for each resource [5], usually assuming linear functions of time.

3   Software Aging Data

We make use of the data reported in [5] and [7] to carry on further analysis using an Artificial Neural Network approach. The collected data is from a Linux web server with an artificial load approaching its maximum optimal load level.

The setup that was used for collecting the data consisted of a server running Apache version 1.3.14 on a Linux platform, and a client connected via an Ethernet local area network. Among the system parameters of the web server monitored during a period of more than 3.5 weeks are the free physical memory and the used swap space. Data were collected during experiments in which the web server was put in a near overload condition indicating the presence of software aging.

4   The Neural Network Approach

4.1   Artificial Neural Networks

ANN is a class of flexible nonlinear models that can discover patterns adaptively from the data. Given an appropriate number of nonlinear processing units, neural

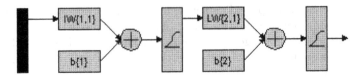

Fig. 1. The implemented MLP Neural Network

networks can learn from experience and estimate any complex functional relationship with high accuracy. Numerous successful ANN applications have been reported in the literature in a variety of fields including pattern recognition and forecasting. For a comprehensive overview of ANN the reader is referred to [8].

4.2   ANN for Mining Patterns in Software Aging

In software aging, we do not have a well defined model describing the aging process that one would like to study. All that is available are measurements of the variables of interest (i.e. time series). Therefore, we propose, in this work, an artificial neural network approach for mining software aging patterns, with the objective of predicting the expected exhaustion patterns of resources like real memory and swap space used. We investigate in this work two ANN based methods for this problem; a univariate time series forecasting method and a multivariate time series empirical modeling method.

4.3   The Proposed Neural Network Structure

Although many types of neural network models have been proposed, the most popular one is the Multi-Layer Perceptron (MLP) feed forward model [13]. A multi layer feed forward network with at least one hidden layer and a sufficient number of hidden neurons is capable of approximating any measurable function [11]. A feed-forward network can map a finite time sequence into the value that the sequence will have at some point in the future [6]. Feed forward ANNs are intrinsically non-linear, non-parametric approximators, which makes them suitable for complex prediction tasks.

For this problem, we choose to use a fully connected, MLP, feed forward ANN with one hidden layer, a logistic activation function as in figure 1, and the back propagation learning algorithm [6].

4.4   Forecasting the Exhaustion of the Apache Server Resources

We use the ANN described above and the data introduced in [5] and [7] to predict the Apache server Free Physical Memory and Swap Space Used performance variables, in order to obtain predictions about possible impending failures due to resource exhaustion. An ANN based univariate time series method is used for forecasting the Swap Space Used and an ANN based non-linear multivariate time series empirical modeling method is used to predict the Free Physical Memory.

This dataset was split into three segments; the first segment is used to train the ANN and the second segment is used to tune the ANN parameters (i.e. number of time lags and number of neurons in the hidden layer) and validation. The third segment is used to measure the ANN generalization performance on data which has not been presented to the NN during parameter tuning.

Forecasting Swap Space Used of the Apache server. The Swap Space Used of the Apache server is forecasted using ANN based univariate time series forecasting. The usage of ANN for time series analysis relies entirely on the data that were observed and is powerful enough to represent any form of time series. ANN can learn even in the case of noisy data and can represent nonlinear time series. For example, Given a series of values of the variable x at time step t and at past time steps $x(t), x(t-1), x(t-2) \cdots x(t-m)$, we look for an unknown function F such that; $X(t+n) = F[x(t), x(t-1), x(t-2) \cdots x(t-m)]$, which gives an $n - step$ predictor of order m for the quantity x.

The ANN sees the time series $X1, \cdots, Xn$ in the form of many mappings of an input vector to an output value [2]. The time-lagged values $x(t), x(t-1)$, $x(t-2) \cdots x(t-m)$ are fed as inputs to the network which once trained on many input-output pairs, gives as output the predicted value for yet unseen x values. The ANN input nodes in this case are the previous lagged observations while the output nodes are the forecast for the future values. Hidden nodes with appropriate non-linear transfer (activation) functions are used to process the information received by the input nodes.

The number of ANN input neurons determine the number of periods the neural network looks into the past when predicting the future. Whereas it has been shown that one hidden layer is generally sufficient to approximate continuous function [8], the number of hidden units necessary is not known in general. To examine the distribution of the ANN main parameters (i.e. number of time lags and number of neurons in the hidden layer), we conducted a number of experiments, where these parameters were systematically changed to explore their effect on the forecasting capabilities. These estimations of the networks most important parameters although rough, allowed us to choose reasonable parameters for our ANN.

The Swap Space Used dataset was collected on a 25-day period with connection rate of 400 per second. We divided the collected data into three segments, one to train the ANN, one for validation, and the third for testing. The testing segment is used to evaluate the forecasting performance of the ANN in predicting the performance parameters values.

The training and forecasting accuracy is measured by Root Mean Square Error (RMSE) and two other common error measures, MAPE and SMAPE.

Mean Absolute Percentage Error (MAPE). MAPE is calculated by averaging the percentage difference between the fitted (forecast) line and the original data:

$$MAPE = \sum_t |e_t/y_t| * 100/n$$

Where y represents the original series and e the original series minus the forecast, and n the number of observations.

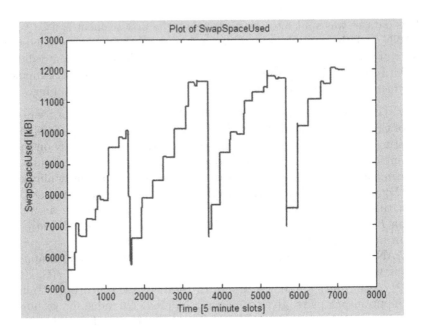

Fig. 2. Swap Space Used

Table 1. Swap Space Used evaluation

Error measures for the predicted data	SMAPE (Symmetric Mean Absolute Percent Error)	MAPE (Mean Absolute Percent Error)	RMSE (Root Mean Square Error)
Non-Parametric Statistical approach	4.313%	4.47%	612.46
ANN approach	0.354%	0.357%	116.68

Symmetric Mean Absolute Percentage Error (SMAPE). SMAPE calculates the symmetric absolute error in percent between the actual X and the forecast F across all observations t of the test set of size n. The formula is

$$MAPE = \frac{1}{n} \sum_{t=1}^{n} \frac{|X_t - F_t|}{(X_t + F_t)/2} * 100$$

Results. Figure 2 shows Swap Space Usage for the Apache server. It is clear that it follows a seasonal pattern and that considerable increases in used swap space occur at fixed intervals.

Table 1 shows the RMSE, MAPE and SMAPE for the forecasts of Swap Space Used of the Apache server for the testing dataset using the MLP described in Figure 1 with 3 input neurons (time lags), 3 neurons in the hidden layer and a sigmoid nonlinear transfer function. As seen in Table 1, the results obtained by the ANN approach are far more accurate than the results obtained by the non-parametric statistical approach reported in [5].

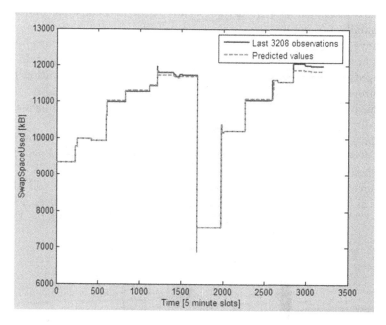

Fig. 3. Swap Space Used results

In Figure 3, we show a plot of the last 3208 observations of the measured SwapSpaceUsed (the testing dataset) and the predicted values obtained by the ANN approach, which shows accurate predictions.

Forecasting Free Physical Memory of Apache Server. In order to model and predict the Apache server Physical Free Memory performance variable, we have developed an ANN based non-linear multivariate time series empirical modeling procedure that involves parameter set reduction and selection, model building and sensitivity analysis.

Parameter set reduction and selection. Since there are 100 different Apache parameters that were monitored in addition to Free Physical Memory, an important question will be which of these parameters are the most important predictors. Some parameters may encode the same information and therefore are redundant and some others may have a trivial or no effect at all on future values of Free Physical Memory. Since parameter set reduction is a subset selection problem, therefore for 100 parameters we have 2 to the power of 100 possible subsets, which is not practical to evaluate.

In order to determine the smallest subset of input parameters which are necessary and sufficient for Free Physical Memory prediction, we have adopted the following approach:

(a) We have excluded 41 parameters because they had constant values during the monitoring period.
(b) We have performed non-linear logistic regression for the remaining 59 parameters in addition to the Free Physical Memory at time $(t-1)$, taking them as inputs to a simple one neuron ANN with a sigmoid activation function (Figure 4).
(c) We have selected seven parameters that had ANN weights values above an arbitrary small threshold (Fig. 5).
(d) We have repeated step (c) above but using a tan-sigmoid activation function and obtained nine parameters that had ANN weight values above the same threshold in step (c).
(e) We have selected the parameters in common between step (c) and (d) which were: si_tcp_tw_bucket, si_tcp_bind_bucket, si_mm_struct, si_files_cache, si_size_1024, udp_socks_high and PhysicalMemoryFree at time $t-1$.

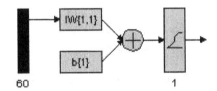

Fig. 4. A single neuron for logistic regression

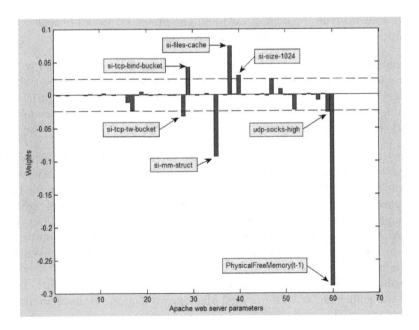

Fig. 5. ANN weights for the Apache web server parameters (Sigmoid activation function)

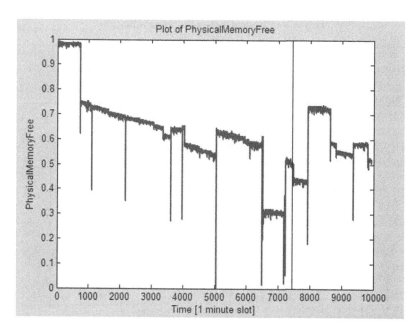

Fig. 6. Physical Free Memory

Emperical model building. Having selected the seven parameters that look more significant in section 4.4 above, we have used them as input nodes for the MLP feed forward ANN of Figure 1. We have used a sigmoid activation function and two neurons in one hidden layer (2 neurons gave the least MAPE, SMAPE and RMSE during validation of this dataset). The output node was selected to be the Free Physical memory. We therefore have formulated the problem as a non-linear multivariate time series model.

Parameter sensitivity analysis. We have conducted sensitivity analysis on the parameters of the ANN model developed in section 4.4 above in order to gain some insight into the type of interactions among the different parameters and the Free Physical Memory and to assess the contribution of each parameter on the predicted value of Free Physical Memory.

We removed one parameter at a time from the input of the developed ANN above and each time we computed the SMAPE, MAPE and RMSE on the testing dataset and recorded the change. We noted that the developed ANN model was particularly sensitive to the Free Physical Memory at time $(t - 1)$ and the si_files_cache which is in accordance with the results reported in [7].

Results. Figure 6 shows a plot over time of the Free Physical Memory of the Apache server that was collected in a 7-day period with a connection rate of 350 per second. The shown irregular utilization pattern can be explained by the fact that the Free Physical Memory cannot be lower than a preset threshold value.

Table 2. Physical Free Memory evaluation

Error measures for the predicted data	SMAPE (Symmetric Mean Absolute Percent Error)	MAPE (Mean Absolute Percent Error)	RMSE (Root Mean Square Error)
ANN approach	1.295%	1.275%	0.01354

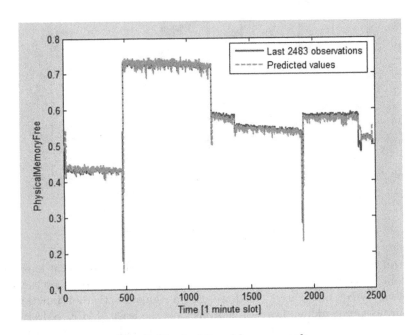

Fig. 7. Physical Free Memory results

If physical memory approaches the lower limit, the system frees up memory by paging [7].

Table 2 shows the RMSE, MAPE and SMAPE for the forecasted Free Physical Memory of the Apache server for the testing dataset using the ANN empirical model.

In Figure 7, we show a plot of the last 2483 observations of the measured physical free memory (the testing dataset) and the predicted values obtained by the ANN empirical model, which shows accurate forecasts. Based on RMSE the obtained results are more accurate than the results reported in [7] obtained from universal basis functions, multivariate linear regression, support vector machines and radial basis functions empirical modeling techniques.

5   Conclusion

In this work we have investigated the use of ANN for mining the software aging patterns in a typical long-running software system: an Apache web server.

ANN based univariate time series forecasting method and ANN nonlinear multivariate time series empirical model were developed to predict swap space used and memory usage that are related to software aging, of an Apache web server subjected to a synthetic load for 25 days. We showed that a Multi-Layer Perceptron (MLP) feed forward ANN is able to accurately predict the future behavior of these performance variables. The results obtained were benchmarked against those reported in the literature that are based on parametric and non-parametric statistical techniques and other empirical modeling techniques and were more accurate.

Future work involves extending the proposed Artificial Neural Network approach to attempt to define an optimal software rejuvenation policy.

Acknowledgements

The authors would like to thank Professor Kishor S. Trivedi, the Hudson Chair in the Department of Electrical and Computer Engineering at Duke University, for valuable discussions during the development of this work and for his review of the manuscript and Michael Grottke for providing the Apache performance dataset.

This work is part of a research project conducted at IBM Center for Advanced Studies in Cairo.

References

1. Bolch, G., Greiner, S., de Meer, H., Trivedi, K.S.: Queueing networks and Markov chains: modeling and performance evaluation with computer science applications. Wiley-Interscience, New York (1998)
2. Chakraborty, K., Mehrota, K., Mohan Chilukuri, K., Ranka, S.: Forecasting the behaviour of multivariate time series using neural networks. Neural Networks 5, 961–970 (1992)
3. Chen, M.Y., Kiciman, E., Fratkin, E., Fox, A., Brewer, E.: Pinpoint: Problem determination in large, dynamic internet services. In: DSN 2002: Proceedings of the 2002 International Conference on Dependable Systems and Networks, Washington, DC, USA, pp. 595–604. IEEE Computer Society, Los Alamitos (2002)
4. Dohi, T., Goseva-Popstojanova, K., Trivedi, K.S.: Analysis of software cost models with rejuvenation. hase, 00:25 (2000)
5. Grottke, M., Li, L., Vaidyanathan, K., Trivedi, K.S.: Analysis of software aging in a web server. IEEE Transactions on Reliability 55(3), 411–420 (2006)
6. Hassoun, M.H.: Fundamentals of Artificial Neural Networks. MIT Press, Cambridge (1995)
7. Hoffmann, G.A., Trivedi, K.S., Malek, M.: A best practice guide to resource forecasting for computing systems. IEEE Transactions on Reliability, 615–628 (2007)
8. Hornik, K., Stinchcombe, M., White, H.: Multilayer feedforward networks are universal approximators. Neural Netw. 2(5), 359–366 (1989)
9. Kolettis, N., Fulton, N.D.: Software rejuvenation: Analysis, module and applications. In: FTCS 1995: Proceedings of the Twenty-Fifth International Symposium on Fault-Tolerant Computing, Washington, DC, USA, p. 381. IEEE Computer Society, Los Alamitos (1995)

10. Ning, M.H., Yong, Q., Di, H., Ying, C., Zhong, Z.J.: Software aging prediction model based on fuzzy wavelet network with adaptive genetic algorithm. In: Proceedings of the 18th IEEE International Conference on Tools with Artificial Intelligence, pp. 659–666 (2006)
11. Siegelmann, H., Sontag Eduardo, D.: Neural nets are universal computing devices. Technical Report SYSCON-91-08, Rugters Center for Systems and Control (1991)
12. Xu, J., You, J., Zhang, K.: A neural-wavelet based methodology for software aging forecasting. In: IEEE International Conference on Systems, Man and Cybernetics, pp. 59–63 (2005)
13. Zhang, G.P., Qi, M.: Neural network forecasting for seasonal and trend time series. European Journal of Operational Research, 501–514 (2005)

Bayesian Classifiers for Predicting the Outcome of Breast Cancer Preoperative Chemotherapy

Antônio P. Braga[1], Euler G. Horta[1], René Natowicz[2], Roman Rouzier[3],
Roberto Incitti[4], Thiago S. Rodrigues[5], Marcelo A. Costa[1],
Carmen D.M. Pataro[1], and Arben Çela[2]

[1] Universidade Federal de Minas Gerais, Depto. Engenharia Eletrônica, Brazil
[2] Université Paris-Est, ESIEE-Paris, France
[3] Hôpital Tenon, Service de gynécologie, France
[4] Institut Mondor de Médecine Moléculaire, Plate-forme génomique, France
[5] Universidade Federal de Lavras, Depto. Ciência da Computação, Brazil

Abstract. Efficient predictors of the response to chemotherapy is an important issue because such predictors would make it possible to give the patients the most appropriate chemotherapy regimen. DNA microarrays appear to be of high interest for the design of such predictors. In this article we propose bayesian classifiers taking as input the expression levels of DNA probes, and a 'filtering' method for DNA probes selection.

1 Introduction

We consider the question of predicting the response of the patients to neoadjuvant chemotherapy (treatment given prior to surgery). A pathologic complete response (PCR) at surgery is correlated with an positive outcome while residual disease (NoPCR) is associated with a negative outcome. An accurate prediction of tumor sensitivity to preoperative chemotherapy is important to avoid the prescription of an inefficient treatment to patients with predicted residual disease while allocating the treatment to all the PCR patients

In [4,5] we have proposed a new method for DNA probes selection. In this paper we investigate the performances of bayesian classifiers taking as input the expression levels of the selected probes. We also compare the results with SVMs [2] and Multi-objective neural networks [6].

2 Previous Studies

The data comes from a clinical trail jointly conducted at the MD Anderson Cancer Center (MDACC) in Houston (USA) and at the Institut Gustave Roussy (IGR) in Villejuif (France) and it was used previously in K. Hess & al. [3]. The dataset was composed of 82 cases in the *training set* and 51 cases in the *test set*. A pathologic complete response (PCR) was defined as no histopathologic evidence of any residual invasive cancer cells in the breast, and a non pathologic complete response as any residual cancer cells after histopathologic study. For

L. Prevost, S. Marinai, and F. Schwenker (Eds.): ANNPR 2008, LNAI 5064, pp. 263–266, 2008.

each patient case, the data was the outcome of the treatment and the expression levels of 22283 DNA probes of an Affymetrix U133A microarray, measured on tumor cells.

Method of probes selection [5,4]. We assigned two sets of expression levels to any probe s, the sets $E_p(s)$ and $E_n(s)$, computed from the training data as follows [5]. Let $m_p(s)$ et $sd_p(s)$ be the mean and standard deviation of the expression levels of probe s for the PCR training cases, and let $m_n(s)$ and $sd_n(s)$ be that of the NoPCR training cases. The set of expression levels of the PCR training cases was defined as the set difference $E_p(s)$,

$$E_p(s) = [m_p(s) - sd_p(s), m_p(s) + sd_p(s)] \setminus [m_n(s) - sd_n(s), m_n(s) + sd_n(s)]$$

and conversely for the NoPCR training cases,

$$E_n(s) = [m_n(s) - sd_n(s), m_n(s) + sd_n(s)] \setminus [m_p(s) - sd_p(s), m_p(s) + sd_p(s)].$$

For any patient case, the individual prediction of a probe was a discrete value in set $\{pcr, nopcr, unspecified\}$: *pcr* if the expression level of patient p lied within the interval $E_p(s)$ and *nopcr* if it lied within $E_n(s)$. Otherwise, the individual prediction value was *unspecified*.

Let $p(s)$ be the number of PCR training cases correctly predicted *pcr* by probe s, and let $n(s)$ be the number of the NoPCR training cases correctly predicted *nopcr* by the probe. The valuation function of the probes was defined so as to favor probes which correctly predicted high numbers of training cases and moreover, whose sets of correctly predicted training cases were 'good' samplings of the training set. To this end, we have considered the ratios $p(s)/P$ and $n(s)/N$ of correctly predicted training cases. The valuation function $v(s)$, $v(s) \in [0,1]$, was defined as:

$$v(s) = 0.5 \times \left(\frac{p(s)}{P} + \frac{n(s)}{N} \right) \tag{1}$$

In our previous studies ([4,5]) we had defined the k-probes majority decision predictor as the set of the k top ranked probes together with the majority decision criterion: for any patient case, when the majority of 'pcr' and 'nopcr' predictions of the k top ranked probes was 'pcr', the patient was predicted to be 'PCR', and when the majority was 'nopcr' the patient was predicted to be 'NoPCR'. In case of tie the patient was predicted 'UNSPECIFIED'. When computing the performances of the predictor, a false negative was a PCR patient case predicted NoPCR or UNSPECIFIED, and conversely for the false positives.

The best k-probes predictor was for $k = 30$ probes. This number of probes was in accordance with the result of a previous study by K. Hess & al. [3].

3   Bayes Classification and Results

In order to obtain the majority classification rule, the marginal distributions of each one of the 30 probes were estimate separately for the two classes, PCR

Table 1. Comparison between all the classifiers on the test set

	Hess et al	Majority decision	Bayes		Bayes discretized		SVM	MOBJ NN
			equal priors	diff. priors	equal priors	diff. priors		
accuracy	0.76	0.86	0.88	0.86	0.88	0.82	0.88	0.88
sensitivity	0.92	0.92	0.84	0.77	0.92	0.38	0.92	0.92
specificity	0.71	0.84	0.89	0.89	0.87	0.97	0.87	0.87
PPV	0.52	0.67	0.73	0.71	0.71	0.83	0.71	0.71
NPV	0.96	0.97	0.94	0.92	0.97	0.82	0.97	0.97

(class C_{+1}) and NoPCR (class C_{-1}). Therefore, for each probe i the distributions $P(x_i|C_{+1})$ and $P(x_i|C_{-1})$ are known in advance. The majority decision rule can be viewed as:

$$\prod_{i=1}^{30} \frac{P(x_i|C_{+1})}{P(x_i|C_{-1})} = \frac{n_{+1s}}{n_{-1s}} > 1 \tag{2}$$

where n_{+1s} and n_{-1s} are, respectively, the number of +1s and -1s in the final discretized vector of the majority decision rule.

So, we can consider the majority rule as a Bayes rule with equal prior probabilities ($P(C_{+1}) = P(C_{-1})$). The effect of this is to *force a linear separation* in the input space. If the priors were used, this linear surface would *bend* in the direction of the minority class, since the NoPCR class is more likely and has larger variance.

The likelihoods $P(\mathbf{x}|C_{+1})$ and $P(\mathbf{x}|C_{-1})$ were estimated by considering independence between the probes, by $P(\mathbf{x}|C_{+1}) = \prod_{i=1}^{30} P(x_i|C_{+1})$ and $P(\mathbf{x}|C_{-1}) = \prod_{i=1}^{30} P(x_i|C_{-1})$, where $P(x_i|C_{+1})$ and $P(x_i|C_{-1})$ are, respectively, the marginal distributions of probe i for classes C_{+1} and C_{-1}.

The Bayesian rule [1] that assigns a pattern to class C_{+1} is given by Equation 3.

$$\frac{\prod_{i=1}^{30} P(x_i|C_{+1}) N p_{+1s}}{\prod_{i=1}^{30} P(x_i|C_{-1}) N p_{-1s}} > 1 \tag{3}$$

The results for Equations 2 and 3, with and without estimated prior probabilities for both discretized and non-discretized input vectors are presented in Table 1. SVMs and Multi-Objective neural networks results [4] are also presented in Table 1.

As can be observed in Table 1, the results of SVM and MOBJ-NN were identical. This can be explained by the low complexity models, bearing linear separation, yielded by these models. The resulting norm of the MOBJ neural network weights was very small, what suggests a strong smoothing effect at the network output. This interpretation is also supported by the identical results achieved by the Bayes rule with equal priors that tends to yield a smooth separation. It is surprising, however, that the Bayes rule with estimated prior probability has performed worse than Bayes with equal priors and the other models. This may be due to the small samples effect or to the independence assumption. Sparsity in the data set and small sample sizes may also favor smoother separating surfaces.

4   Conclusion

The trade-off between bias and variance [1] of a model is usually achieved by complexity control, re-sampling or data-set partition strategies like cross-validation [1]. SVMs and the MONJ-NN embody complexity control in their formulations, what makes as believe that the results achieved with these models point out to limit performance indexes that may be achieved with the current data set. This assumption is reinforced by the fact that the results, obtained by different training strategies, are very close to each other. For such a small data set, representativeness in data set partition may be difficult to achieve, so we rely on the results presented on Table 1 as reliable estimations of the separating boundaries of the generator functions. We believe also that the use or not of estimated priors in the Bayes/majority decision rule is a model design decision and is not against any statistical principle. For our problem, the smoother decision surface yielded results that are quite close to those obtained with the complexity control models.

Acknowledgments. The authors would like to thank CAPES, COFECUB and CNPq for the support.

References

1. Bishop, C.M.: Pattern Recognition and Machine Learning. Springer, Heidelberg (2006)
2. Cortes, C., Vapnik, V.: Support vector networks. Machine Learning 20, 273–279 (1995)
3. Hess, K.R., Anderson, K., Symmans, W.F., Valero, V., Ibrahim, N., Mejia, J.A., Booser, D., Theriault, R.L., Buzdar, A.U., Dempsey, P.J., Rouzier, R., Sneige, N., Ross, J.S., Vidaurre, T., Gomez, H.L., Hortobagyi, G.N., Pusztai, L.: Pharmacogenomic predictor of sensitivity to preoperative chemotherapy with paclitaxel and fluorouracil, doxorubicin, and cyclophosphamide in breast cancer. Journal of Clinical Oncology 24(26), 4236–4244 (2006)
4. Natowicz, R., Braga, A.P., Incitti, R., Horta, E.G., Rouzier, R., Rodrigues, T.S., Costa, M.A.: A new method of dna probes selection and its use with multi-objective neural networks for predicting the outcome of breast cancer preoperative chemotherapy. In: European Symposium on Neural Networks (ESANN 2008) (Accepted, 2008)
5. Natowicz, R., Incitti, R., Charles, B., Guinot, P., Horta, E.G., Pusztai, L., Rouzier, R.: Prediction of the outcome of preoperative chemotherapy in breast cancer by dna probes that convey information on both complete and non complete responses. BMC Bioinformatics (Accepted, 2008)
6. Teixeira, R.A., Braga, A.P., Takahashi, R.H.C., Saldanha, R.R.: Improving generalization of mlps with multi-objective optimization. Neurocomputing (35), 189–194 (2000)

Feature Ranking Ensembles for Facial Action Unit Classification

Terry Windeatt and Kaushala Dias

Centre for Vision, Speech and Signal Proc (CVSSP), University of Surrey,
Guildford, Surrey, United Kingdom GU2 7XH
t.windeatt@surrey.ac.uk

Abstract. Recursive Feature Elimination RFE combined with feature-ranking is an effective technique for eliminating irrelevant features. In this paper, an ensemble of MLP base classifiers with feature-ranking based on the magnitude of MLP weights is proposed. This approach is compared experimentally with other popular feature-ranking methods, and with a Support Vector Classifier SVC. Experimental results on natural benchmark data and on a problem in facial action unit classification demonstrate that the MLP ensemble is relatively insensitive to the feature-ranking method, and simple ranking methods perform as well as more sophisticated schemes. The results are interpreted with the assistance of bias/variance of 0/1 loss function.

1 Introduction

Consider a supervised learning problem, in which many features are suspected to be irrelevant. To ensure good generalisation performance dimensionality needs to be reduced, otherwise there is the danger that the classifier will specialise on features that are not relevant for discrimination, that is the classifier may over-fit the data. It is particularly important to reduce the number of features for small sample size problems, where the number of patterns is less than or of comparable size to the number of features [1]. To reduce dimensionality, features may be extracted (for example Principal Component Analysis PCA) or selected. Feature extraction techniques make use of all the original features when mapping to new features but, compared with feature selection, are difficult to interpret in terms of the importance of original features.

Feature selection has received attention for many years from researchers in the fields of pattern recognition, machine learning and statistics. The aim of feature selection is to find a feature subset from the original set of features such that an induction algorithm that is run on data containing only those features generates a classifier that has the highest possible accuracy [2]. Typically with tens of features in the original set, an exhaustive search is computationally prohibitive. Indeed the problem is known to be NP-hard [2], and a greedy search scheme is required. For problems with hundreds of features, classical feature selection schemes are not greedy enough, and filter, wrapper and embedded approaches have been developed [3].

Although feature-ranking has received much attention in the literature, there has been relatively little work devoted to handling feature-ranking explicitly in the

L. Prevost, S. Marinai, and F. Schwenker (Eds.): ANNPR 2008, LNAI 5064, pp. 267–279, 2008.
© Springer-Verlag Berlin Heidelberg 2008

context of Multiple Classifier System (MCS). Most previous approaches have focused on determining feature subsets to combine, but differ in the way the subsets are chosen. The Random Subspace Method (RSM) is the best-known method, and it was shown that a random choice of feature subset, (allowing a single feature to be in more than one subset), improves performance for high-dimensional problems. In [1], forward feature and random (without replacement) selection methods are used to sequentially determine disjoint optimal subsets. In [4], feature subsets are chosen based on how well a feature correlates with a particular class. Ranking subsets of randomly chosen features before combining was reported in [5].

In this paper an MLP ensemble using Recursive Feature Elimination RFE [12] is experimentally compared for different feature-ranking methods. Ensemble techniques are discussed in Section 2, and feature-ranking strategies in Section 3. The datasets, which include a problem in face expression recognition, are described in Section 4, with experimental results in Section 5.

2 Ensembles, Bootstrapping and Bias/Variance Analysis

In this paper, we assume a simple parallel Multiple Classifier System (MCS) architecture with homogenous MLP base classifiers and majority vote combiner. A good strategy for improving generalisation performance in MCS is to inject randomness, the most popular strategy being Bootstrapping. An advantage of Bootstrapping is that the Out-of-Bootstrap (OOB) error estimate may be used to tune base classifier parameters, and furthermore, the OOB is a good estimator of when to stop eliminating features [6]. Normally, deciding when to stop eliminating irrelevant features is difficult and requires a validation set or cross-validation techniques.

Bootstrapping is an ensemble technique which implies that if μ training patterns are randomly sampled with replacement, $(1-1/\mu)^{\mu} \cong 37\%$ are removed with remaining patterns occurring one or more times. The base classifier OOB estimate uses the patterns left out of training, and should be distinguished from the ensemble OOB. For the ensemble OOB, all training patterns contribute to the estimate, but the only participating classifiers for each pattern are those that have not been used with that pattern for training (that is, approximately thirty-seven percent of classifiers). Note that OOB gives a biased estimate of the absolute value of generalisation error [7], but for tuning purposes the estimate of the absolute value is not important [8]. Bagging, that is Bootstrapping with majority vote combiner, and Boosting (Section 3.3) are probably the most popular MCS methods.

The use of Bias and Variance for analysing multiple classifiers is motivated by what appears to be analogous concepts in regression theory. The notion is that averaging a large number of classifiers leads to a smoothing out of error rates. Visualisation of simple two-dimensional problems appears to support the idea that Bias/Variance is a good way of quantifying the difference between the Bayes decision boundary and the ensemble classifier boundary. However, there are difficulties with the various Bias/Variance definitions for 0/1 loss functions. A comparison of Bias/Variance definitions [9] shows that no definition satisfies all properties that would ideally be expected for 0/1 loss function. In particular, it is shown that it is impossible for a single definition to satisfy both zero Bias and Variance for Bayes classifier, and additive Bias and Variance decomposition of error (as in regression theory).

Also, the effect of bias and variance on error rate cannot be guaranteed. It is easy to think of example probability distributions for which bias and variance are constant but error rate changes with distribution, or for which reduction in variance leads to increase in error rate [9] [11]. Besides these theoretical difficulties, there is the additional consideration that for real problems the Bayes classification needs to be known or estimated. Although some definitions, for example [10], do not require this, the consequence is that the Bayes error is ignored.

In our experiments, we use Breiman's definition [11] which is based on defining Variance as the component of classification error that is eliminated by aggregation. Patterns are divided into two sets, the Bias set B containing patterns for which the Bayes classification disagrees with the aggregate classifier and the Unbias set U containing the remainder. Bias is computed using B patterns and Variance is computed using U patterns, but both Bias and Variance are defined as the difference between the probabilities that the Bayes and base classifier predict the correct class label. Therefore, the reducible error (what we have control over) with respect to a pattern is either assigned to Bias or Variance, an assumption that has been criticised [9]. However, this definition has the nice property that the error of the base classifiers can be decomposed into additive components of Bayes error, Bias and Variance.

3 Feature-Ranking and RFE

RFE is a simple algorithm [12], and operates recursively as follows:

1) Rank the features according to a suitable feature-ranking method
2) Identify and remove the r least ranked features

If $r \geq 2$, which is usually desirable from an efficiency viewpoint, this produces a feature subset ranking. The main advantage of RFE is that the only requirement to be successful is that at each recursion the least ranked subset does not contain a strongly relevant feature [13]. In this paper we use RFE with MLP weights, SVC weights (Section 3.1), and noisy bootstrap (Section 3.2).

The issues in feature-ranking can be quite complex, and feature relevance, redundancy and irrelevance has been explicitly addressed in many papers. As noted in [13] it is possible to think up examples for which two features may appear irrelevant by themselves but be relevant when considered together. Also adding redundant features can provide the desirable effect of noise reduction.

One-dimensional feature-ranking methods consider each feature in isolation and rank the features according to a scoring function $Score(j)$ where $j=1...p$ is a feature, for which higher scores usually indicate more influential features. One-dimensional functions ignore all $p-1$ remaining features whereas a multi-dimensional scoring function considers correlations with remaining features. According to [3] one-dimensional methods are disadvantaged by implicit orthogonality assumption, and have been shown to be inferior to multi-dimensional methods that consider all features simultaneously. However, there has not been any systematic comparison of single and multi-dimensional methods in the context of ensembles.

In this paper, the assumption is that all feature-ranking strategies use the training set for computing ranking criterion (but see Section 5 in which the test set is used for best case scenario). In Sections 3.1-3.4 we describe the ranking strategies that are compared in Section 5, denoted as *rfenn, rfesvc (Section 3.1) rfenb (Section 3.2) boost (Section 3.3) and SFFS, 1dim (Section 3.4)*. Note that SVC, Boosting and statistical ranking methods are well-known so that the technical details are omitted.

3.1 Ranking by Classifier Weights (*rfenn, rfesvc*)

The equation for the output O of a single output single hidden-layer MLP, assuming sigmoid activation function S is given by

$$O = \sum_j S(\sum_i x_i W_{ij}^1) * W_j^2 \tag{1}$$

where i,j are the input and hidden node indices, x_i is input feature, W^1 is the first layer weight matrix and W^2 is the output weight vector. In [14], a local feature selection gain w_i is derived form equation (1)

$$w_i = \sum_j \left| W_{ij}^1 * W_j^2 \right| \tag{2}$$

This product of weights strategy has been found in general not to give a reliable feature-ranking [15]. However, when used with RFE it is only required to find the least relevant features. The ranking using product of weights is performed once for each MLP base classifier. Then individual rankings are summed for each feature, giving an overall ranking that is used for eliminating the set of least relevant features in RFE.

For SVC the weights of the decision function are based on a small subset of patterns, known as support vectors. In this paper we restrict ourselves to the linear SVC in which linear decision function consists of the support vector weights, that is the weights that have not been driven to zero.

3.2 Ranking by Noisy Bootstrap (*rfenb*)

Fisher's criterion measures the separation between two sets of patterns in a direction w, and is defined for the projected patterns as the difference in means normalised by the averaged variance. FLD is defined as the linear discriminant function for which $J(w)$ is maximized

$$J(w) = \frac{\left| w^T S_B w \right|}{\left| w^T S_W w \right|} \tag{3}$$

where, S_B is the between-class scatter matrix and S_W is the within-class scatter matrix (Section 3.4). The objective of FLD is to find the transformation matrix w^* that maximises $J(w)$ in equation (3) and w^* is known to be the solution of the following eigenvalue problem $S_B - S_W \Lambda = 0$ where Λ is a diagonal matrix whose elements are the eigenvalues of matrix $S_W^{-1} S_B$. Since in practice S_W is nearly always singular, dimensionality reduction is required. The idea behind the *noisy bootstrap* [16] is to estimate the noise in the data and extend the training set by re-sampling with simulated noise. Therefore, the number of patterns may be increased by using a re-sampling rate greater than 100 percent. The noise model assumes a multi-variate Gaussian distribution with zero mean and diagonal covariance matrix, since there are generally insufficient number of patterns to make a reliable estimate of any correlations between features. Two parameters to tune are the noise added γ and the sample to feature ratio *s2f*. We set for our experiments $\gamma = 0.25$ and *s2f = 1* [17].

3.3 Ranking by Boosting *(boost)*

Boosting, which combines with a fixed weighted vote is more complex than Bagging in that the distribution of the training set is adaptively changed based upon the performance of sequentially constructed classifiers. Each new classifier is used to adaptively filter and re-weight the training set, so that the next classifier in the sequence has increased probability of selecting patterns that have been previously misclassified. The algorithm is well-known and has proved successful as a classification procedure that 'boosts' a weak learner, with the advantage of minimal tuning. More recently, particularly in the Computer Vision community, Boosting has become popular as a feature selection routine, in which a single feature is selected on each Boosting iteration [18]. Specifically, the Boosting algorithm is modified so that, on each iteration, the individual feature is chosen which minimises the classification error on the weighted samples [19]. In our implementation, we use Adaboost with decision stump as weak learner.

3.4 Ranking by Statistical Criteria *(1dim, SFFS)*

Class separability measures are popular for feature-ranking, and many definitions use S_B and S_W (equation (3)) [20]. Recall that S_W is defined as the scatter of samples around respective class expected vectors and S_B as the scatter of the expected vectors around the mixture mean. Although many definitions have been proposed, we use trace($S_W^{-1} * S_B$), a one-dimensional method.

A fast multi-dimensional search method that has been shown to give good results with individual classifiers is Sequential Floating Forward Search (SFFS). It improves on (plus l – take away r) algorithms by introducing dynamic backtracking. After each forward step, a number of backward steps are applied, as long as the resulting subsets are improved compared with previously evaluated subsets at that level. We use the implementation in [21] for our comparative study.

4 Datasets

The first set of experiments use natural benchmark two-class problems selected from [22] and [23] and are shown in Table 1. For datasets with missing values the scheme suggested in [22] is used. The original features are normalised to mean 0 std 1 and the number of features increased to one hundred by adding noisy features (Gaussian std

Table 1. Benchmark Datasets showing numbers of patterns, continuous and discrete features and estimated Bayes error rate

DATASET	#pat	#con	#dis	%error
cancer	699	0	9	3.1
card	690	6	9	12.8
credita	690	3	11	14.1
diabetes	768	8	0	22.0
heart	920	5	30	16.1
ion	351	31	3	6.8
vote	435	0	16	2.8

0.25). All experiments use random training/testing splits, and the results are reported as mean over twenty runs. Two-class benchmark problems are split 20/80 (20% training, 80% testing) 10/90, 5/95 and use 100 base classifiers.

The second set of experiments addresses a problem in face expression recognition, which has potential application in many areas including human-computer interaction, talking heads, image retrieval, virtual reality, human emotion analysis, face animation, biometric authentication [24]. The problem is difficult because facial expression depends on age, ethnicity, gender, and occlusions due to cosmetics, hair, glasses. Furthermore, images may be subject to pose and lighting variation. There are two approaches to automating the task, the first concentrating on what meaning is conveyed by facial expression and the second on categorising deformation and motion into visual classes. The latter approach has the advantage that the interpretation of facial expression is decoupled from individual actions. In FACS (facial action coding system) [25], the problem is decomposed into forty-four facial action units (e.g. *au1* inner brow raiser). The coding process requires skilled practitioners and is time-consuming so that typically there are a limited number of training patterns. These characteristics make the problem of face expression classification relevant and suitable to the feature-ranking techniques proposed in this paper.

The database we use is Cohn-Kanade [26], which contains posed (as opposed to the more difficult spontaneous) expression sequences from a frontal camera from 97 university students. Each sequence goes from neutral to target display but only the last image is *au* coded. Facial expressions in general contain combinations of action units (*aus*), and in some cases *aus* are non-additive (one action unit is dependent on another). To automate the task of *au* classification, a number of design decisions need to be made, which relate to the following a) subset of image sequences chosen from the database b) whether or not the neutral image is included in training c) image resolution d) normalisation procedure e) size of window extracted from the image, if at all f) features chosen for discrimination, g) feature selection or feature extraction procedure h) classifier type and parameters, and i) training/testing protocol. Researchers make different decisions in these nine areas, and in some cases are not explicit about which choice has been made. Therefore it is difficult to make a fair comparison with previous results.

We concentrate on the upper face around the eyes, (involving *au1, au2, au4, au5, au6, au7*) and consider the two-class problem of distinguishing images containing inner brow raised (*au1*), from images not containing *au1*. The design decisions we made were

a)   all image sequences of size 640 x 480 chosen from the database
b)   last image in sequence (no neutral) chosen giving 424 images, 115 containing *au1*
c)   full image resolution, no compression
d)   manually located eye centres plus rotation/scaling into 2 common eye coordinates
e)   window extracted of size 150 x 75 pixels centred on eye coordinates
f)   Forty Gabor filters [18], five special frequencies at five orientations with top 4 principle components for each Gabor filter, giving 160-dimensional feature vector
g)   Comparison of feature selection schemes described in Section 3
h)   Comparison of MLP ensemble and Support Vector Classifier
i)   Random training/test split of 90/10 and 50/50 repeated twenty times and averaged

With reference to b), some studies use only the last image in the sequence but others use the neutral image to increase the numbers of *non-aus*. Furthermore, some researchers consider only images with single *au*, while others use combinations of *aus*. We consider the more difficult problem, in which neutral images are excluded and images contain combinations of *aus*. With reference to d) there are different approaches to normalisation and extraction of the relevant facial region. To ensure that our results are independent of any eye detection software, we manually annotate the eye centres of all images, and subsequently rotate and scale the images to align the eye centres horizontally. A further problem is that some papers only report overall error rate. This may be mis-leading since class distributions are unequal, and it is possible to get an apparently low error rate by a simplistic classifier that classifies all images as *non-au1*. For the reason we report area under ROC curve, similar to [18].

5   Experimental Evidence

The purpose of the experiments is to compare the various feature-ranking schemes described in Section 3, using an MLP ensemble and a Support Vector Classifier. The SVC is generally recognised to give superior results when compared with other single classifiers. A difficulty with both MLPs and SVCs is that parameters need to be tuned. In the case of SVC, this is the kernel and regularisation constant C. For MLP ensemble, it is the number of hidden nodes and number of training epochs. There are other tuning parameters for MLPs, such as learning rate but the ensemble has been shown to be robust to these parameters [8]. When the number of features is reduced, the ratio of the number of patterns to features is changing, so that optimal classifier parameters will be varying. In general, this makes it a very complex problem, since theoretically an optimisation needs to be carried out after each feature reduction. To make a full comparison between MLP and SVC, we would need to search over the full parameter space, which is not feasible. For the two-class problems in table 1, we compare linear SVC with linear perceptron ensemble. We found that the differences

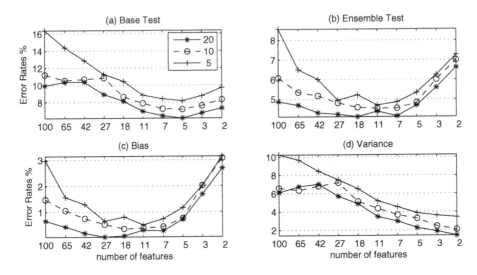

Fig. 1. Mean test error rates, Bias, Variance for RFE perceptron ensemble with Cancer Dataset 80/20, 10/90. 5/95 train/test split

between feature selection schemes were not statistically significant (McNemar test 5% [27]), and we show results graphically and report the mean over all datasets.

Random perturbation of the MLP base classifiers is caused by different starting weights on each run, combined with bootstrapped training patterns, Section 2. The experiment is performed with one hundred single hidden-layer MLP base classifiers, using the Levenberg-Marquardt training algorithm with default parameters. The feature-ranking criterion is given in equ. (2). In our framework, we vary the number of hidden nodes, and use a single node for linear perceptron. We checked that results were consistent for Single layer perceptron (SLP), using absolute value of orientation weights to rank features.

In order to compute bias and variance we need to estimate the Bayes classifier for the 2-class benchmark problems. The estimation is performed for 90/10 split using original features in Table 1, and a SVC with polynomial kernel run 100 times. The polynomial degree is varied as well as the regularisation constant. The lowest test error found is given in Table 1, and the classifications are stored for the bias/variance computation. All datasets achieved minimum with linear SVC, with the exception of 'Ion' (degree 2).

Figure 1 shows RFE linear MLP ensemble results for 'Cancer' 20/80, 10/90, 5/95 which has 140, 70, 35 training patterns respectively. With 100 features the latter two splits give rise to small sample size problem, that is number of patterns less than number of features [1]. The recursive step size for RFE is chosen using a logarithmic scale to start at 100 and finish at 2 features. Figure 1 (a) (b) show base classifier and ensemble test error rates, and (c) (d) the bias and variance as described in Section 2. Consider the 20/80 split for which Figure 1 (a) shows that minimum base classifier error is achieved with 5 features compared with figure (b) 7 features for the ensemble. Notice that the ensemble is more robust than base classifiers with respect to noisy

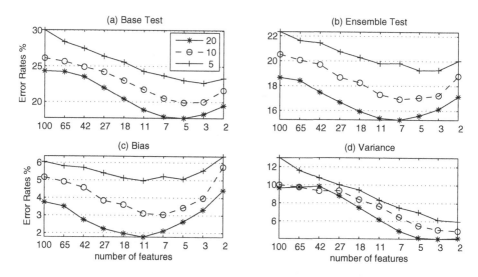

Fig. 2. Mean test error rates, Bias, Variance for RFE MLP ensemble over seven 2-class Datasets 80/20, 10/90. 5/95 train/test split

features. In fact, Figure 1 (c) shows that bias is minimised at 27 features, demonstrating that the linear perceptron with bootstrapping benefits (in bias reduction) from a few extra noisy features. Figure 1 (d) shows that Variance is reduced monotonically as number of features is reduced, and between 27 and 7 features the Variance reduction more than compensates for bias increase. Note also that according to Breiman's decomposition (Section 2), (c) + (d) + 3.1 (Bayes) equals (a).

Figure 2 shows RFE linear MLP ensemble mean test error rates, bias and variance over all seven datasets from table 1. On average, the base classifier achieves minimum error rate at 5 features and the ensemble at 7 features. Bias is minimised at 11 features and Variance at 3 features. For the 5/95 split there appears to be too few patterns to reduce bias, which stays approximately constant as features are reduced. Note that for SVC (not shown) the error is due entirely to bias, since variance is zero.

The comparison for various schemes defined in Section 3 can be found in Table 2. It may be seen that the ensemble is fairly insensitive to the ranking scheme and the linear perceptron ensemble performs similarly to SVC. In particular, the more sophisticated schemes of SFFS and Boosting are slightly worse on average than the simpler schemes. Although the 1-dimensional method (Section 3.4) is best on average for 20/80 split, as number of training patterns decreases, performance is slightly worse than RFE methods. We also tried MLP base classifier with 8 nodes 7 epochs which was found to be the best setting without added noisy features [8]. The mean ensemble rate for 20/80, 10/90 5/95 was 14.5%,15.7%, 17.9% respectively the improvement due mostly to 'ion' dataset which has a high bias with respect to Bayes classifier.

To determine the potential effect of using a validation set with a feature selection strategy, we chose SVC plus SFFS with the unrealistic case of full test set for tuning. The mean ensemble rate for 20/80, 10/90 5/95 was 13.3%, 14.0%, 15.0% for SVC

Table 2. Mean best error rates (%)/number of features for seven two-class problems (20/80) with five feature-ranking schemes (Mean 10/90, 5/95 also shown)

	perceptron-ensemble classifier					SVC-classifier				
	rfenn	rfenb	1dim	SFFS	boost	rfesvc	rfenb	1dim	SFFS	boost
diab	24.9/2	25.3/2	25.3/2	25.8/2	25.6/2	24.5/3	24.8/5	24.9/2	25.3/2	25.3/2
credita	16.5/5	15.7/3	14.6/2	15.6/2	15.5/2	15.7/2	15.1/2	14.6/2	15.4/2	15.1/2
cancer	4/7	4/5	4.1/5	4.4/3	4.9/7	3.7/7	3.7/7	3.8/11	4.2/5	4.5/7
heart	21/27	21/18	21/11	23/5	23/18	20/18	20/11	20/18	22/7	24/18
vote	5.5/5	5.3/7	5.6/18	5.7/2	5.5/2	4.8/2	4.8/2	4.7/2	4.3/3	4.7/2
ion	18/11	16.7/3	14.8/3	15.8/3	18.1/2	15/11	15.9/7	15.3/5	17.9/5	19.5/5
card	15.7/7	15/2	14.7/2	16.9/2	14.8/2	15.5/2	14.8/2	14.5/2	16.6/2	14.5/2
Mean20/80	15.1	14.6	14.2	15.4	15.4	14.2	14.2	13.9	15.1	15.3
Mean10/90	16.3	16.3	16.6	18.0	17.6	15.5	15.7	15.8	17.5	17.3
Mean5/95	18.4	18.5	20.0	21.3	21.3	17.0	17.7	18.4	20.3	20.7

Table 3. Mean best error rates (%)/number of features for au1 classification 90/10 with five feature ranking schemes

MLP-ensemble classifier					SVC-classifier				
rfenn	rfenb	1dim	SFFS	boost	rfesvc	rfenb	1dim	SFFS	boost
10.0/28	10.9/43	10.9/43	12.3/104	11.9/43	11.6/28	12.1/28	11.9/67	13.9/67	12.4/43

and 13.5%, 14.1%, 15.4% for MLP. We also repeated *rfenn* without Bootstrapping, showing that although variance is lower, bias is higher and achieved 15.7%, 17.6%, 20.0% respectively, demonstrating that Bootstrapping has beneficial effect on performance.

Table 3 shows feature-ranking comparison for *au1* classification from the Cohn-Kanade database as described in Section 4. It was found that lower test error was obtained with non-linear base classifier and Figure 3 shows test error rates, using an MLP ensemble with 16 nodes 10 epochs. The minimum base error rate for 90/10 split is 16.5% achieved for 28 features, while the ensemble is 10.0% at 28 features. Note that for 50/50 split there are too few training patterns for feature selection to have much effect. Since class distributions are unbalanced, the overall error rate may be mis-leading, as explained in Section 4. Therefore, we show the true positive rate in Figure 3 c) and area under ROC in Figure d). Note that only 71% of *au1s* are correctly recognised. However, by changing the threshold for calculating the ROC, it is clearly possible to increase the true positive rate at the expense of false negatives. Nevertheless, it is believed that the overall ensemble rate of 10% is among the best for *au1* on this database (recognising the difficulty of making fair comparison as explained in Section 4). We did try SVC for degree 2,3,4 polynomials with C varying,

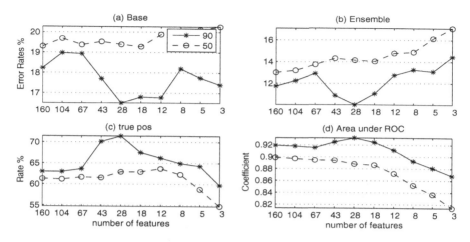

Fig. 3. Mean test error rates, True Positive and area under ROC for RFE MLP ensemble for au1 classification 90/10. 50/50 train/test split

but did not improve on degree 1 results. The results are not presented but the performance of SVC was very sensitive to regularisation constant C, which makes it difficult to tune and we did not try different kernels.

6 Discussion

There is conflicting evidence over whether an SVC ensemble gives superior results compared with single SVC, but in [28] it is claimed that an SVC ensemble with low bias classifiers gives better results. However, it is not possible to be definitive, without searching over all kernels and regularisation constants C. In our experiments, we chose to consider only linear SVC, and found the performance to be sensitive to C. In contrast, the ensemble is relatively insensitive to number of nodes and epochs [8], and this is an advantage of the MLP ensemble. However, we believe it is likely that we could have achieved comparable results to MLP ensemble by searching over different kernels and values of C for SVC.

The feature-ranking approaches have been applied to a two-class problem in facial action unit classification. The problem of detecting action units is naturally a multi-class problem, and the intention is to employ multi-class approaches that decompose the problem into two-class problems, such as Error-Correcting Output Coding (ECOC) [29].

7 Conclusion

A bootstrapped MLP ensemble, combined with RFE and product of weights feature-ranking, is an effective way of eliminating irrelevant features. The accuracy is comparable to SVC but has the advantage that the OOB estimate may be used to tune parameters and to determine when to stop eliminating features. Simple feature-ranking techniques, such as 1-dimensional class separability measure or product of MLP weights plus RFE, perform at least as well as more sophisticated techniques such as multi-dimensional methods of SFFS and Boosting.

References

1. Skuruchina, M., Duin, R.P.W.: Combining feature subsets in feature selection. In: Oza, N., Polikar, R., Roli, F., Kittler, J. (eds.) Proc. 6th Int. Workshop Multiple Classifier Systems, Seaside, Calif. USA, June 2005. LNCS, pp. 165–174. Springer, Heidelberg (2005)
2. Kohavi, R., John, G.H.: Wrappers for feature subset selection. Artificial Intelligence Journal, special issue on relevance 97(1-2), 273–324 (1997)
3. Guyon, I., Elisseeff, A.: An introduction to variable and feature selection. Journal of Machine Learning Research 3, 1157–1182 (2003)
4. Oza, N., Tumer, K.: Input Decimation ensembles: decorrelation through dimensionality reduction. In: Kittler, J., Roli, F. (eds.) Proc. 2nd Int. Workshop Multiple Classifier Systems, Cambridge, UK. LNCS, pp. 238–247. Springer, Heidelberg (2001)
5. Bryll, R., Gutierrez-Osuna, R., Quek, F.: Attribute bagging: improving accuracy of classifier ensembles by using random feature subsets. Pattern Recognition 36, 1291–1302 (2003)
6. Windeatt, T., Prior, M.: Stopping Criteria for Ensemble-based Feature Selection. In: Proc. 7th Int. Workshop Multiple Classifier Systems, Prague, May 2007. LNCS, pp. 271–281. Springer, Heidelberg (2007)
7. Bylander, T.: Estimating generalisation error two-class datasets using out-of-bag estimate. Machine Learning 48, 287–297 (2002)
8. Windeatt, T.: Accuracy/ Diversity and Ensemble Classifier Design. IEEE Trans Neural Networks 17(5), 1194–1211 (2006)
9. James, G.: Variance and Bias for General Loss Functions. Machine Learning 51(2), 115–135 (2003)
10. Kong, E.B., Dietterich, T.G.: Error- Correcting Output Coding corrects Bias and Variance. In: 12th Int. Conf. Machine Learning, San Francisco, pp. 313–321 (1995)
11. Breiman, L.: Arcing Classifiers. The Annals of Statistics 26(3), 801–849 (1998)
12. Guyon, I., Weston, J., Barnhill, S., Vapnik, V.: Gene selection for cancer classification using support vector machines. Machine Learning 46(1-3), 389–422 (2002)
13. Yu, L., Liu, H.: Efficient feature selection via analysis of relevance and redundancy. Journal of Machine Learning Research 5, 1205–1224 (2004)
14. Hsu, C., Huang, H., Schuschel, D.: The ANNIGMA-wrapper approach to fast feature selection for neural nets. IEEE Trans. System, Man and Cybernetics-Part B:Cybernetics 32(2), 207–212 (2002)
15. Wang, W., Jones, P., Partridge, D.: Assessing the impact of input features in a feedforward neural network. Neural Computing and Applications 9, 101–112 (2000)
16. Efron, N., Intrator, N.: The effect of noisy bootstrapping on the robustness of supervised classification of gene expression data. In: IEEE Int. Workshop on Machine Learning for Signal Processing, Brazil, pp. 411–420 (2004)
17. Windeatt, T., Prior, M., Effron, N., Intrator, N.: Ensemble-based Feature Selection Criteria. In: Proc. Conference on Machine Learning Data Mining MLDM2007, Leipzig, July 2007, pp. 168–182 (2007) ISBN 978-3-940501-00-4

18. Bartlett, M.S., Littlewort, G., Lainscsek, C., Fasel, I., Movellan, J.: Machine learning methods for fully automatic recognition of facial expressions and facial actions. In: IEEE Conf. Systems, Man and Cybernetics, October 2004, vol. 1, pp. 592–597 (2004)
19. Silapachote, P., Karuppiah, D.R., Hanson, A.R.: Feature Selection using Adaboost for Face Expression Recognition. In: Proc. Conf. on Visualisation, Imaging and Image Processing, Marbella, Spain, September 2004, pp. 84–89 (2004)
20. Fukunaga, K.: Introduction to statistical pattern recognition. Academic Press, London (1990)
21. Heijden, F., Duin, R.P.W., Ridder, D., Tax, D.M.J.: Classification, Parameter Estimation and State Estimation. Wiley, Chichester (2004)
22. Prechelt, L.: Proben1: A set of neural network Benchmark Problems and Benchmarking Rules, Tech Report 21/94, Univ. Karlsruhe, Germany (1994)
23. Merz, C.J., Murphy, P.M.: UCI repository of machine learning databases (1998), http://www.ics.uci.edu/~mlearn/MLRepository.html
24. Fasel, B., Luettin, J.: Automatic facial expression analysis: a survey. Pattern Recognition 36, 259–275 (2003)
25. Tian, Y., Kanade, T., Cohn, J.F.: Recognising action units for facial expression analysis. IEEE Trans. PAMI 23(2), 97–115 (2001)
26. Kanade, T., Cohn, J.F., Tian, Y.: Comprehenive Database for facial expression analysis. In: Proc. 4th Int. Conf. automatic face and gesture recognition, Grenoble, France, pp. 46–53 (2000)
27. Dietterich, T.G.: Approx. statistical tests for comparing supervised classification learning algorithms. Neural Computation 10, 1895–1923 (1998)
28. Valentini, G., Dietterich, T.G.: Bias-Variance Analysis for Development of SVM-Based Ensemble Methods. Journal of Machine Learning Research 4, 725–775 (2004)
29. Windeatt, T., Ghaderi, R.: Coding and Decoding Strategies for Multi-class Learning Problems. Information Fusion 4(1), 11–21 (2003)

Texture Classification with Generalized Fourier Descriptors in Dimensionality Reduction Context: An Overview Exploration

Ludovic Journaux[1], Marie-France Destain[1], Johel Miteran[2], Alexis Piron[1], and Frederic Cointault[3]

[1] Unité de Mécanique et Construction, Faculté Universitaire des Sciences Agronomiques de Gembloux, Passage des Déportés 2, B-5030 Gembloux, Belgique
[2] Lab. Le2i, Université de Bourgogne, BP 47870, 21078 Dijon Cedex, France
[3] ENESAD UP GAP, 26 Bd Dr Petitjean BP 87999, 21079 Dijon Cedex, France
journaux.l@fsagx.ac.be

Abstract. In the context of texture classification, this article explores the capacity and the performance of some combinations of feature extraction, linear and nonlinear dimensionality reduction techniques and several kinds of classification methods. The performances are evaluated and compared in term of classification error. In order to test our texture classification protocol, the experiment carried out images from two different sources, the well known Brodatz database and our leaf texture images database.

Keywords: Texture classification, Motion Descriptors, Dimensionality reduction.

1 Introduction

For natural images the texture is a fundamental characteristic which plays an important role in pattern recognition and computer vision. Thus, texture analysis is an essential step for any image processing applications such as medical and biological imaging, industrial control, document segmentation, remote sensing of earth resources. A successful classification or segmentation requires an efficient feature extraction methodology but the major difficulty is that textures in the real world are often not uniform, due to changes in orientation, scale, illumination conditions, or other visual appearance. To overcome these problems, numerous approaches are proposed in the literature, often based on the computation of invariants followed by a classification method as in [1]. In the case of a large size texture image, these invariants texture features often lead to very high-dimensional data, the dimension of the data being in the hundreds or thousands. Unfortunately, in a classification context these kinds of high-dimensional datasets are difficult to handle and tend to suffer from the problem of the "curse of dimensionality", well known as "Hughes phenomenon" [2], which cause inaccurate classification. One possible solution to improve the classification performance is to use Dimensionality Reduction (DR) techniques in order to transform high-dimensional data into a meaningful representation of reduced

L. Prevost, S. Marinai, and F. Schwenker (Eds.): ANNPR 2008, LNAI 5064, pp. 280–291, 2008.

dimensionality. Numerous studies have aimed at comparing DR algorithms, usually using synthetic data [3 , 4] but less for natural tasks as in [5] or [6].

In this paper, considering one family of invariants called Motion descriptors (MD) or Generalized Fourier Descriptors (GFD) [7] which provide well-proven robust features in complex areas of pattern recognition (faces, objects, forms) [8], we propose to compare in 6 classification methods context, the high-dimensional original features datasets extracted from two different textures databases to reduced textures features dataset obtained by 11 DR methods. This paper is organized as follows : section 2 presents the textured images databases, review the definition of invariants features used for classification methods which are also quickly described. In section 3 we propose a review of Dimensionality Reduction techniques and the section 4 presents the results allowing to compare the performances of some combinations of feature extraction, dimensionality reduction and classification. The performances are evaluated and compared in term of classification error.

2 Materials and Methods

2.1 Textures Images Databases

In order to test our texture classification protocol, the experiment carried out images from two different sources:

■ The well known Brodatz textures dataset [9] adapted from the Machine Vision Group of Oulun University and first used by Valkealahti [10]. The dataset is composed of 32 different textures (Fig. 1). The original images are grey levels images with a 256×256 pixels resolution. Each image has been cropped into 16 disjoint 64x64 samples. In order to evaluate scale and rotation invariance, three additional samples were generated per original sample (90° degrees rotation, 64×64 scaling, combinations of rotation and scaling). Finally, the set contains almost 2048 images, 64 samples per texture.

Fig. 1. The 32 Brodatz textures used in the experiments

■ The second textures images used in this study have been provided by the Matters and Materials laboratory at the "Free University of Brussels" for agronomic application. They are grey levels images acquired with a scanning electron microscope (SEM) and representing different kinds of leaf surfaces coming from six leaf plant species (Fig. 2). Thus, the image database contains 6 classes of leaf textures images. For each class 150 to 200 images have been acquired. Each image consists of a 100 μm scale image, with a resolution of 512×512 pixels adapting the scale to our biological application (1242 textures images in six classes).

Fig. 2. The six classes of leaf texture images

2.2 Texture Characterisation Using Generalized Fourier Descriptors (GFD)

The GFD are defined as follows. Let f be a square summable function on the plane, and $\hat{f}$ its Fourier transform:

$$\hat{f}(\xi) = \int_{\mathbb{R}^2} f(x) \exp\left(-j\xi x\right) dx. \tag{1}$$

If $\left(\lambda, \theta\right)$ are polar coordinates of the point ξ, we shall denote again $\hat{f}\left(\lambda, \theta\right)$ the Fourier transform of f at the point $\left(\lambda, \theta\right)$. Gauthier et al. [7] defined the mapping D_f from $\mathbb{R}_+$ into $\mathbb{R}_+$ by

$$D_f\left(\lambda\right) = \int_0^{2\pi} \left|\hat{f}(\lambda, \theta)\right|^2 d\theta. \tag{2}$$

So, D_f is the feature vector (the GFD) which describes each texture image and will be used as an input of the supervised classification method and be reduced by DR methods.

Motion descriptors, calculated according to equation (2), have several properties useful for object recognition : they are translation, rotation and reflexion-invariant [7, 8].

2.3 Classification Methods

Classification is a central problem of pattern recognition [11] and many approaches to solve it have been proposed such as connectionist approach [12] or metrics based methods, k-nearest neighbours (k-nn) and kernel-based methods like Support Vector Machines (SVM) [13], to name the most common. In our experiments, we want to evaluate the average performance of the dimensionality reduction methods and one basic feature selection method applied on the GFD features. In this context, we have chosen and evaluated six efficient classification approaches coming from four classification families: The boosting (adaboost) family [14] using three weak classifiers, (Hyperplan, Hyperinterval and Hyperrectangle), the Hyperrectangle (Polytope) method [15], the Support Vector Machine (SVM) method [13, 16] and the connectionist family with a Multilayers perceptron (MLP) [17]. We have excluded the majority of neural networks methods due to the high variability of textures from natural images; Variability which included an infinite number of samples required for the learning step (Kind of leaves, growth stage, pedo-climatic conditions, roughness, hydration state,...). In order to validate the classification performance and estimate the average error rate for each classification method, we performed 20 iterative experiments with a 10-fold cross validation procedure.

3 Dimensionality Reduction Methods

The GFD provide features that are of great potential in pattern recognition as it was shown by Smach et al. in [8]. Unfortunately, these high dimensional datasets are however difficult to handle, the information is often redundant and highly correlated with one another. Moreover, data are also typically large, and the computational cost of elaborate data processing tasks may be prohibitive. Thus, to improve the classification performance it is well interesting to use Dimensionality Reduction (DR) techniques in order to transform high-dimensional data into a meaningful representation of reduced dimensionality. At this time of our work, we selected a dozen of DR methods. However, it is important to note that works employing recent approaches as it could be find in [18] are being finalized (another distance, topology or angle preservation methods like Kernel Discriminant Analysis, Generative Topographic Mapping, Isotop, Conformal Eigenmaps,...).

3.1 Estimating Intrinsic Dimensionality

Let $\mathbf{X} = (\mathbf{x}_1, ..., \mathbf{x}_n)^T$ be the $n{\times}m$ data matrix. The number n represents the number of images examples contained in each texture dataset, and m the dimension of the vector $\mathbf{x}_i$, which his the vector corresponding to the discrete computing of the D_f (from eq. (2)). We have in our case $n=2048$ and $m=32$ for Brodatz textures database and $n=1034$ and $m=254$ for plants leaf textures database. This dataset represent respectively 32 and 6 classes of textures surfaces.

Ideally, the reduced representation has a dimensionality that corresponds to the intrinsic dimensionality of the data. One of our working hypotheses is that, though data points (all texture image) are points in $\mathbb{R}^m$, there exists a p-dimensional manifold $\mathcal{M} = (\mathbf{y}_1, ..., \mathbf{y}_n)^T$ that can satisfyingly approximate the space spanned by the data points. The meaning of "satisfyingly" depends on the dimensionality reduction technique that is used. The so-called intrinsic dimension (ID) of $\mathbf{X}$ in $\mathbb{R}^m$ is the lowest possible value of p $(p{<}m)$ for which the approximation of $\mathbf{X}$ by $\mathcal{M}$ is reasonable. In order to estimate the ID of our two datasets, we used a geometric approach that estimates the equivalent notion of fractal dimension [19]. Using this method, we estimated and fixed the intrinsic dimensionality of our two datasets as being $p=5$.

3.2 Review of DR Methods

DR methods can be classified according to three characteristics:

   -   **Linearity : DR can be Linear or nonlinear.** This describes the type of transformation applied to the data matrix, mapping it from $\mathbb{R}^m$ to $\mathbb{R}^p$.

- **Scale analysis : DR can be Local *or* global.** This reflects the kind of properties the transformation does preserve. In most nonlinear methods, there is a compromise to be made between the preservation of local topological relationships between data points, or of the global structure of $\mathbf{X}$.
- **Metric : Euclidean *or* geodesic.** This defines the distance function used to estimate whether two data points are close to each other in $\mathbb{R}^m$, and should consequently remain close in $\mathbb{R}^p$, after the DR transformation.

In this context, we retained 11 methods based on these various criteria, 3 are linear methods and 8 are nonlinear. In order to complete this review of dimensionality reduction methods, we opposed them to one classical feature selection method. This comparison will show which approaches are the most relevant.

3.2.1 Linear Methods

3.2.1.1 Principal Components Analysis. Principal Components Analysis (PCA, [11]) is the best known DR method. PCA finds a linear transformation for keeping the subspace that has largest variance. PCA aims at solving the following problem: given p<m, find an orthonormal basis $< u_1, u_2, ..., u_p >$ that minimizes the so-called reconstruction error:

$$J_{PCA}(\mathbf{X}, p) = \sum_{i=1}^{n} \|\mathbf{x}_i - \mathbf{y}_i\|^2, \qquad (3)$$

It can be shown that J_{PCA} is minimized for the u_i being the eigenvectors of the covariance matrix of $\mathbf{X}$. In practice, it is implemented using singular value decomposition. PCA is linear, global and Euclidean technique.

3.2.1.2 Second-Order Blind Identification. Second Order Blind Identification (SOBI) [20] relies only on stationary second-order statistics that are based on a joint diagonalization of a set of covariance matrices. The set X is assimilated to a set of signals $X_i(t)$ and the p features of the destination space we are searching are assimilated to a fixed number of original sources $s_i(t)$. Each $X_i(t)$ is assumed to be an instantaneous linear mixture of n unknown components (sources) $s_i(t)$, via the unknown mixing matrix A.

$$X(t) = As(t) \qquad (5)$$

This algorithm can be described by the following steps (more details on SOBI algorithm can be found in [20]) : (1) Estimate the sample covariance matrix $R_x(0)$ and compute the whitening matrix W with $R_x(0) = E(X(t).X^*(t))$. (2) Estimate the covariance matrices $R_z(\tau)$ of the whitened process $z(t)$ for fixed lag times τ. (3) Jointly diagonalize the set $\{R_z(\tau_j)\, /\, j = 1, ..., k\}$, by minimizing the criterion

$$J(M, V) = \sum_{k=1,...,n} \left(\sum_{i \neq j=1,...,n} \left| (V^t M_{i,j} V \right|^2 \right) \qquad (6)$$

where M is a set of matrices in the form $M_k = VD_k V$, where V is a unitary matrix, and D_k is a diagonal matrix. (4) Determinate an estimation $\hat{A}$ of the mixing matrix A such as $\hat{A} = W^{-1}$. (5) Determinate the source matrix and then extracting the p components. SOBI is a linear, global and Euclidean method.

3.2.1.3 Projection Pursuit (PP). This projection method [21] is based on the optimization of the gradient descent. Our algorithm uses the Fast-ICA procedure [22] that allows estimating the new components one by one by deflation. The symmetric decorrelation of the vectors at each iteration was replaced by a Gram-Schmidt orthogonalization procedure. When p components $w_1, ..., w_p$ have been estimated, the fix point algorithm determines w_{p+1}. After each iteration, the projections $w_{p+1}^T w_j w_j (j = 1, ..., p)$ of the p precedent estimated vectors are subtracted from w_{p+1}. Then, w_{p+1} is re-normalized:

$$w_{p+1} = w_{p+1} - \sum_{j=1}^{p} w_{p+1}^T w_j w_j = \frac{w_{p+1}}{\sqrt{w_{p+1}^T w_{p+1}}} \tag{7}$$

The algorithm stops when p components have been estimated. Projection Pursuit is linear, global and Euclidean.

3.2.2 Nonlinear Methods : Global Approaches

3.2.2.1 Sammon's Mapping (Sammon). Sammon's mapping is a projection method that tries to preserve the topology of the set of data (neighbourhood) in preserving distances between points [23]. To evaluate the preservation of the neighbourhood topology, we use the following stress function

$$J_{sam} = \frac{1}{\sum_{i,j=1}^{n} d_{i,j}^m} \left(\sum_{i,j=1}^{n} \frac{(d_{i,j}^m - d_{i,j}^p)^2}{d_{i,j}^m} \right) \tag{8}$$

Where $d_{i,j}^m$ and $d_{i,j}^p$ are the distances between points i^{th} and j^{th} points, in $\mathbb{R}^m$ and $\mathbb{R}^p$.

This function, minimized by a gradient descent, allows adapting the distances in the projection space at best as distances in the initial space. Sammon's mapping is a nonlinear, global, and Euclidean method.

3.2.2.2 Isometric Feature Mapping (Isomap). Isometric Feature Mapping (Isomap) [24] estimates the geodesic distance along the manifold using the shortest path in the nearest neighbours' graph. It then looks for a low-dimensional representation that approximates those geodesic distances in the least square sense (which amounts to MDS). It consists of three steps: (1) Build $D_m(X)$, the all-pairs distance matrix. (2) Build a graph from X (k nearest neighbours). For a given point x_i in $\mathbb{R}^m$, a neighbour is either one of the K nearest data points from x_i or one for which $d_{ij}^m < \varepsilon$. Build the all-pairs geodesic distance matrix $\Delta_m(X)$, using Dijkstra's all-pairs shortest path algorithm. (3) Use classical MDS to find the transformation from $\mathbb{R}^m$ to $\mathbb{R}^p$ that minimizes

$$J_{ISOMAP}(X,p) = \sum_{i,j}^{n} (\delta_{ij}^m - \delta_{ij}^p)^2 \tag{9}$$

Isomap is nonlinear, global and geodesic.

3.2.2.3   Kernel Methods (K-PCA, K-Isomap). Recently, several well-known algorithms for dimensionality reduction of manifolds have been developed in a new way, taking the kernel machine viewpoint [25, 26]. We retain here the two most known : the kernel-PCA (K-PCA) [27] and the kernel Isomap (K-Isomap) [28]. The non-linearity is introduced via a mapping of the data from the input space $\mathbb{R}^m$ to a feature space $\mathcal{F}$. Projection methods (PCA or Isomap) are then performed in this new feature space. This feature space is expressed by a kernel K in terms of a Mercer Kernel function [29]. More details on K-PCA and K-Isomap algorithm can be found respectively in [27] and [28]. These methods are nonlinear and global, K-PCA is Euclidean and K-Isomap is geodesic.

3.2.3   Nonlinear Methods: Local Approaches

3.2.3.1   Local Linear Embedding (LLE). The LLE algorithm [30] estimates the local coordinates of each data point in the basis of its nearest neighbours, then looks for a low-dimensional coordinate system that has about the same expansion. The 3 steps are: (1) Find the neighbourhood graph (see steps 1 and 2 of Isomap). (2) Compute the weights W_{ij} that best reconstruct $\mathbf{x}_i$ from its neighbours, thus minimizing the reconstruction error, $\|\mathbf{x}_i - \hat{\mathbf{x}}_i\|$, where $\hat{\mathbf{x}}_i = \sum_j W_{ij}\mathbf{x}_j \approx \mathbf{x}_i$. (3) Compute vectors $\mathbf{y}_i$ in $\mathbb{R}^p$ reconstructed by the weights W_{ij}. Solve for all $\mathbf{y}_i$ simultaneously.

$$\mathbf{y}_i \approx \sum_j W_{ij}\mathbf{y}_j \tag{10}$$

LLE is nonlinear, local and Euclidean. It finds the local affine structure of the data manifold, and identifies the manifold by joining the affine patches.

3.2.3.2   Laplacian Eigenmaps (Laplacian). Similar to LLE, Laplacian Eigenmaps find a low-dimensional data representation by preserving local properties of the manifold [31]. The three steps of the algorithm are the following: (1) Build the non-oriented symmetric neighbourhood graph. (2) Associate a positive weight W_{ij} to each link of the graph. These weights can be constant ($W_{ij} = 1/k$), or exponentially decreasing ($W_{ij} = \exp\left(-\|\mathbf{x}_i - \mathbf{x}_j\|^2 / \sigma^2\right)$). (3) Obtain the final coordinates $\mathbf{y}_i$ of the points in $\mathbb{R}^p$ by minimizing the cost function

$$J_{LE} = \sum_{ij} \left(W_{ij} \left\| \mathbf{y}_i - \mathbf{y}_j \right\|^2 / \sqrt{D_{ii} D_{jj}} \right) \tag{11}$$

where D is the diagonal matrix $D_{ii} = \sum_j W_{ij}$. The minimum of the cost function is found with the eigenvectors of the Laplacian matrix: $L = I - D^{-\frac{1}{2}} - WD^{-\frac{1}{2}}$. LE is a nonlinear, local, Euclidean method.

3.2.3.3   Curvilinear Components Analysis (CCA). CCA is an evolution of the nonlinear Multidimensional Scaling (MDS) and Sammon's mapping algorithms [32]. Instead of the optimization of a reconstruction error, CCA and the related Curvilinear Distance Analysis (CDA) aim at preserving of the so-called distance matrix while projecting data onto a lower dimensional manifold.

Let $D_m(X)$ be the $n^2 \times n^2$ matrix of distances between pairs of points in **X**

$$D_m(X) = (d_{ij}^m), \text{ where } d_{ij}^m = \left\| \mathbf{x}_i - \mathbf{x}_j \right\| \tag{12}$$

After DR transformation to $\mathbb{R}^p$, we also have

$$D_p(X) = (d_{ij}^p), \text{ where } d_{ij}^p = \left\| \mathbf{y}_i - \mathbf{y}_j \right\| \tag{13}$$

As with PCA, the $\mathbf{y}_i$ are the transformed approximations of the $\mathbf{x}_i$. CCA tries to find the best suitable transformation, minimizing

$$J_{CCA}(X, p) = \sum_{i,j=1}^{n} (d_{ij}^m - d_{ij}^p)^2 F(d_{ij}^p), \tag{14}$$

Where F is a decreasing, positive function. It acts as a weighting function, giving more importance to the preservation of small distances. In practice, J_{CCA} is minimized using stochastic gradient descent and vector quantization to limit the optimization to a reduced set of representative points. CCA is nonlinear, local and Euclidean.

CDA is a refinement of CCA [4], minimizing

$$J_{CDA}(X, p) = \sum_{i,j=1}^{n} (\delta_{ij}^m - d_{ij}^p)^2 F(d_{ij}^p), \tag{15}$$

Where δ_{ij}^m measures the geodesic distance between $\mathbf{x}_i$ and $\mathbf{x}_j$, approximated by the shortest path distance along a neighbourhood graph. CDA is nonlinear, local and geodesic.

3.2.4 Feature Selection Method

Parameter selection with an exhaustive search is impractical due to the large amount of possible feature subsets. To select the 5 best parameters, we use sequential forward selection (SFS) [33]. The criteria function is the average correct classification rate over all classes, obtained by quadratic discriminant analysis (QDA) on all observations. The QDA approach was chosen because it is not dependant on parameters other than the observations and that the goal is not to compute the optimal classification rate but a measure of the feature subsets efficiencies. At the end of the process, the 5 best features have been selected.

4 Results

In order to compare the classification performance and estimate the average error rate for each classification method, we performed 20 iterative experiments with a 10-fold cross validation procedure. In the case of SVM, we used the gaussian kernel, for which we tuned the determined the optimum value of :

$$K(\mathbf{x},\mathbf{y}) = e^{\left(-\frac{\|\mathbf{x}-\mathbf{y}\|^2}{\sigma}\right)} \tag{16}$$

Table 1. Classification results on the Brodatz dataset (% error rate)

	Methods	Boosting			Hyperrectangle	SVM	MLP
		Hyperplan	Hyperinterval	Hyperrectangle			
linear	Original features	17,26	12,2	22,5	15,5	2,65	41,4
	Selection	21,3	19,7	13,4	8,3	3,06	15,3
	PCA	23,4	18,5	13,2	7,4	8,4	11,2
	SOBI	46,6	27,1	25,7	24,8	10,46	16,4
	PP	84	82	69	75	61,4	73,0
nonlinear	Sammon	23,8	21,2	12,9	15,6	7,8	13,3
	Isomap	23,4	19,5	12,9	7,3	6,55	12,1
	LLE	22,5	23,4	15,3	8	4,5	15,9
	CCA	23,7	20,7	13,9	9,1	5,7	18,2
	CDA	22,6	20	15,3	7,4	3,9	15,6
	Laplacian	16,7	11,8	14	5,56	1,20	10,2
	K-PCA	23,6	19,1	14,2	7,2	6,65	11,8
	K-Isomap	21,3	17,7	15,1	6,13	1,9	9,6

In the case of Brodatz texture dataset (Table 1), regarding to the classification error using the original feature space, the best result are obtained using SVM (e=2.65%). In this case, the backpropagation algorithm of the MLP seems to converge to a local minimum and not to the global one. The use of a second order optimization method, such as BFGS or Levenberg-Marquart method [34] could overcome this problem. All the other methods give poorer results (from 12.2% to 22.5%). Their performances are generally improved by DR: the optimum error is obtained combining Laplacian Eigenmaps and SVM (e=1.2%, i.e. the error is divided by a factor 2). The K-Isomap combination with SVM gives some similar results. One can note that the use of kernel in DR methods generally improve performances compared to original ones (Isomap vs K-Isomap, PCA vs K-PCA). In the family of fast decision methods, the best result is obtained using Hyperrectangle also combined with Laplacian Eigenmaps.

These results are generally confirmed by the experiments performed using the plants Leaf dataset (Table 2), although the original dimensional space is significantly higher than in the previous case (254 vs 32) and the number of classes is lower (6 vs 32). In this case, the gain factor is 1.18 (comparing SVM using original feature space, and SVM combined with Laplacian Eigenmaps).

Table 2. Classification results on the Plants leaf dataset (% error rate)

	Methods	Boosting			Hyperrectangle	SVM	MLP
		Hyperplan	Hyperinterval	Hyperrectangle			
linear	*Original features*	6,52	3,3	16,87	27,61	1,47	35,7
	Selection	18,19	14,97	**3,7**	**10,5**	5,71	**9,7**
	PCA	7,64	3,94	**8,62**	**9,66**	2,35	**11,9**
	SOBI	15,29	4,99	**9,56**	**13,2**	4,8	**15,8**
	PP	87,2	85,58	87,45	84,54	82	81,2
nonlinear	Sammon	26,9	25,84	**10,89**	**10,1**	5,48	**13**
	Isomap	7,2	5,12	**4,25**	**7,8**	2,28	**11,2**
	LLE	22,86	17,87	**7,81**	**8,29**	1,96	**14,1**
	CCA	31,07	17,47	**5,23**	**9,98**	2,89	**16,2**
	CDA	34,13	7,6	**4,83**	**9,75**	1,92	**15,4**
	Laplacian	**5,2**	**2,5**	10,38	**8,1**	1,25	**7,8**
	K-PCA	7,05	13,2	**11,75**	**11,51**	1,86	**13,9**
	K-Isomap	6,8	3,9	**11,43**	**6,3**	1,31	**10,2**

In order to classify the DR methods, we computed the average rank of each method for both datasets (Table 3). Laplacian Eigenmaps and K-Isomap are the best ranked, but the standard PCA (linear) is still a good compromise between computation time and performances.

Table 3. Average rank mean for each classification results for the two dataset

Methods	Boosting			Hyperrectangle	SVM	MLP	*rank means*
	Hyperplan	Hyperinterval	Hyperrectangle				
Laplacian	1	1	7	2	1	1,5	*2,25*
K-Isomap	3	3	9	1,5	2	2	*3,42*
Isomap	6	6	1,5	3	7,5	4,5	*4,75*
PCA	7	4	4,5	5,5	9,5	4	*5,75*
Selection	6	8	2,5	8,5	8	4,5	*6,25*
K-PCA	6,5	6,5	9	6,5	6,5	5,5	*6,75*
Original data	2	2	11,5	11	3	12	*6,92*
CDA	9	7,5	6	5,5	5	8,5	*6,92*
LLE	7	11	7,5	5,5	6	8,5	*7,58*
CCA	10,5	9,5	4,5	8	8	11	*8,58*
Sammon	10,5	11	5,5	9,5	10,5	6	*8,83*
SOBI	9,5	8,5	9,5	11,5	11	10	*10,00*
PP	13	13	13	13	13	13	*13,00*

Moreover, it is interesting to note that DR methods allow to minimize the number of support vectors needed for the decision function of SVM (Table 4). For Laplacian Eigenmaps, the gain is 29% in the case of Brodatz dataset and 47% in the case of Plants leaf dataset. Since the computation time of the SVM decision function depends linearly of this number, the process is accelerated. This is particularly true using PCA, since it is not always necessary to update the PCA transformation during the classification step.

Table 4. Number of support vectors needed for the decision function of SVM

Dataset	Original data	Selection	CCA	SOBI	PCA	K-PCA	CDA	LLE	Lapl	K-iso	Iso	Sam	PP
Brodatz	1545	1120	1006	1025	1035	1050	1059	1078	1095	1155	1163	1189	1467
Plants leaf	504	209	363	172	232	332	419	291	267	296	328	272	1090

5 Conclusion

In this paper we proposed a comparison of DR methods combined with several classification methods, in the context of texture classification of natural images using GFD. We used the powerful Generalized Fourier Descriptors which have interesting properties such as translation, rotation and reflexion invariants.

In any case, the SVM classifier outperforms all other classification methods using the original feature space. However, we experimentally demonstrated that some DR methods still improve final classification performances, and we proposed a rank classification of these methods. The best DR methods are the Laplacian Eigenmaps and K-Isomap, even if the standard PCA is still a good compromise between computation time and performances. In any case, the use of DR methods allows to minimize the number of support vectors, thus optimizing the computational cost of the final decision step.

In our future work, we will apply this comparison review to multispectral textures images for which the original dimensional space is higher and for which the correlation between spectral bands are often very important.

References

1. Arivazhagan, S., Ganesan, L., Priyal, S.P.: Texture classification using Gabor wavelets based rotation invariant features. Pattern Recognition Letters 27, 1976–1982 (2006)
2. Hughes, G.F.: On the mean accuracy of statistical pattern recognizers. IEEE Transactions on Information Theory 14, 55–63 (1968)
3. Aldo Lee, J., Archambeau, C., Verleysen, M.: Locally Linear Embedding versus Isotop. In: ESANN 2003 proceedings, Bruges (Belgium), pp. 527–534 (2003)
4. Aldo Lee, J., Lendasse, A., Verleysen, M.: Nonlinear projection with curvilinear distances: Isomap versus curvilinear distance analysis. Neurocomputing 57, 49–76 (2004)
5. Journaux, L., Foucherot, I., Gouton, P.: Reduction of the number of spectral bands in Landsat images: a comparison of linear and nonlinear methods. Optical Engineering 45, 67002 (2006)
6. Niskanen, M., Silven, O.: Comparison of dimensionality reduction methods for wood surface inspection. In: QCAV 2003 proceedings, Gatlinburg, Tennessee, USA, pp. 178–188 (2003)
7. Gauthier, J.-P., Bornard, G., Silbermann, M.: Harmonic analysis on motion groups and their homogeneous spaces. IEEE Transactions on Systems, Man and Cybernetics 21, 159–172 (1991)
8. Lemaître, C., Smach, F., Miteran, J., Gauthier, J.-P., Atri, M.: A comparative study of motion descriptors and Zernike moments in color object recognition. In: proceeding of International Multi-Conference on Systems, Signal and Devices. IEEE, Hammamet, Tunisia (2007)
9. Brodatz, P.: Textures: A Photographic Album for Artists and Designers. Dover, New York (1966)
10. Valkealahti, K., Oja, E.: Reduced multidimensional cooccurrence histograms in texture classification. IEEE Transactions on Pattern Analysis and Machine Intelligence 20, 90–94 (1998)
11. Duda, R.O., Hart, P.E., Stork, D.G.: Pattern Classification, 2nd edn. (2001)

12. Bishop, C.M.: Neural Networks for Pattern Recognition. Oxford University Press, Oxford (1995)
13. Vapnik, V.: Statistical learning theory. John Wiley & sons, inc., Chichester (1998)
14. Schapire, R.E.: The strenght of weak learnability. Machine Learning 5, 197–227 (1990)
15. Miteran, J., Gorria, P., Robert, M.: Geometric classification by stress polytopes. Performances and integrations. Traitement du signal 11, 393–407 (1994)
16. Abe, S.: Support Vector Machines for Pattern Classification. Springer, Heidelberg (2005)
17. Rumelhart, D.E., McClelland, J.L., Group, a.t.P.R.: Parallel Distributed Processing, vol. 1. MIT Press, Cambridge (1986)
18. Aldo Lee, J., Verleysen, M.: Nonlinear Dimensionality Reduction. Springer, Heidelberg (2007)
19. Camastra, F., Vinciarelli, A.: Estimating the Intrinsic Dimension of Data with a Fractal-Based Method. IEEE Transactions on Pattern Analysis and Machine Intelligence 24, 1404–1407 (2002)
20. Belouchrani, A., Abed-Meraim, K., Cardoso, J.F., Moulines, E.: A blind source separation technique using second order statistics. IEEE Transactions on signal processing 45, 434–444 (1997)
21. Friedman, J.H., Tukey, J.W.: A projection pursuit algorithm for exploratory data analysis. IEEE Transactions on computers C23, 881–890 (1974)
22. HyvÄarinen, A.: Fast and Robust Fixed-Point Algorithms for Independent Component Analysis. IEEE Transactions on Neural Networks 10, 626–634 (1999)
23. Sammon, J.W.: A nonlinear mapping for data analysis. IEEE Transactions on Computers C-18, 401–409 (1969)
24. Tenenbaum, J.B., de Silva, V., Langford, J.C.: A Global Geometric Framework for Nonlinear Dimensionality Reduction. Science 290, 2319–2323 (2000)
25. Shawe-Taylor, J., Cristianini, N.: Kernel methods for pattern analysis. Cambridge University Press, Cambridge (2004)
26. Ham, J., Lee, D.D., Mika, S., Schölkopf, B.: A kernel view of the dimensionality reduction of manifolds. In: 21th ICML 2004, Banff, Canada, pp. 369–376 (2004)
27. Schölkopf, B., Smola, A.J., Müller, K.-R.: Nonlinear component analysis as a kernel eigenvalue problem. Neural Computation 10, 1299–1319 (1998)
28. Choi, H., Choi, S.: Robust kernel Isomap. Pattern Recognition 40, 853–862 (2007)
29. Schölkopf, B., Burges, J.C.C., Smola, A.J.: Advances in Kernel Methods - Support Vector Learning. MIT Press, Cambridge (1999)
30. Roweis, S.T., Saul, L.K.: Nonlinear dimensionality reduction by locally linear embedding. Science 290, 2323–2326 (2000)
31. Belkin, M., Niyogi, P.: Laplacian eigenmaps for dimensionality reduction and data representation. Neural Computation 15, 1373–1396 (2003)
32. Demartines, P., Hérault, J.: Curvilinear Component Analysis: A self-organizing neural network for nonlinear mapping of data sets. IEEE Transactions on neural networks 8, 148–154 (1997)
33. Kittler, J.: Feature set search algorithms. In: Noordhoff, S. (ed.) Pattern Recognition and Signal Processing. Chen, H., pp. 41–60 (1978)
34. Fletcher, R.: Practical Methods of Optimization. John Wiley & Sons, Chichester (2000)

Improving Features Subset Selection Using Genetic Algorithms for Iris Recognition

Kaushik Roy and Prabir Bhattacharya

Concordia Institute For Information Systems Engineering (CIISE)
Concordia University, Montreal, QC, Canada H3G 1M8
{kaush_ro,prabir}@ciise.concordia.ca

Abstract. In this paper, we propose an iris recognition method based on genetic algorithms (GA) to select the optimal features subset. The iris data usually contains huge number of textural features and a comparatively small number of samples per subject, which make the accurate iris patterns classification challenging. Feature selection scheme is used to identify the most important and irrelevant features from extracted features set of relatively high dimension based on some selection criterions. The traditional feature selection schemes require sufficient number of samples per subject to select the most representative features sequence; however, it is not always practical to accumulate a large number of samples due to some security issues. In this paper, we propose GA to improve the feature subset selection by combining valuable outcomes from multiple feature selection methods. The main objective of GA is to achieve a balance among the recognition rate, the false accept rate, the false reject rate and the selected features subset size. This paper also motivates and introduces the use of Gaussian Mixture Model for iris pattern classification. The proposed technique is computationally effective with the recognition rates of 97.81 % and 96.23% on the ICE (Iris Challenge Evaluation) and the WVU (West Virginia University) iris datasets respectively.

Keywords: Biometrics, Gaussian mixture model, genetic algorithms, collarette area localization.

1 Introduction

The popularity of the iris biometric has grown considerably over the past three to four years. The iris has been known as a biometric for some time [1, 5]. However, it has gained substantial attention to both the research community and governmental organizations recently. Five crucial factors that influenced the increased interest in the iris biometric are as follows: 1) unique structure of iris; 2) stability of iris pattern throughout the person's lifetime; 3) public acceptance; 4) new user-friendly capture devices with broad improved capabilities; and 5) a wide range of applications. As a result, a large number of new iris encoding and processing techniques have been developed over this short period of time [1]. Based on the technology developed by Daugman [2], iris scans have been used in several international airports for the rapid processing of passengers through the immigration who have pre-registered their iris

L. Prevost, S. Marinai, and F. Schwenker (Eds.): ANNPR 2008, LNAI 5064, pp. 292–304, 2008.

images. Iris technology has also been widely used in several countries for various security purposes (and also by the United Nations High Commission for refugees). A new technology development project for iris recognition namely, the *Iris Challenge Evaluation* (ICE) has been conducted by the National Institute of Standards and Technology (NIST) [10]. While most of the literatures are focused on preprocessing of iris images [1], recently, there have been important new directions identified in iris biometric research. These include optimal feature selection and iris pattern classification.

The optimal features set selection from a feature sequence with a relative high dimension has become an important factor in the field of iris recognition [3]. The conventional feature selection techniques (e.g., *Principal components analysis, Independent components analysis, Singular valued decomposition* etc.) require sufficient number of samples per subject to select the most representative features sequence. However, it is not realistic to accumulate a large number of samples due to some security issues. Moreover, different feature selection algorithm based on various theoretical arguments may produce different results on the same data set [15]. This makes selecting the optimal features subset for a data set difficult. In this paper, we emphasize on the utilization of the useful information from different feature selection methods to select the most important features subset and also to improve the classification accuracy. We propose Genetic algorithms (GA) to select the significant features subset by combining the multiple feature selection criteria. The proposed approach provides the convenient way of selecting a better feature subset based on the performance of the different feature selection schemes, and this approach is regarded as independent of the inductive learning algorithm used to build the classifier. To evaluate the proposed scheme, support vector machines (SVM)-recursive feature elimination (RFE), k-NN, T-statistics, and entropy-based methods are used to provide the candidate features for the selection of features subset using GA.

In this paper, we also introduce a new iris-subject model based on the Gaussian mixture model (GMM). GMM is used to take into account the interpersonal and intrapersonal variations due to occlusion occurred by the eyelids and the eyelashes and also due to changing light conditions, head tilt etc. The GMM is also applied to satisfy several security requirements with high matching accuracy based on the variation of the Gaussian mixture components.

Fig. 1. Samples of iris images form ICE and WVU datasets

2 Iris Image Preprocessing

The iris is surrounded by various non-relevant regions such as the pupil, the sclera, the eyelids, and also has some noise that include the eyelashes, the eyebrows, the reflections and the surrounding skin [5]. In order to isolate the iris, pupil, and collarette boundaries from digital eye's image, we use an efficient approach proposed in our previous work in [16]. Though collarette region is less affected by the eyelids and the eyelashes, there are few cases where this region is occluded by the eyelids and the eyelashes [16]. These noisy regions are required to be eliminated in order to improve the performance, and this approach is also illustrated in [16]. Fig. 2 shows the localized iris images. We use the rubber sheet model to normalize or unwrap the isolated collarette area [2]. Fig. 3 shows the unwrapping procedure. Since the normalized iris image has relatively low contrast and may have non-uniform intensity values due to the position of the light sources, a local intensity based histogram equalization technique is applied to enhance the contrast of the quality of the normalized iris image to improve the subsequent recognition accuracy. Fig. 3 also shows the effect of enhancement on the unwrapped iris image.

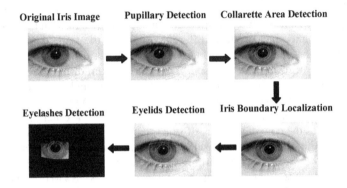

Fig. 2. Iris Image preprocessing on ICE dataset

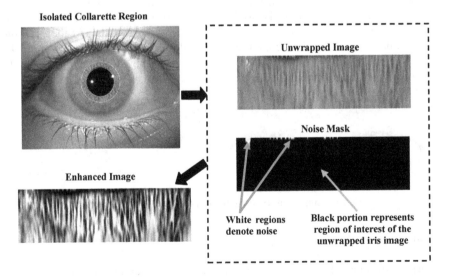

Fig. 3. Unwrapping and enhancement of an iris image on WVU dataset

3   Feature Extraction

Gabor filters based methods have been widely used as feature extractor in computer vision, especially for the texture analysis [2, 16, 17]. However, one weakness of the Gabor filter in which the even symmetric filter will have a DC component whenever the bandwidth is larger than one octave. To overcome this disadvantage, a type of Gabor filter known as log-Gabor filter, which is Gaussian on a logarithmic scale, can be used to produce zero DC components for any bandwidth. The log-Gabor function more closely reflects the frequency response for the task of analyzing natural images and is consistent with measurement of the mammalian visual system. The log-Gabor filters are obtained by multiplying the radial and the angular components together where each even and odd symmetric pair of log-Gabor filters comprises a complex log-Gabor filter at one scale. The frequency response of a log-Gabor filter is given as

$$G(f) = \exp\left(-\left(\log\left(f/f_0\right)\right)^2 / 2\left(\log\left(\sigma/f_0\right)\right)\right) \tag{1}$$

where f_0 is the centre frequency, and σ provides the bandwidth of the filter. In order to extract the discriminating features from the normalized collarette area, the normalized pattern is convolved with 1D log-Gabor filters [16].

4   Feature Subset Selection Using Genetic Algorithms

In this paper, we propose GA to select the prominent features based on the outcomes of the four feature selection algorithms, namely: the Entropy-based approach, k-NN based method, T-statistics and the SVM-RFE approach. Usually, the feature selection algorithms can be divided into two categories: the filter approach and wrapper approach based on whether the selection method is performed independently of the learning algorithm used to construct the classifier. If the feature selection is done independently of the learning algorithm, the technique is referred as the filter approach. Otherwise, it is referred to as a wrapper approach [15]. Several feature selection schemes produce different results on the same data set because of the feature redundancy, interactions and correlations between features, and the biases in the selection or ranking criteria. In order to obtain the most significant feature subset from the different feature selection algorithms, we use a hybrid approach as shown in Fig. 4.

   We adopt GA to combine multiple feature selection criteria to find the optimal subset of informative features. The GA searches the pool of hypotheses (denoted as population) consisting of complex interaction parts. Each hypothesis or individual of the current population is evaluated based on the specific fitness function. A new population is generated by applying genetic operations like selection, mutation and crossover. In this paper, we select sets of features by utilizing four feature selection algorithms instead of using all features set from the original extracted iris features sequence to form the collection of candidate features called the *feature pool*. The selection of features subset from these feature selection algorithm can be subjective to their performance. In order to choose the sets of feature selected by several feature

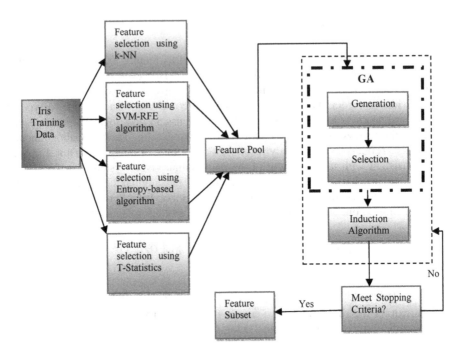

Fig. 4. Feature Selection Procedure using GA (Hybrid approach)

selection algorithms instead of using all the features from the original data set, we deploy four existing feature selection algorithms, two filters (entropy-based, T-statistics) approaches and two wrapper (SVM-RFE, k-NN) approaches to form the feature pool. We apply each algorithm to the extracted features sequence and generate a ranking of those features. Given a ranking of features, we pick a number of top ranked features from each algorithm and provide these top-ranked features into the feature pool. Here, we briefly describe four feature selection algorithms.

In entropy-based method, entropy is lower for orderly configurations and higher for disorderly configurations. Therefore, when an irrelevant feature is eliminated, the entropy is reduced more than that for a relevant feature. This algorithm ranks the features in descending order of the entropies after removing each feature one at a time. We can estimate the entropy measure of a data set of N instances as follows:

$$E = -\sum_{i=1}^{N}\sum_{j=1}^{N}\left(d_{ij} \times \log d_{ij} + \left(1 - d_{ij}\right) \times \log\left(1 - d_{ij}\right)\right) \qquad (2)$$

Where, $d_{ij} = \exp^{-\alpha \times ED_{ij}}$ and $\alpha = -\ln 0.5 / \overline{AD}$.

Here, d_{ij} denotes the similarity between two instances x_i and x_j, ED_{ij} is the Euclidean distance between the two, and $\overline{AD}$ is the average distance among the instances. This approach is used for unsupervised data since no class information is required.

In [15], the SVM-RFE has been used for selecting the genes that are relevant for cancer classification problem. Here, we adopt this approach to find top ranked iris features form the extracted features sequence. The idea is to eliminate one worst feature (i.e., the one that modified the objective function *Obj* least after being eliminated) at one time. This method is based on backward sequential selection.

$$Obj = \|w\|^2 / 2 \tag{3}$$

$$w = \sum_{i=1}^{N_s} \alpha_i y_i x_i \tag{4}$$

Where N_s denotes the number of support vectors that are defined to be the training samples with $0 < \alpha_i \leq C$. C is the penalty parameter for the error term. x_i and y_j are the data instance and its class label respectively. The modification of *Obj* is approximated by *Optimal Brain Damage* (OBD) algorithm so that

$$\Delta Obj(i) = (\Delta w_i)^2 \tag{5}$$

w_i^2 is considered as the ranking criteria. The iterative procedure of RFE is as follows:

- The SVM is trained with training data.
- The ranking criterion is measured for all features.
- Then the feature with smallest ranking criterion is eliminated.
- The procedure is stopped when all the features are ranked.

In T-statistics based feature selection approach, each sample is labeled with $\{1,-1\}$. The mean, $\mu_i^1 (\mu_i^{-1})$ and the standard deviation, $\delta_i^1 (\delta_i^{-1})$ are calculated for the samples labeled as 1 (-1) or each feature, f_i. Then a score $T(f_i)$ is obtained as follows

$$T(f_i) = \left. |\mu_i^1 - \mu_i^{-1}| \middle/ \sqrt{\frac{(\delta_i^1)^2}{n_1} + \frac{(\delta_i^{-1})^2}{n_{-1}}} \right. \tag{6}$$

where $n_1 (n_{-1})$ denotes the number of samples labeled as 1 (-1). In order to make decision, the features with highest scores are considered as the most distinctive features.

In k-NN based feature selection, a direct method based on nonparametric feature subset selection evaluation is applied. The evaluation technique denoted as 'leave-one-out (LOO)' method has been used. The main idea of LOO method is as follows:

- Design the decision rule using N-1 samples of the total N samples.
- Apply decision rule to the one remaining sample.

- This process is repeated for all partitions of size N-1 for the design sample set and size one for the test
- Estimate the probability of error by the ratio of the test samples incorrectly classified to the total number of samples classified.

The k-nearest neighbour (k-NN) has been used as non parametric classification technique in the evaluation procedure. Overall the feature selection procedure is given as below:

1. Apply k-NN as the classifier.
2. Use the LOO test for recognition rate estimation.
3. Select the first feature that has the highest LOO recognition rate among all features.
4. Select the feature, among all unselected features, together with the selected features that gives the highest recognition rate.
5. Repeat the previous process until enough number of features has been selected, or until the recognition rate is good enough.

Each individual represents a feature subset. In this subsection, we present the choice of a representation for encoding the candidate solutions to be manipulated by the genetic algorithms, and each individual in the population represents a candidate solution to the feature subset selection problem. If n be the total number of features available to represent the patterns to be classified, the individual (chromosome) is represented by a binary vector of dimension, n. If a bit is a 1, it means that the corresponding feature is selected; otherwise the feature is not selected. This is the simplest and most straightforward representation scheme [3]. In this paper, we propose the following fitness function based on the nature of our problem:

$$Fitness = W_1.(1 - RR) + W_2.FAR + W_3.FRR + W_4.\left(\frac{FeatureSize}{TotalNumberOfFeatures}\right) \qquad (7)$$

Where W_1, W_2, W_3 and W_4 are constant weighting parameters which reflect the relative importance between *Recognition Rate* (RR), *False Accept Rate* (FAR), *False Reject Rate* (FRR) and *Feature Size*. The genetic algorithm is independent of the inductive learning algorithm used by the classifier. In this paper, we use asymmetrical SVM classifier as an induction algorithm in the experiments to separate the cases of false accepts and false reject [12]. We use Roulette wheel selection to probabilistically select individuals from a population for latter breeding. The probability of selecting an individual ind_{ij} is estimated as;

$$p(ind_i) = F(ind_i)\Big/\sum_{i=1}^{p} F(ind_i) \qquad (8)$$

The probability that an individual will be selected is proportional to its own fitness and is inversely proportional to the fitness of the other competing hypothesis in the current population. Here, we use single point crossover, and each individual has a probability, P_n to mutate. The number of n bits is randomly selected to be flipped in every mutation stage.

5 Iris Pattern Classification Using Multi-class Gaussian Mixture Model

In this paper, we propose Gaussian mixture model (GMM) to accurately classify the iris pattern. We apply GMM to address the following two important issues. First issue is the significant inter and intra personal variation and second issue is to obtain the required *false accept* and *false reject* rates with a high recognition accuracy to meet several security demands by changing the number of Gaussian mixtures. In the following subsections, we briefly discuss the form of GMM. A detailed discussion on GMM can be found at [14].

5.1 Model Description

A Gaussian mixture model is s weighted sum of M component densities and can be described by the following equation

$$p(\bar{x} \mid \lambda) = \sum_{i=1}^{M} p_i b_i(\bar{x})$$

(9)

Where, $\bar{x}$ denotes the D-dimensional random vector, $b_i(\bar{x}), i = 1, 2, \ldots, M$, are the component densities and $p_i, i = 1, \ldots, M$, are the mixture weights. Each component density is a D-variate Gaussian function of the following form [14]

$$b_i(\vec{x}) = \frac{1}{(2\pi)^{D/2} |\Sigma_i|^{1/2}} \exp\left\{ -\frac{1}{2} (\vec{x} - \vec{\mu_i})' \Sigma_i^{-1} (\vec{x} - \vec{\mu_i}) \right\}$$

(10)

Here, μ_i is the mean vector and $\sum_i$ is the covariance matrix. The mixture weight satisfies the constraint $\sum_{i=1}^{M} p_i = 1$. Therefore, the Gaussian mixture density is parameterized by the mean vectors, covariance matrix and mixture weights from all the component weights. The parameters can be represented by the following equation [14]

$$\lambda = \{p_i, \mu_i, \Sigma_i\} i = 1, 2, \ldots, M$$

(11)

For iris recognition, each subject is represented by a GMM and is denoted by the model, λ.

5.2 Estimation of Maximum Likelihood Parameters

Given a training sample from a subject, the main objective of the person model training is to estimate the parameters of the GMM, λ, that best matches the distribution of the training feature vectors. The popular maximum likelihood estimation (ML) is used to estimate the parameters of a GMM. The idea is to find the model parameters that maximize the likelihood of the GMM provided the training data is given. If T denotes the sequence of training vectors $X = \{\vec{x}_1, \ldots\ldots, \vec{x}_T\}$, the GMM likelihood can be defined as

$$p(X \mid \lambda) = \prod_{t=1}^{T} p(\vec{x}_t \mid \lambda) \qquad (12)$$

However, this expression is a non linear function of the parameters λ and direct maximization is not possible. Therefore, ML parameters estimation can be obtained iteratively by using a special case of the expectation-maximization algorithm [14]. The basic idea of the EM algorithm is to begin with an initial model λ, then a new model $\overline{\lambda}$ is estimated from the initial model such that $p(X \mid \overline{\lambda}) > p(X \mid \lambda)$. The new model becomes the initial model for the next iteration and the process is repeated until some convergence threshold is reached.

5.3 Subject Identification

For iris recognition, a group of *subjects, S= { 1, 2, ..., S}* represented by GMM's $\lambda_1, \lambda_2, \ldots, \lambda_s$. The objective is to find the person model which has the maximum a posteriori probability for a given observation sequence.

Formally,

$$\hat{S} = \arg\max_{1 \le k \le S} P_r(\lambda_k \mid X) = \arg\max_{1 \le k \le S} \frac{p(X \mid \lambda_k) P_r(\lambda_k)}{p(X)} \qquad (13)$$

Let us consider the equally likely subjects (i. e., $P_r(\lambda_k) = 1/S$) and, it is also assumed that $P(X)$ is the same for all subjects, the classification simplifies to

$$\hat{S} = \arg\max_{1 \le k \le S} P_r(\lambda_k \mid X) \qquad (14)$$

By using independence between observations, the iris recognition system computes

$$\hat{S} = \arg\max_{1 \le k \le S} \sum \log p(\vec{x}_t \mid \lambda_k) \qquad (15)$$

Where, $p(\vec{x}_t \mid \lambda_k) = \sum_i^M p_i b_i(\vec{x})$, the Gaussian mixture density which is weighted sum of M components as given in (9).

6 Experimental Results

We conduct the experimentation on two iris data sets namely, the ICE (Iris Challenge Evaluation) dataset created by the University of Notre Dame, USA, [10] and the WVU (West Virginia University) dataset [11]. The ICE database consists of left and right iris images for experimentation. We consider only the left iris images in our experiments. There are 1528 left iris images corresponding to the 120 subjects in our

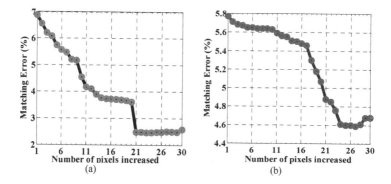

Fig. 5. Matching error vs. number of pixels increased on (a) ICE and (b) WVU datasets

experiments. We also evaluated the performance of the proposed iris recognition scheme on the WVU dataset [11]. The WVU iris dataset has a total of 1852 iris images from 380 different persons. The number of iris images for each person ranges from 3-6 in this database. Since the number of samples from most iris research is limited, cross-validation procedure is commonly used to evaluate the performance of a classifier. In k-fold cross validation, the data is divided into k subsets of (approximately) equal size. We train the classifier k times, each time leaving out one of the subsets from training, but using only the omitted subset to compute the classification accuracy. LOO cross-validation (LOOCV) is a special case of k-fold cross-validation where k equals the sample size. LOOCV is used for ICE dataset, and for WVU dataset, we use 3-fold cross-validation to obtain the training accuracy for GA.We evaluate the success rate for the proposed method on the ICE and WVU datasets by detecting the pupil boundary and the collarette area. The obtained success rates are 98.80% and 97.95% for the ICE and WVU data sets respectively. From the experimental results, it is found that a reasonable recognition accuracy is achieved when the collarette area is isolated by increasing the previously detected radius value of the pupil up to a certain number of pixels. A rapid drop of matching error from 3.61% to 2.48% is observed in Fig.5 (a) for the case of ICE data set when the pixel value is increased from 20 to 21. Therefore, we choose to increase the pupil radius up to 23 pixels because a stable matching accuracy of 97.54% is achieved in this case. From Fig. 5(b) it is found that if we increase the pixel values up to 26 we obtain the highest matching accuracy of 95.53% for WVU data set. Fig. 6 shows the accuracy of the feature subsets with a different number of top-ranked features from the four feature selection algorithms on two data sets. Fig.6 (a) shows that SVM-RFE achieves the better accuracy than the other feature selection methods used in this paper with a subset of 600 top-ranked features. In Fig. 6(b), we can see that SVM-RFE also find the better accuracy among the four algorithms with the 800-top ranked features. Therefore, after obtaining the top-ranked features subset from the SVM-RFE algorithm on both of the two data sets, we input them to the feature pool used by the GA. In order to select the optimum features for the improvement of the matching accuracy, GA involves running the genetic process for several generations. We

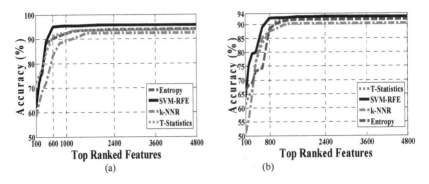

Fig. 6. Accuracy vs. top ranked features on (a) ICE (b) WVU datasets

Table 1. The Selected Values of the Arguments of GA for ICE and WVU Datasets

Parameters	ICE Dataset	WVU dataset
Population Size	120 (the scale of iris sample)	380 (the scale of iris sample)
Length of chromosome code	600 (selected dimensionality of top ranked feature sequence)	800 (selected dimensionality of top-ranked feature sequence)
Crossover probability	0.40	0.89
Mutation probability	0.008	0.007
Number of generation	130	80
Weighting Parameters	W_1= 2000, W_2 =150 W_3= 10, W_4 =1000	W_1= 3500, W_2 =100 W_3= 10, W_4 =2000

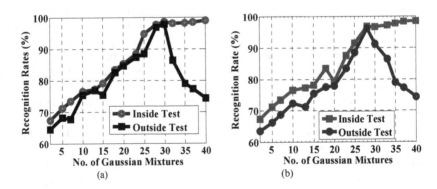

Fig. 7. Recognition accuracy vs. No. of Gaussian mixtures for (a) ICE and (b) (WVU) data sets

conduct several experimentations, and the arguments of the GA are set as shown in Table 1. From experimentation, we find that the proposed GA scheme achieves the highest accuracy of 97.60% at the generation 90 with reduced features subset of 520 for the ICE dataset. Based on the experimentation, we also find that at the generation

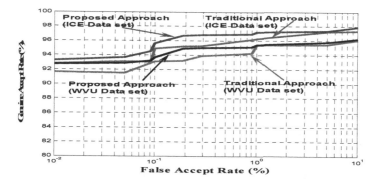

Fig. 8. ROC curve shows the comparison between GAR (%) and FAR (%) for the traditional approach with the complete iris information and proposed approach with collarette information

of 60, the highest accuracy of 95.95% is obtained with the reduced features subset of 680 on the WVU data set. Fig 7 shows the classification accuracy vs. number of Gaussian mixture components. From this figure, we can see that highest accuracy of 97.80% is achieved when the number of Gaussian components is 32 on the ICE data set. For WVU data set, highest recognition accuracy of 96.20% is found at 28 mixture components. In Fig. 8, ROC curve shows how the Genuine Accept Rate (GAR) changes with the False Accept Rate (FAR). It is found form this figure that the proposed approach performs better than the traditional approach with the complete iris information for both of the data sets.

7 Conclusions

In this paper, we mainly focus on the feature subset selection of iris data based on GA. The proposed GA incorporates four feature selection criterions, namely: the SVM-RFE, the k-NN, the T-statistics, and the entropy-based methods to find the subset of informative texture features that can improve the analysis of iris data. The experimental results show that the proposed method is capable of finding feature subsets with a better classification accuracy and/or smaller size than each single individual feature selection algorithm does. This paper also introduces the use of GMM as an iris patterns classifier. The experimental results indicate that the proposed iris recognition scheme with GMM approach can be applied to a wide range of security-related application fields with encouraging recognition rates.

References

1. Schuckers, S.A.C., Schmid, N.A., Abhyankar, A., Dorairaj, V., Boyce, C.K., Hornak, L.A.: On techniques for angle compensation in nonideal iris recognition. IEEE Trans. SMC-B 37(5), 1176–1190 (2007)
2. Daugman, J.: Demodulation by complex-valued wavelets for stochastic pattern recognition. Internat. J. Wavelets, Multi-Res. and Info. Processing 1, 1–17 (2003)

3. Deb, K.: Multi-Objective Optimization using Evolutionary Algorithms. J. Wiley Ltd., West Sussex (2004)
4. He, X., Shi, P.: An efficient iris segmentation method for recognition. In: Internat. Conf. on Adv. Pattern Recog. LNCS, vol. 3687, pp. 120–126. Springer, Heidelberg (2005)
5. Ma, L., Tan, T., Wang, Y., Zhang, D.: Efficient iris recognition by characterizing key local variations. IEEE Trans. Image Processing 13, 739–750 (2004)
6. Son, B., Won, H., Kee, G., Lee, Y.: Discriminant iris feature and support vector machines for iris recognition. In: Internat. Conf. on Image Processing, vol. 2, pp. 865–868 (2004)
7. Sung, H., Lim, J., Park, J., Lee, Y.: Iris Recognition using collarette boundary localization. In: Proc. of Internat. Conf. on Pattern Recog., vol. 4 (2004)
8. Vapnik, V.N.: Statistical Learning Theory. J. Wiley Ltd., New York (1998)
9. http://www.csie.ntu.edu.tw/~cjlin/libsvm
10. http://iris.nist.gov/ICE/
11. Iris Dataset obtained from West Virginia University (WVU), http://www.wvu.edu/
12. Ding, P., Chen, Z., Liu, Y., Xu, B.: Asymmetrical support vector machines and application in speech processing. In: IEEE Internat. Conf. on Acousts, Speech, and Signal Process, vol. 1, pp. 73–76 (2002)
13. Oliveira, L.S., Sabourin, R.F., Bortolozzi, C.Y., Suen, C.Y.: Feature selection using multiobjective genetic algorithms for handwritten digit recognition. Internat. Conf. on Pattern Recog. 1, 568–571 (2002)
14. Reynolds, D.A., Rose, R.C.: Robust text-independent speaker identification using Gaussian mixture models. IEEE Trans. On speech and audio process 3(1), 72–83 (1995)
15. Tan, F., Fu, X., Zhang, Y., Bourgeois, A.G.: Improving feature subset selection using a genetic algorithm for microarray gene expression data. IEEE congress on evolutionary computation, 2529–2534 (2006)
16. Roy, K., Bhattacharya, P.: Iris Recognition Based on Collarette Region and Asymmetrical Support Vector Machines. In: Kamel, M., Campilho, A. (eds.) ICIAR 2007. LNCS, vol. 4633, pp. 854–865. Springer, Heidelberg (2007)
17. Roy, K., Bhattacharya, P.: Iris Recognition Using Support Vector Machine. In: IAPR Internat. Conf. on Biometric Authentication. LNCS, vol. 3882, pp. 486–492. Springer, Heidelberg (2006)

Artificial Neural Network Based Automatic Face Model Generation System from Only One Fingerprint

Seref Sagiroglu[1] and Necla Ozkaya[2]

[1] Gazi University, Engineering and Architecture Faculty, Computer Engineering Department,
06570 Ankara, Turkey
[2] Erciyes University, Engineering Faculty, Computer Engineering Department,
38039, Kayseri, Turkey
ss@gazi.edu.tr, neclaozkaya@erciyes.edu.tr

Abstract. Biometrics technology has received increasingly more attention during the last three decades. Since the performance of biometric systems has reached a satisfactory level for applications, a number of biometric features have been deeply studied, tested and successfully deployed in applications. Relationships among biometric features have not been studied so far. This study focuses on analysing the existence of any relationships among fingerprints and faces. For doing that an intelligent system based on artificial neural networks for generating face models including *eyes, nose, mouth, ears* and *face border* from only one fingerprint with the errors among 2.0-12.9 % was developed. Experimental results have shown that there are close realitionships among fingerprints and faces and it is possible to generate faces from only one fingerprint image without knowing any information about faces. Although the proposed system is an initial study and it is still under development, the results are very encouraging and promising for the future developments and applications.

Keywords: Biometrics, artificial neural network, intelligent systems, fingerprint identification, face recognition.

1 Introduction

Biometrics is a science that extracts the physical or behavioral parameters of individuals with the aim of identification. Several personal biological characteristics such as fingerprint, face, iris, voice, or hand geometry are now used in biometric systems that are more reliable to identify people than traditional methods based on features that we have (key, card) or we know (password). Because a biometric system relies on specific characteristics of a particular person to register an identity. Therefore, it can differentiate between an authorized person and fraudulent impostor. So, any system assuring reliable person identification must necessarily involve a biometric component. Recently a number of biometric features have been successfully applied to the applications including information security, law enforcement, surveillance, forensics, smart cards, access control, time/place control points and computer networks etc. Furthermore various biometric devices and complete systems

L. Prevost, S. Marinai, and F. Schwenker (Eds.): ANNPR 2008, LNAI 5064, pp. 305–316, 2008.

that provide business and benefit management solutions that using biometric based person identification systems have been produced and they are commercially available. In spite of all these developments in biometrics, there is no study on investigating relationships among the biometric features or obtaining one feature from another. It should be emphasized that most of the works in biometrics have been focused on how to improve the accuracy and processing time of the biometric systems, to design the more intelligent systems, and to develop more effective and robust techniques and algorithms [1], [2].

The aim of our study is intended to develop an automatic and intelligent system capable of generating the face of a person from only one fingerprint of the same person without having any priori knowledge about his or her face. In order to achieve that, an artificial neural network based intelligent face model generation system from only one fingerprint (fingerprint to face system: F2FS) has been developed and introduced in this study.

This paper is organized as follows. Section 2 briefly describes background information on biometrics, automatic fingerprint authentication systems (AFASs) and face recognition systems (FRSs). Section 3 basically introduces artificial neural networks (ANNs). Section 4 highlights the novelty of the proposed technique, introduces basic information, notation and performance metrics related to the F2FS and explains the various steps of the new approach. Section 5 demonstrates our experimental results. Finally, the proposed work is concluded and discussed in Section 6.

2 Background

In general, a biometric system gets a biometric data from a person, extracts a feature set from the acquired data, and compares this feature set against the template feature set in the database [3]. A biometric system works in four modes depending on the application status [4]: the enrolment, the verification, the identification and the screening. *The enrolment* is responsible for scanning, categorization and registration of the biometric characteristics. All other modes use the biometric data that was acquired to the system in the enrolment mode. In *the verification,* a person desired to be identified by submitting to the system a claim to an identity, usually via a magnetic card, login name, smart card etc., and the system either rejects or accepts the submitted claim of the identity at the end [5]. Commercial applications, such as physical access control, computer network logon, electronic data security, ATMs, credit-card purchases, cellular phones, personal digital assistants, medical records management and distance learning are samples of the verification applications [2], [4]. In *the identification,* the system identifies a person's identity without the person having to claim an identity or it fails if the person is not enrolled in the system database. The input and the output of the system are just a biometric feature and a combination of a list of identities and the scores indicating the similarity among two biometric features, respectively [5]. Welfare-disbursement, national ID cards, border control, voter ID cards, driver's license, criminal investigation, corpse identification, parenthood determination, missing children identification are from typical identification applications [2], [4]. In *the screening,* the results of determination

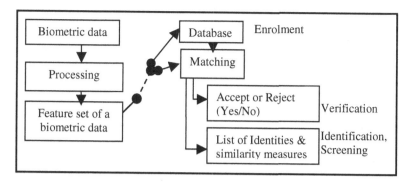

Fig. 1. A generic biometric system

whether a person belongs to a watch list of identities or not is displayed. Security at airports, public events and other surveillance applications are some of the screening examples [4], [6]. A generic biometric system is given in Fig. 1.

It is expected that a biometric system always takes the correct decision when a biometric feature is presented to the system. However, in practice a biometric system can make two basic types of errors: false match rate (FMR) and false non-match rate (FNMR) [1]. These errors generally used to show the accuracy and performance of the system in the literature. Nevertheless it is more informative to report the system accuracy in terms of a Receiver Operating Characteristic (ROC) curve that shows the system performance at all operating points [6].

2.1 Automatic Fingerprint Authentication Systems

Fingerprint is the most widely used biometric feature due to its uniqueness, immutability, reliability, permanence and universality [7]. It has a ridge-valley structure, core and delta points called singular points, end points and bifurcations called minutiaes. These structures are given in Fig. 2. Many approaches to AFASs have been presented in the literature [1], [2], [5], [7]-[19]. Yet, it is still an actively researched area.

The AFASs might be broadly classified as being *minutiae-based, correlation-based* and *image-based* systems [8]. A good survey about these techniques was given in [1]. *The minutiae-based approaches* rely on the comparisons for similarities and differences of the local ridge attributes and their relationships to make a personal

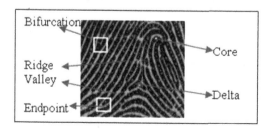

Fig. 2. Ridge-valley structure and features of a fingerprint

identification [9]-[11]. They attempt to align two sets of minutiaes from two fingerprints and count the total number of matched minutiaes [4]. If minutaes and their parameters are computed relative to the singular points which are highly stable, rotation, translation and scale invariant, then these minutiaes will also become rotation, translation and scale invariant [5], [12]-[14]. Core points are the points where the innermost ridge loops are at their steepest. Delta points are the points from which three patterns deviate [13], [15], [16]. The general methods to detect the singular points are poincare-based methods [17], intersection-based methods [13] or filter-based methods [18]. Main steps of the operations in the minutiae-based AFASs are summarized as follows: selecting the image area, detecting the singular points, enhancing, improving and thinning the fingerprint image, extracting the minutiae points and calculating their parameters, eliminating the false minutiaes, properly representing the fingerprint images with their feature sets, recording the feature sets into a database, matching the feature sets, test and evaluating the system [19]. The results of these processes are given in Fig. 3, respectively. The performance of the minutiae-based techniques relies on the accuracy of all these processes. Especially the feature extraction and the use of sophisticated matching techniques to compare two minutiae sets often more affect the performance.

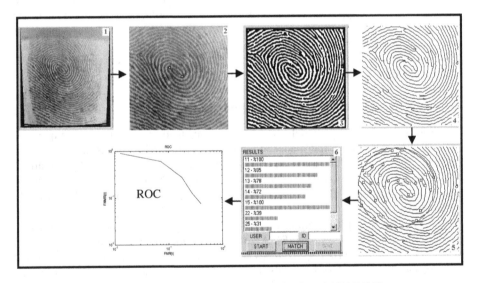

Fig. 3. Main operation steps of a minutiae-based AFAS [19]

In the correlation-based AFASs, global patterns of the ridges and valleys are compared to determine if the two fingerprints align. The template and query fingerprint images are spatially correlated to estimate the degree of similarity between them. The performance of the correlation-based techniques is affected by non-linear distortions and noise present in the image. In general, it has been observed that minutiae-based techniques perform better than correlation-based ones [20]. In *the image-based approaches,* the decision is made using the features that are extracted directly from the raw image that might be the only viable choice when image quality is too low to allow reliable minutiae extraction [8].

2.2 Face Recognition Systems

Face is probably the most common biometric characteristic used by humans to make personal recognition. Face recognition is an active area of research with several applications ranging from static to dynamic [9]. So, face recognition technology is well advanced. In general, a FRS consists of three main steps. These steps cover detection of the faces in a complicated background, localization of the faces followed by extraction of the features from the face regions and finally identification or verification tasks [21]. Face detection and recognition process is really complex and difficult due to numerous factors affecting the appearance of an individual's facial features such as 3D pose, facial expression, hair style, make-up, etc. [22]. In addition to these varying factors, lighting, background, scale, noise and face occlusion and many other possible factors make these tasks even more challenging [21]. The most popular approaches to face detection and recognition are based on either the location and shape of the facial attributes, such as the faces, eyebrows, nose, lips and chin and their spatial relationships or the overall analysis of the face images that represents a face as a weighted combination of a number of canonical faces [4], [21], [23]. Also many effective and robust methods for the face recognition have been proposed [2], [9], [21]-[25]. They are categorized as follows: Knowledge-based methods encode human knowledge of what constitutes a typical face. Feature invariant methods aim to find structural features that exist even when the pose, viewpoint or lighting conditions vary to locate faces. In template matching based methods several standard patterns of a face are used to describe the face as a whole or the facial features separately. Appearance-based methods operate directly on images or appearances of the face objects and process the images as two-dimensional holistic patterns [23].

3 Artificial Neural Networks

ANNs have been applied to solve many problems [26], [30]. Learning, generalization, less data requirement, fast computation, ease of implementation and software and hardware availability features have made the ANNs very attractive for many applications [27], [28]. These fascinating features have also made them popular in biometrics as well [23], [24], [28]-[30]. Multilayered perceptron (MLP) is one of the most used ANN architectures. Because of the MLP structure can be trained by many learning algorithms, it has been applied to a variety of problems successfully in the literature. The MLP structure consists of three layers: input, output and hidden layers. One or more hidden layers might be used. The neurons in the input layer can be treated as buffers and distribute x_i input signals to the neurons in the hidden layer. The output of the each neuron y_j in the hidden layer is obtained from sum of the multiplication of all input signals x_i and weights w_{ji} that follow all these input signals. The sum can be calculated as a function of y_j and can be expressed as:

$$y_j = f\left(\sum w_{ji} x_i\right) \tag{1}$$

where f can be a simple threshold function, a hyperbolic tangent or a sigmoid function. The outputs of the neurons in other layers are calculated in the same way. The weights are adapted with the help of a learning algorithm according to the error that can be calculated by subtracting the ANN output from the desired output. The ANNs might be trained with many different learning algorithms [28].

4 Proposed ANN Based Intelligent Face Generation System

Fingerprint and face recognition topics have received significantly increased attention due to possessing the merits of their reliability, performance and high accuracy. The proposed F2FS generates the face of a person from only one fingerprint of the same person without having any information about his or her face. It is thought that it will be a very interesting innovation to biometrics.

Implementation steps of the F2FS to establish a relationship among fingerprints and faces (Fs&Fs) can be mentioned as follows:

1. A database consisted of Fs&Fs was established.
2. Feature sets of Fs&Fs were obtained.
3. Training and test data sets were established for ANN application.
4. Suitable ANN parameters were selected.
5. Randomly selected 80 of 120 data sets covering pairs of Fs&Fs were used to train the ANN based F2FS.
6. Feature sets of test sets covering only fingerprints of remaining 40 people from the database were used to test the system.
7. In order to evaluate the accuracy of the F2FS, the test results were compared with their desired values against to a variety of state-of-the-art methods [1].

The proposed F2FS has constructed by appropriately combining a data enrolment module, a feature extraction module, an ANN module, an evaluation module and a face reconstruction module. The first module of the system which is called the data enrolment module helps store biometric data of individuals into the biometric system database. During this process, Fs&Fs of an individual have been captured to produce a digital representation of the characteristics. Two types of biometric data are used in this study: Fs&Fs. The second module of the system extracts discriminative feature sets from the acquired data. Extracting local and global feature sets of the fingerprints, which include fingerprint singularities, minutiae points and their parameters, is achieved. Similarly feature sets of the faces were obtained. The ANN module is used to analyze the existence of any relationship among Fs&Fs. This part of the system was implemented with the help of 3-layered ANN structure that is trained with the scaled conjugate gradient (SCG) algorithm. The SCG algorithm is an ANN training algorithm that adjusts the weights and biases of an ANN structure according to its learning strategy. The SCG algorithm is based on conjugate directions. The details of SCG algorithm can be found in the references [31] and [32]. Sigmoid transfer function was used in the proposed study for generating the output of each neuron used in the structure. The ANN module has only a hidden layer with 200 neurons. The block diagram of the ANN Module is given in Fig. 4.

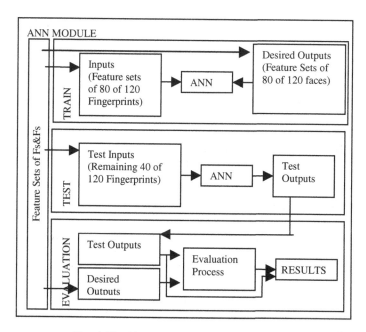

Fig. 4. The block diagram of the ANN Module

The ANN module is the most critical and important module of the system. Because, all modules of the system except the ANN module are on duty, either in pre-processing or post-processing of the main process that is done by the ANN module. So, if we investigate this module deeply, we can explain the working principles of the F2FS properly.

The ANN module operates its task in two stages: the training and the testing. In the training process, randomly selected input-output image sets covering Fs&Fs belonging to the same people were used. The ANN structure and the training parameters were determined for achieving the training stage accurately. The training process is started with applying a person's fingerprint feature set to the system as an input and same person's face feature set as the desired output. The system achieves the training process with these feature sets according to the learning algorithm and the ANN parameters. Both fingerprint and face feature sets are required in the training. Only fingerprint feature sets are used in test. These fingerprints are unknown biometric data to the F2FS.

Producing the faces as close to the real one as possible is critical for this study. The traditional metrics of an ordinary biometric system are no longer appropriate to characterize the performance of the F2FS. In addition to the FMR-FNMR and the ROC curve representations, the results of the system were evaluated to give more prices perceptions to the researchers by considering the following metrics: mean squared error (MSE), sum squared error (SSE), average correlation, absolute percent error (APE) and mean APE [33]. In addition to these numerical evaluations, a visual evaluation platform was also created from face features by drawing the results of desired and actual outputs together. Moreover, another visual evaluation platform was also established by drawing the results of actual and desired outputs on the involved real face images of test people.

Consequently, for a more objective comparison, the performance and accuracy of the system have been evaluated and presented on the basis of the combination of these metrics for illustrating the qualitative properties of the proposed methods as well as a quantitative evaluation of their performance. The face reconstruction module facilitates the evaluation process, simplifies to understand the results and presents to the users to evaluate the results perceptionally. To achieve all these processes easily and efficiently, an automatic system has been proposed and a graphical interface was designed to achieve the results and the metrics in the expected form.

5 Experimental Results

The concept of generating faces from only one fingerprint is a novel and challenging idea to biometrics technology. The proposed ANN based face generation system from only one fingerprint that was discussed in previous section is implemented. A dedicated software has been developed to conduct the experiments easily and efficiently. In the experiments, a multimodal database having Fs&Fs belonging to 120 people was established. The index finger of the right hand was used because of being the most used finger in AFASs.

In the training processes, 80 of 120 image sets were randomly selected from the database. The remaining 40 of 120 fingerprint images were used in the test. The experimental image sets used in test processes contain only fingerprint images of the test people and these data sets are unknown data for the system. The face images of the test people were also used for evaluation of the system's performance. The inputs and the outputs of the system were vectors sized 298 and 148, respectively. These vectors were the feature sets of fingerprints and faces, respectively. The feature vectors of fingerprints were computed using a SDK developed by Neurotechnologija. The reason of this preference was to prove the system's success with a known method for the F2FS. This software is known as an effective, robust and reliable AFAS in the field of biometrics and it uses a minutiae-based algorithm. Detailed explanation of its algorithms, detailed information of fingerprint feature sets and its storage format are given in [34].

To get the feature sets of the faces a feature-based approach has been selected from the face recognition literature and used by modifying fundamentally [35]. It can be explained the reason of this preference is that it is used a minutiae-based approach to get the feature sets of the fingerprints. Actually minutiae-based approaches rely on the physical features of the fingerprints. Therefore it is reasonable that the feature sets of both Fs&Fs should be obtained in the same way. So, a feature-based approach was used to get the feature sets of the faces. A template was used for faces at the beginning to provide appropriate features to the ANN model.

In order to evaluate the performance of the system effectively, we have benchmarked our system against to the extra metrics in addition to the traditional evaluation metrics of biometric systems that include FMR-FNMR representation and ROC curve of the test results. The metrics MSE and SSE were computed before rescaling, while the other metrics MSE, SSE, average APE and average correlation

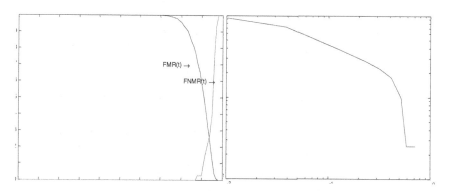

Fig. 5. FMR-FNMR representation and ROC curve of the test results

were calculated after rescaling as 8.8583e-004, 5.2441, 132.2764, 7.8308e+005, 5.43656 and 0.993698, respectively. FMR-FNMR representation and ROC curve of the test results are given in Fig. 5.

These results indicated that the proposed system performed the tasks with high similarity measures to the desired values. For the purpose of more realistic and visual evaluation, 12 of 40 desired and achieved test results were drawn on the same platform as shown in Fig. 6. The same test results were shown on the real face images as given in Fig. 7. It needs to be emphasized that because of the page limitation, only 12 of 40 test results are given. However it is possible to show the overall system performance graphically for the all test results. The APE values belonging to all test results were demonsrated in Fig. 8.

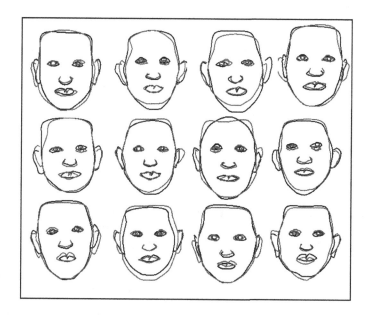

Fig. 6. Combined the test faces achieved from the F2FS and their desired values

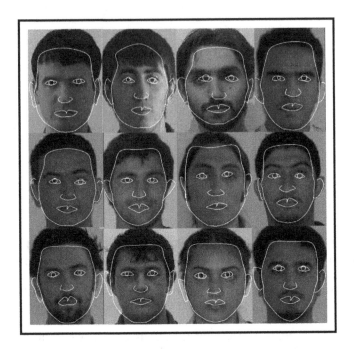

Fig. 7. Test faces obtained from the system are drawn on the real face images

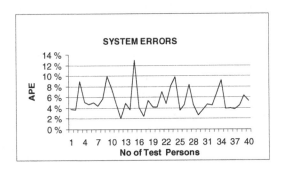

Fig. 8. APE values for all test results

Besides the numerical results indicated the system success clearly, graphical results also confirmed this success as well. As it can be seen from Figures 5-8, the system is very successful in achieving the faces according to each standard metric defined in the literature or extra metrics presented by us. Based on the observations, the fundamental novelty and diversity of the proposed approach over the most other studies in biometrics was to investigate the relationships between fingerprints and faces. In addition, the proposed system can be used for an intelligent and efficient translator that effectively converts a person's fingerprint to the face of the same person without any information about his or her face.

6 Conclusions and Future Work

In this study, the existence of relationships among biometric features was experimentally illustrated. In addition, it is demonstrated that it is possible to achieve an unknown biometric feature from a known biometric feature, successfully.

It has a very simple structure. It is constructed by appropriately combining a data enrolment module, a feature extraction module, an ANN module, a test & evaluation module and a face reconstruction module. In order to achieve the experiments easily and effectively, these modules were combined in the F2FS for generating the face models from fingerprints without any need of face information. Because of establishing the relationship among Fs&Fs, the structure of the ANN module plays an important role in the system. The difficulty faced in the study was to explain the relationships among Fs&Fs mathematically.

Finally it is hoped that this approach would lead to create new concepts, research areas, and especially new applications in the field of biometrics.

References

1. Maio, D., Maltoni, D., Jain, A.K., Prabhakar, S.: Handbook of fingerprint recognition. Springer, New York (2003)
2. Jain, L.C., Halici, U., Hayashi, I., Lee, S.B., Tsutsui, S.: Intelligent biometric techniques in fingerprint and face recognition. CRC press, New York (1999)
3. Jain, A.K., Ross, A., Prabhakar, S.: An introduction to biometric recognition. IEEE Trans. on Circuits and Systems for Video Technology 14(1), 4–19 (2004)
4. Jain, A.K., Ross, A., Pankanti, S.: Biometrics: a tool for information security. IEEE Trans. on Information Forensics and Security, 1(2), 125–143 (2006)
5. Jain, A.K., Hong, L., Pankanti, S., Bolle, R.: An identity authentication system using fingerprints. Proceedings of the IEEE 85(9), 1365–1388 (1997)
6. Jain, A.K., Pankanti, S., Prabhakar, S., Hong, L., Ross, A., Wayman, J.L.: Biometrics: A Grand Challenge. In: Proceedings of the Int. Conf. on Pattern Recognition, Cambridge, UK, August, vol. II, pp. 935–942 (2004)
7. Kovács-Vajna, Z.M.: A fingerprint verification system based on triangular matching and dynamic time warping. IEEE Trans. Pattern Anal. Mach. Intell. 22(11), 1266–1276 (2000)
8. Lumini, A., Nanni, L.: Two-class Fingerprint matcher. Pattern Recognition 39(4), 714–716 (2006)
9. Hong, L., Jain, A.: Integrating faces and fingerprints for personal identification. IEEE Trans. Pattern Analysis and Machine Intelligence 20(12), 1295–1307 (1998)
10. Jain, A.K., Hong, L., Bolle, R.: On-line fingerprint verification. IEEE Trans. on Pattern Analysis and Machine Intelligence 19(4), 302–314 (1997)
11. Zhou, J., Gu, J.: Modeling orientation fields of fingerprints with rational complex functions. Pattern Recognition 37(2), 389–391 (2004)
12. Hsieh, C.T., Lu, Z.Y., Li, T.C., Mei, K.C.: An Effective Method To Extract Fingerprint Singular Point. In: The Fourth Int. Conf./Exhibition on High Performance Computing in the Asia-Pacific Region, pp. 696–699 (2000)
13. Rämö, P., Tico, M., Onnia, V., Saarinen, J.: Optimized singular point detection algorithm for fingerprint images. In: Int. Conf. on Image Processing, pp. 242–245 (2001)

14. Zhang, Q., Yan, H.: Fingerprint classification based on extraction and analysis of singularities and pseudo ridges. Pattern Recognition 11, 2233–2243 (2004)
15. Wang, X., Li, J., Niu, Y.: Definition and extraction of stable points from fingerprint images. Pattern Recognition 40(6), 1804–1815 (2007)
16. Li, J., Yau, W.Y., Wang, H.: Combining singular points and orientation image information for fingerprint classification. Pattern Recognition 41(1), 353–366 (2008)
17. Kawagoe, M., Tojo, A.: Fingerprint pattern classification. Pattern Recognition 17(3), 295–303 (1984)
18. Nilsson, K., Bigun, J.: Localization of corresponding points in fingerprints by complex filtering. Pattern Recognition Lett. 24, 2135–2144 (2003)
19. Ozkaya, N., Sagiroglu, S., Wani, A.: An intelligent automatic fingerprint recognition system design. In: 5th Int. Conf. on Machine Learning and App., pp. 231–238 (2006)
20. Ross, A., Jain, A.K., Reisman, J.: A Hybrid Fingerprint Matcher. Pattern Recognition 36(7), 1661–1673 (2003)
21. Cevikalp, H., Neamtu, M., Wilkes, M., Barkana, A.: Discriminative common vectors for face recognition. IEEE Trans. on Pattern Analysis and Machine Intelligence 27(1), 4–13 (2005)
22. Bouchaffra, D., Amira, A.: Structural Hidden Markov Models for Biometrics: Fusion of Face and Fingerprint. Special Issue of Pattern Recognition Journal, Feature Extraction and Machine Learning for Robust Multimodal Biometrics (Article in press, 2007) (available online)
23. Li, S.Z., Jain, A.K.: Handbook of Face Recognition. Springer, New York (2004)
24. Yang, M.H., Kriegman, D.J., Ahuja, N.: Detecting faces in images: a survey. IEEE Trans. on Pattern Analysis and Machine Intelligence 1(24), 34–58 (2002)
25. Zhao, W., Chellappa, R., Phillips, P.J., Rosenfeld, A.: Face recognition: a literature survey. ACM Computing Surveys 35, 399–459 (2003)
26. Haykin, S.: Neural Networks: A Comprehensive Foundation. Macmillan College Publishing Company, New York (1994)
27. Sagiroglu, S., Beşdok, E., Erler, M.: Artificial intelligence applications in Engineering I: artificial neural networks. Ufuk publishing, Kayseri, Turkey (2003)
28. Sagar, V.K., Beng, K.J.A.: Hybrid Fuzzy Logic And Neural Network Model For Fingerprint Minutiae Extraction. In: Int. Conf. on Neural Netw., pp. 3255–3259 (1999)
29. Nagaty, K.A.: Fingerprints classification using artificial neural networks: a combined structural and statistical approach. Neural Networks 14, 1293–1305 (2001)
30. Maio, D., Maltoni, D.: Neural network based minutiae filtering in fingerprints. In: 14th Int. Conf. on Pattern Recognition, pp. 1654–1658 (1998)
31. The Mathworks, Accelerating the Pace of Engineering and Science (2008), http://www.mathworks.com/access/helpdesk/help/toolbox/nnet/nnet.html?/access/helpdesk/help/toolbox
32. Moller, M.F.: A Scaled Conjugate Gradient Algorithm for Fast Supervised Learning. Neurall Networks 6, 525–533 (1993)
33. Novobilski, A., Kamangar, F.A.: Absolute percent error based fitness functions for evolving forecast models. In: FLAIRS Conf., pp. 591–595 (2001)
34. Biometrical and Artificial intelligence Technologies (2008), http://www.neurotechnologija.com/vf_sdk.html
35. Cox, I.J., Ghosn, J., Yianilos, P.N.: Feature-Based Face Recognition Using Mixture Distance. Computer Vision and Pattern Recognition, 209–216 (1996)

Author Index

Lecture Notes in Artificial Intelligence (LNAI)